D0621856

The Art of Voice Acting

The Art of

Voice

Acting

fourth edition

The Craft and Business
of Performing for Voiceover

James R. Alburger

AMSTERDAM • BOSTON • HEIDELBERG • LONDON
NEW YORK • OXFORD • PARIS • SAN DIEGO
SAN FRANCISCO • SINGAPORE • SYDNEY • TOKYO
Focal Press is an imprint of Elsevier

ELSEVIER

Focal Press is an imprint of Elsevier
30 Corporate Drive, Suite 400, Burlington, MA 01803, USA
The Boulevard, Langford Lane, Kidlington, Oxford, OX5 1GB, UK

© 2011 James R. Alburger. Published by Elsevier Inc. All rights reserved.

No part of this publication may be reproduced or transmitted in any form or by any means, electronic
or mechanical, including photocopying, recording, or any information storage and retrieval system,
without permission in writing from the publisher. Details on how to seek permission, further
information about the Publisher's permissions policies and our arrangements with organizations such
as the Copyright Clearance Center and the Copyright Licensing Agency, can be found at our
website: www.elsevier.com/permissions.

This book and the individual contributions contained in it are protected under copyright by the
Publisher (other than as may be noted herein).

Notices
Knowledge and best practice in this field are constantly changing. As new research and experience
broaden our understanding, changes in research methods, professional practices, or medical
treatment may become necessary.

Practitioners and researchers must always rely on their own experience and knowledge in evaluating
and using any information, methods, compounds, or experiments described herein. In using such
information or methods they should be mindful of their own safety and the safety of others, including
parties for whom they have a professional responsibility.

To the fullest extent of the law, neither the Publisher nor the authors, contributors, or editors, assume
any liability for any injury and/or damage to persons or property as a matter of products liability,
negligence or otherwise, or from any use or operation of any methods, products, instructions, or
ideas contained in the material herein.

Library of Congress Cataloging-in-Publication Data
Alburger, James R., 1950-
   The art of voice acting : the craft and business of performing for voice-over/ James R. Alburger. -- 4th
      p. cm.
   Includes bibliographical references and index.
   ISBN 978-0-240-81211-3
   1. Television announcing--Vocational guidance. 2. Radio announcing--Vocational guidance. 3. Voice-ov
   4. Television advertising--Vocational guidance. 5. Radio advertising--Vocational guidance. I. Title.
   PN1992.8.A6A42 2010
   791.4502'8023--dc22
                                    2010023399

British Library Cataloguing-in-Publication Data
A catalogue record for this book is available from the British Library.

For information on all Focal Press publications
visit our website at www.elsevierdirect.com

10 11 12 13   5 4 3 2 1

Printed in the United States of America

Working together to grow
libraries in developing countries

www.elsevier.com | www.bookaid.org | www.sabre.org

ELSEVIER   BOOK AID   Sabre Foundation
              International

This edition is dedicated…

To my business partner and the other half of The Voiceover Team, Penny Abshire. Thank you for your positive attitude; your expertise as a performance coach, director, and voice actor; and your dedication to maintaining the highest of standards in everything you do.

And…

To the many extremely talented voice actors, past and present, who have made their mark in the world of voiceover, and who freely share their knowledge and experience. Your support and encouragement of those students who have the passion for performing voiceover, and are serious, about mastering the skills of this challenging business is truly inspirational.

Voice acting is the performing craft of creating believable characters, in interesting relationships, telling compelling stories, using only the spoken word.

Mastering and applying the skills of voice acting will take your personal and professional communication to an entirely new level of effectiveness.
James R. Alburger
www.jamesalburger.com

Contents

CD Index

Foreword

By Beau Weaver

One of the things that distinguishes human beings from our fellow creatures, is that we are animals who tell stories. In this digital age, the warm, personal connection of a flesh-and-blood person telling that story is more important than ever. At the heart of it, that is what voice acting is all about. Marice Tobias, one of the most sought-after Hollywood voice acting coaches, calls voiceover "storytelling with a point of view." However simple this might seem, it's much harder than it looks, which is why this book is so valuable.

Across the modern media landscape, there are many different places where the interpretive talents of the professional voice actor are used. From old-school national radio and television commercial campaigns, trailers for feature films, documentary narration, to more contemporary audio books, website opens, interactive flash animations and mini-documentary webisodes, each genre and subgenre is actually it's own very separate and distinct world. Each has very different players, conventions and rules of engagement.

There may appear to be a unified field called "voiceover work" but it's actually many different disciplines, with very little overlap. It's much like professional sports, where all the players are athletes, but most specialize in only one sport, with precious few able to master more than one game. This book will help the beginner and journeyman alike to untangle the many related, but completely distinct, "games" in this business.

And as the subtitle of this book points out, voice acting is an art, but it is also a business. While it is still true that on the east and west coasts, most national work is handled through talent agents and managers, gone are the days when a voiceover actor could simply play the role of artist. Today, increasingly, to be successful one must have much in common with the entrepreneur, wearing the hat of both actor and manager of your own career. Few people possess the sensibilities of both actor and marketer, but both are necessary to compete in what is now a global marketplace.

Fortunately, James Alburger understands how to wear both of these hats, and has broken down each area in great depth for those who are new to the field and for veterans who are looking for ways to move their craft, and business to the next level.

In his book *Outliers*, author Malcolm Gladwell points out that his own research shows that in order to master any field of endeavor, from computer programming to playing guitar to basketball, at least 10,000 hours of practice are required. This has proven true in my own experience. A few years back, I knew a self-taught artist, who was invited to teach a beginner-level class in drawing at a community college. Though he was a great illustrator, he had no idea how to teach others. So he surprised himself when he heard these words come out of his own mouth on day one of his first class. He said: "Well, it's going to take you about 10,000 drawings before you do anything that's worth a damn... so, I guess you'd better get started."

The Art of Voice Acting is a great place to start.

Beau Weaver
www.spokenword.com

ABOUT BEAU WEAVER:

Beau Weaver is one of the West Coast's perennial A-list voiceover artists, narrating trailers for feature films, network television promos, syndicated television shows, cable network documentaries, national radio and television commercial campaigns, and imaging for major market television affiliates and radio stations. He has had starring roles in animated series such as *Superman* and *The Fantastic Four* and narrates several popular documentary series for NatGeo, Discovery Channel, and Animal Planet.

Preface
James R. Alburger

When I first began teaching voiceover workshops in 1997, I never dreamed that my workshop notes would evolve into the most popular, and what is considered by many the most comprehensive, book on the craft and business of voiceover.

Upon completing the first edition, I knew I had only scratched the surface, but the thought that there would be a fourth edition never crossed my mind! Fortunately, with each new edition I have been able to expand the content to cover more and more aspects of this fascinating type of work.

The business of voiceover is one that is in a constant state of flux. Although the fundamental performing techniques may be consistent, there are trends and performing styles that are constantly changing. As voice actors, we must keep up with the trends and maintain our performing skills in order to keep the work coming in. But the reality is, there's actually a lot more to it than that.

What you hold in your hands is a manual for working in the business of voiceover that will take you from the fundamentals of performing to the essentials of marketing... and everything in-between.

Most books on voiceover talk about interpretation—how to deliver phrases or analyze a script, some actually teach "announcing," and some are more about the author than the craft of voiceover.

This book is different! This book was written with the intention of giving you a solid foundation in both the craft *and* business of voiceover. Within these pages you'll find dozens of tools and techniques that are essential for success in this area of show business—some of which you won't find anywhere else. With this book, you will learn exactly how to use these tools, not just in voiceover, but in everything you do. Unlike some other books on voiceover, I don't focus on how I did it, or go into boring stories of my voiceover career. No... I'll show you how *you* can do it! Every story and every technique you'll read in this book is here for a reason—to teach you exactly how some aspect of this craft and business works, and how you can make it work for you.

The tools and techniques are just that—tools and techniques. Without understanding how to use them, they are little more than words on the page. But once you learn how to use a few of these tools, you'll discover that they can be used to improve relationships, get more customers, resolve problems, close more sales, make you a better actor, improve your public speaking skills, and so on, and so on. You won't use every performance tool all the time, and some of the tools may not work for you at all. That's fine. Find the tools and techniques that *do* work for you, take them, and make them your own. Create your own unique style.

This is a book about how you can bring your personal life experience to every message you present and, by using a few simple techniques, communicate more effectively than you can imagine. When you apply voice-acting techniques, you are communicating on an emotional level with your audience. Your message has power and impact that would not be there otherwise. Everything you experience in life holds an emotion that can be used to make you more effective as a voice actor. And even if you never intend to stand in front of a microphone in a recording studio, you can still use what you learn here to become a more effective communicator.

Performing voiceover is much like performing music: There is a limited number of musical notes, yet there is an almost unlimited variety of possibilities for performing those notes. The same is true with a voiceover script. Words and phrases can be delivered with infinite variety, subtlety, and nuance. A voiceover performance is, indeed, very similar to the way a conductor blends and balances the instruments of the orchestra. Your voice is your instrument, and this book will give you the tools to create a musical performance. You might think of voiceover as a "Symphony of Words," or "Orchestrating Your Message," both catch phrases I've used to describe the results of what we do as voice actors.

Acknowledgments

This fourth edition of my book would not have been possible without generous support and help from so many people and companies who work in the world of voiceover every day. As you read through these pages, you will see names, web site links, and other references to the many individuals who have supported my efforts with their contributions. Please join me in thanking them for their willingness to share their knowledge and experience.

A very special thank you goes to my coaching, creative, and business partner, Penny Abshire. As a skilled coach and brilliant copywriter, you're the best! Thank you for your contributions and keen editing eye.

My goal with this book, as it is with my workshops, seminars, and other products, is to provide the best possible training in this craft that I possibly can. With the assistance of those who have contributed to this book, it is my hope that your journey into the world of voiceover is enjoyable and productive. If I help you on your journey, please let me know.

Introduction

"You should be doing commercials!"

"You've got a great voice!"

"You should be doing cartoons!"

If anyone has ever said any of these things to you—and you have, even for an instant, considered his or her suggestions—this may be just the book you need! If you simply enjoy making up funny character voices or sounds, or enjoy telling stories and jokes, this book will show you how to do it better and more effectively. If you need to make presentations as part of your job, this book will definitely give you a new insight into reaching your audience. If you are involved in any line of work for which you need to communicate any sort of message verbally to one or more individuals, this book will help you make your presentation more powerful and more memorable.

This book is about acting and performing, but a kind of acting that is not on a stage in front of thousands of people. In fact, with this kind of acting you rarely, if ever, see your audience or receive any applause. This is a kind of acting in which you will create illusions and believable images in the mind of the audience—a listening audience who might never see you, but who may remember your performance for many years.

This is a book about acting and performing for voiceover. Even though the focus here is on developing your talent for working in the world of voiceover, the skills and techniques you will learn can be applied to any situation in which you want to reach and motivate an audience on an emotional level.

Voiceover!

The term can be inspiring or intimidating. It can conjure up visions of a world of celebrity and big money. True, that can happen, but as you will learn in the pages that follow, the business of voiceover is just that—a business. It is a business that can be lots of fun and it can be a business that is, at times, very challenging work. And it can be both at the same time!

Voiceover is also an art! It is a highly specialized craft with skills that must be developed. The voiceover performer is an actor who uses his or her voice to create a believable character. The business of voiceover might, more accurately, be called the business of *voice acting*. It is most definitely a part of show business.

I'll be perfectly honest with you right from the beginning. Working as a voice actor is not for everyone. It requires an investment of time, energy, persistence, and money to get started. And, perhaps, just a bit of luck. As the saying goes in show business: An overnight success is the result of 20 years of study and paying dues.

However… if you love to play, have the desire to learn some acting skills, can speak clearly, read well, don't mind the occasional odd working hours, don't take things too seriously, have a good attitude, can motivate yourself to be in the right place at the right time, and are willing to do what is necessary to develop your skills and build your business, this type of work may be just right for you. In addition, as I mentioned earlier, the skills and techniques of voice acting can be applied to any situation in which you want your audience to connect emotionally with the message you are delivering. These skills are not limited to radio and TV commercials.

This book shows you the steps to take to learn the performing skills necessary to be successful as a voice talent. It also has the information you need to get your demo produced and into the hands of those who will hire you. Study these pages and you will get a solid foundation that you can build on to achieve lasting success in the business of voiceover.

You *don□t* have to be in Los Angeles, New York, or Chicago to find voiceover work. Work is available everywhere. You *do* need to have the right attitude, the right skills, and a high-quality, professionally produced presentation of your talents, or the casting people won't even give you a second look (or listen). If you master the techniques explained in this book, you will be able to present yourself like a pro—even if you have never done anything like this before.

Notice that I refer to the voiceover artist as voice talent, voiceover performer, or voice actor—never as "announcer." This is because that's exactly what you are—a performer (or actor) telling a story to communicate a message to an audience on an emotional level.

"Voiceover" is the common reference for all areas of performing in which the actor is not seen. However, this term misrepresents what the work actually entails. It tends to place the focus on the voice, when the real emphasis needs to be on the performance. *Announcers* read and often focus their energy on the sound of their voice, striving to achieve a certain "magical" resonance. Effective voice acting shifts the focus from the voice to the emotional content of the message. This requires knowledge, skill, and a love of performing. Focusing on your performance, instead of on the sound of your voice, helps you become more conversational, more real and more believable. In other words, this type of work is not about your voice… it's about what you can *do* with your voice.

This is why acting is such an important aspect of good voiceover work. Talking *to* your audience conversationally is much better than talking *at* them as an uninterested, detached speaker. The best communication closes the gap between the audience and the performer and frames the performance with a mood that the audience can connect with emotionally. Many people have "great pipes" or wonderfully resonant voices. But it takes much more than a good voice to be an effective communicator or voice actor. In fact, a good voice isn't even necessary—most people have a voice that is perfectly suitable for voice acting. What is necessary are the knowledge and skill to use your voice dramatically and effectively as part of a voiceover performance.

As with most businesses today, the business of voiceover is constantly evolving. The applications for voiceover work are growing every day and changing trends may require new or modified performing techniques. This revised, fourth edition has been expanded to include even more techniques, new scripts, more "tricks of the trade," and lots of Internet resources. Still more information and resources can be found at **www.voiceacting.com** and **www.voiceacting.com/aovaextras**.

I began my adventure through the world of sound and voice acting when I taught myself to edit music at the age of 12. I've worked for several radio stations creating hundreds of commercials as engineer, performer, writer, and producer. I have also performed professionally for more than four decades as a stage and close-up magician. I put my ideas about performing magic to music in my first book, *Get Your Act Together— Producing an Effective Magic Act to Music*, which became a standard in the magic community. However, it was when I worked as a recording engineer in Hollywood that I began to realize what voice acting was all about.

In the nearly four decades since then, I have directed some of the top voice talent in the country, I have been honored as a recipient of 11 Emmy Awards[1] for sound design, and have received numerous awards for creative commercial production. I teach voice-acting workshops and seminars; speak professionally on how to improve performing skills and the effective use of radio advertising; and operate my own business as a voice actor, sound designer, and performance coach. But this book is not about me... it's about giving you the tools you'll need to succeed in voiceover.

As you read the pages that follow, I promise to be straightforward and honest with you. You will find techniques and tricks of the trade that you cannot find anywhere else. For those of you considering a move into the business of voiceover, you will learn what it takes to be successful. If you simply want to learn new ways to use your voice to communicate effectively, you will find a wealth of information within these pages.

I wish you much success—and please let me know when you land your first national commercial or big contract as a result of using the techniques in this book.

---

[1] National Academy of Television Arts and Sciences, Southwestern Regional Emmys awarded for outstanding sound design for television promos and programs.

The Art of Voice Acting

**James R. Alburger
and Penny Abshire**

For your continued training in the craft and business of voiceover, we invite you to consider The VoiceActing Academy Art of Voice Acting workshops, seminars, and personalized coaching.

We also invite you to join our VoiceActing Academy Conductor's Club and subscribe to our VoiceActing News & Information Blog.

You'll receive:

- Articles on the craft and business of voiceover
- Interviews with top voiceover professionals
- Access to free conference calls
- Dates and locations of our workshops and training events
- And much, much, more

And, when you're ready to produce your voiceover demo, we're here to help.

For more information, please visit **www.voiceacting.com**.

VoiceActing Academy™

Changing lives one voice at a time
www.voiceacting.com

1

What Is Voice Acting?

We live in an age of information and communication. We are bombarded with messages of all types 24 hours a day. From 30-second commercials to hour-long infomercials; from documentary films, to video games; from telemarketing sales messages to corporate presentations; and thousands of others. Much of our time is spent assimilating and choosing to act—or not act—on the information we receive.

It is well-known among marketing and communications specialists that there are only two ways to communicate a message: intellectually and emotionally. Of these, the most effective is to connect on an emotional, often unconscious level. This involves drawing the listener (or viewer) into a story or creating a dramatic or emotional scene that the listener can relate to; in short, effective communication in really excellent storytelling. And the best storytellers create vivid imagery through a combination of interpretation, intonation, attitude, and the incorporation of a variety of acting skills. This is exactly what we do as voice actors! The voiceover (VO) performer, in fact, can be more accurately referred to as a voice actor.

The Many Roles of the Voice Actor

Voice actors play a very important role in entertainment, marketing, sales, and delivery of information. It is the voice actor's job to play a role that has been written into the script. To effectively play the role and thus sell the message, the performer must, among other things, be able to quickly determine how to best communicate the message using nothing more than the spoken word. Chapters 5 through 10 cover these subjects in detail. For the moment, you only need to know that this type of work requires more from you than simply reading words off a page.

The purpose of voice acting is to get "off the page" with your message. Make it real—connect with your audience emotionally—and make the message memorable in the mind of the listener.

1

Types of Voiceover Work

You can only do voiceover in one of two ways: either as a business or as a hobby. Both are completely valid approaches to voiceover, and this book will help in either case. However, the real focus on this book is in terms of how to work in voiceover as a business.

When most people think of voiceover, they think of radio and TV commercials. These are only a small part of the business of voiceover. There is actually much more to it.

Let's begin with a simplified definition of voiceover. *Voiceover* can be defined as *any recording or performance of one or more unseen voices for the purpose of communicating a message*. The voiceover is the spoken part of a commercial, program, or other announcement that you hear, but do not see the person speaking. It could be anything from a phone message to a television commercial, sales presentation, instructional video, movie trailer, feature film, or documentary narration. It may be nothing more than a single voice heard on the radio or over a public address system. The production may include music, sound effects, video, animation, or multiple voices. In most cases, the message is selling something, providing information, or asking the listener to take some sort of action.

You hear voiceover messages many times every day, and you are probably not even aware of it. Here are just some of the many types of voiceover work that require talented performers, like you.

RADIO—On-air, Commercials, Promo

There are three basic categories of radio voiceover work:

- **The radio DJ**—This is a specialized job that requires a unique set of skills. Most radio DJs are not considered to be voice actors.
- **Promo & Imaging**—Most radio promos are produced in-house, using station staff. Outside talent will be used for station imaging.
- **Commercials**—Most commercials are produced outside the station by advertising agencies. However, many radio stations will produce local commercials for their clients, usually with station staff.

TELEVISION—On-air, News, Commercial, Promo, Programs

Television stations use voice talent in three ways:

- **Promotions Department**—Handles the station's on-air promotion, including VOCs (voice over credits), promos, and marketing. Voice talent may be on staff, on contract, or booked through an agent.
- **Production Department**—This department is responsible for the production of commercials, programs, sales presentations and other productions. Voice talent may be on staff or hired per project.

- **News and Sales Departments**—News reporters handle their own voiceover for news stories. Sales will usually use on-staff VO talent or work through the Production Department.

Most TV stations have an established pool of voiceover talent on staff or readily available. Staff announcers may come to the station to record on a daily basis, or, more often, the copy may be emailed or faxed to the talent a day or so before it is needed. The voice actor records it in their home studio and delivers the track to the station for production, or it may be recorded at a local recording studio. Some TV stations are equipped with *ISDN* or Internet technology that allows for a live, high-quality recording of a voiceover performer in another city, or across the country.

CORPORATE/INDUSTRIAL—Training, Web Learning, Marketing

There are literally thousands of locally produced audio and video presentations recorded each year for the business community. Here are just a few examples of corporate and/or industrial voiceover work:

- **Telephony**—Messages-on-hold are what you hear while on hold. Also includes voice prompts (IVR) and outgoing messages.
- **In-store Offers**—Usually these are part of the background music program played over a store's speaker system while you shop.
- **Sales and Marketing Presentations**—Video presentations that are designed to attract clients and promote vendors or products. Talent could be either on-camera or voiceover. You will often find these videos as ongoing product demos in department stores or shopping mall kiosks.
- **Convention and/or Trade Show Presentations**—These are similar to sales and marketing presentations, but usually target potential buyers at a convention or trade show. These are usually video presentations.
- **Training and Instructional**—As the name implies, these projects are designed to train personnel on anything from company policies and procedures, to the proper use of equipment. Most corporate presentations are rarely seen by the general public.
- **Web Learning**—The Internet has opened up an entirely new world of voiceover opportunities for online training and education.

ANIMATION—Cartoons, Anime, Video Games

This is a very specialized area of voiceover work. It's definitely not for everyone, and it can be difficult to break into. Good animation voice actors can usually perform a wide range of character voices and have many years of acting experience. Most animation voiceover work is done in Los Angeles, while anime and video game work is done in many cities.

CD-ROM AND MULTIMEDIA—Games, Training, Marketing

This market for voiceover talent developed as a result of the explosion of computer-based CD-ROM games and instructional software. Some software manufacturers produce audio tracks for these products entirely in house, while others are produced by outside production companies.

FILM—Looping, ADR, Narration

Looping (recreating the background crowd ambience of a scene) and ADR (Automatic, or Automated Dialog Replacement) are specialized areas of voiceover work that require a high level of acting ability and often a talent for mimicking other voices. Film narration is common for documentaries, instructional, and marketing programs.

AUDIO BOOKS—Entertainment

Recordings of books and magazines fall into two basic categories: commercial audio books for sale and recorded books or magazines for the visually impaired. Audio books of best-selling novels are often read by a celebrity to make the recording more marketable. However, there is a growing market for audio book projects that use unknown voice talent. Recorded books and magazines for the visually impaired may be produced locally by any number of service organizations or radio stations. The pay is usually minimal or nonexistent (you volunteer).

Most reading services prefer their "readers" to deliver their copy in a somewhat flat tone. There may be several people reading chapters from a book over a period of days. To maintain a degree of continuity in the "reading," the readers are generally asked to avoid putting any emotional spin or dramatic characterization into their reading. This type of work is excellent for improving reading skills and acquiring the stamina to speak for long periods of time, but it limits your opportunities to develop characterization and emotional or dramatic delivery skills. Check your local white pages under Blind Aids and Services or contact your local PBS radio station and ask about any reading services they might provide.

INTERNET STREAMING AUDIO—RSS, MP3, Web Learning

The introduction of the iPod® by Apple Computer created opportunities for streaming audio podcasts on the Internet. Basically, a podcast is an RSS (Really Simple Syndication) streaming audio program that is designed to synchronize with subscriber's computers. It is commonly used to provide news feeds and other audio content that is automatically downloaded to a computer for listening later, at a convenient time. Many podcasts are also available online in the form of downloadable MP3 files.

Podcasting gives anyone with the ability to record audio on their computer an opportunity to record their opinions and original creations or performances for the world to hear. Many podcast programs are recorded by people who are not trained in the craft and performance of voiceover.

Web learning, or Internet-based, online training, is becoming very popular with many businesses that need to efficiently train a large number of people on a small budget.

THE ESSENTIALS

Regardless of the type of voiceover work you do, there are several basic requirements:

- **A decent speaking voice:** The days of the "Golden Pipes" are history! Voice acting is *not* about your voice—it's about what you can *do* with your voice.
- **Excellent reading skills:** All voiceover work requires excellent reading skills. There is no memorization in voiceover work.
- **An ability to act and take direction:** You must be able to change your delivery and interpretation at the whim of the director.
- **Passion:** You must be willing to spend the time, energy, and money necessary to develop your acting and business skills, and market and promote your talent.

The Difference between "Voiceover" and "Voice Acting"

If you accept the definition of voiceover as being anything in which you hear the voice but don't see the performer, then, in the strictest sense, anyone who can speak can do voiceover. But that doesn't mean that anyone who can speak has the ability or skill to work professionally as a voice talent. If you've ever recorded an outgoing message on your answering machine, you've done a form of "voiceover." But does that mean you can do voiceover professionally? Probably not. I frequently receive demos from self-proclaimed "voiceover artists" who have just completed a class and produced a demo in an attempt to break into the business. Some of these individuals have some raw talent, but unfortunately, most have not polished their skills or honed their craft to a point where they can effectively compete as voice talent. It is rare that I receive a demo that demonstrates a level of professionalism that shows me the individual has made the transition from "doing voiceover" to performing as a "voice actor."

"So, what's the difference?" I hear you ask.

There are several factors that differentiate simple voiceover and professional voice acting. Among them are: competent training, acting ability, interpretive skill, dedication, business acumen, and computer skills. But the real difference can be summed up in a single word:

Believability

You can listen to the radio or watch TV any hour of any day and hear commercials that literally make you cringe. If you analyze the performance, most of these "bad" commercials have several things in common: they sound flat and lifeless, with every sentence sounding the same; they sound like the script is being read; the performers sound like they are shouting with no clear focus as to who they are speaking to; the performer is talking "at" and not "to" the listener, or they are trying to talk to everyone listening at the same time; there is absolutely nothing compelling about the delivery or the message. In short, the performance lacks "believability."

Here's a simple way to determine if it's "voiceover" or "voice acting." A "voiceover" performance has at least one or more of the following:

- Often "read-y" or "announcer-y" (sounds like reading the script).
- Content is information-heavy, primarily intellectual, often with many featured items, and with little or no emotional content.
- The goal of the message is to "sell" the listener on something, and this attitude of "selling" comes through in the performance.
- The overall effect of the message is to create "listener tune-out."
- Delivery of the message may, in some way, actually damage or reduce credibility of the advertiser.

A "voice acting" performance has ALL of the following characteristics:

- The performer creates a believable and real character in conversation with the listener.
- The message is primarily emotional, with a clearly defined focus.
- The goal of the message is to "tell a story" that the listener can relate to on an emotional level—often coming from a place of helping the listener in some way, rather than "selling."
- The overall effect of the message is one of keeping the listener's attention and creating a memorable moment.

Using the preceding definition, there is certainly a place for voiceover, and if done properly it can be quite effective. But good voiceover is done within the context of a larger performance or is designed for a very specific purpose, and presented by a very specific character. The best "voiceover" work is performed from a foundation of "voice acting."

Voice acting is about creating real and believable characters in real and believable situations that listeners can relate to and be motivated by. To do this, a performer must be able to reach the audience on an emotional level.

In other words, voice acting is about creating compelling characters in interesting relationships.

We communicate on an emotional level every day in a completely natural manner. But when we work from a script, we suddenly flounder: the words are not ours and the life behind those words is not ours. It's not as easy as it may appear to get "off the page" and speak from a written script in a manner that is natural, real, and conversational. In order for us to speak those words from the point of view of a real and believable character, we must momentarily forget who we are and become that character. That's why it is important to master basic acting techniques.

Learning basic "voiceover" techniques for reading and interpreting a script is a good start. But don't stop there. If performing with your voice is something you love to do, keep studying: take acting and improvisation classes; study commercials and analyze what the professionals are doing to create character and make their scripted words sound real; learn how to take direction; read every book on this craft you can get your hands on; visit talent websites and listen to the demos to learn what works and what doesn't; watch television programs about acting and theater, and finally... never stop learning.

Even if you are an experienced actor, you need to know that the disciplines of "voice acting" are different from stage, film, or TV. In all other forms of acting, your lines are committed to memory and you have time to internalize, understand, and develop your character. In voice acting, you may have only a few minutes to create a believable character, find the voice, and deliver a compelling performance as you read from a script.

Voice acting is creative, fun and potentially lucrative—if you know what you are doing and have the patience to master the necessary skills! In some circles, the term "voice acting" is used to refer to the niche area of voiceover work for anime. However, if you look closely at what anime voice actors are doing, you will discover that they are creating (or in some cases attempting to create) what will ideally be perceived as real and believable characters. And that is exactly what we need to be doing as we voice a radio commercial, corporate narration, or audio book.

To be a successful voice actor, learn how to be natural, confident, real, and most of all... believable.

Breaking into the Business of Voiceover

For the balance of this book, the terms "voiceover" and "voice acting" will be used interchangeably." This book is about the acting craft behind the business of voiceover work, so although "voice acting" is a more accurate term, you'll usually see the term "voiceover." I'll refer to the performer as either a voice actor, voiceover performer, or voice talent.

Most people think voiceover work is easy. You have probably even said to yourself after listening to a commercial, "I can do that!" For some people, it is easy. For most, though, voiceover—just like theatrical acting—is an ongoing learning process. In our VoiceActing Academy, Art of Voice Acting Workshops, it is not uncommon for someone, after only the first or second class, to say "Oh, my! I had no idea there was this much to voiceover! This really isn't about just reading a script!"

Even experienced professionals will tell you that voiceover work is far more difficult than on-camera or on-stage acting. There is no memorization and the advantages of props, scenery, wardrobe, makeup, and lighting are not available to the voice actor. The drama, comedy, emotions, or subtext of a message must be communicated solely through the spoken word. This requires a tremendous amount of focus and concentration, plus an ability to make quick changes in midstream. Prior acting experience is an advantage, but the essential performing skills can be picked up as you go, so don't let a lack of experience stop you. If you can use your imagination, tell a story with vivid imagery, and take direction, you can do voiceover.

One of the greatest misconceptions is that you need a certain type of voice to do voiceover. You do not need a "good" voice, or "announcer" voice. You do need a voice that is easily understood. If your voice has a unique quality or sound, you can use that to your advantage, especially for animation work. But a unique voice quality can also become a limitation if that is the only thing you do. You may find you are better suited to one particular type of voiceover work—corporate/industrial, for example. If that's the case, you can focus on marketing yourself for that type of work. Still, you should consider other types of voice work when the call comes.

Variety is an important aspect of voiceover performing. By variety I mean being able to use your voice to convey a wide range of attitudes, delivery styles, personality, interpretation, energy, and emotions. These are the characteristics of your voice presentation that will allow you to effectively tell a story that contains a message. And communicating a message is what working as a voice actor is all about.

Many people think that because they can do lots of impersonations or make up crazy character voices, they can do voiceover work. Vocal versatility is certainly valuable; however, success in the world of voiceover also takes focus, discipline, and an ability to act.

So, just how do you learn voiceover performing skills, break in, and get yourself known as a voice actor? There is no simple answer to this question. To be successful, you should learn everything you can about acting, communication, and marketing. In this business, an old adage, "It's not what you know, but who you know," is also very true. Getting voiceover work is largely a numbers game—a game of networking and making yourself known in the right circles. To be successful you cannot be shy. Let every person you meet know what you do! But you must also possess both the performing and business skills that qualify you as a professional, and that is what this book is really about! Or, you can do voiceover as a hobby!

2

The Best-Kept Secret

Let's face it—if everyone were equally good at every job, there would be no need for résumés or auditions. Fortunately, in this world, every person has uniquely different talents, abilities, and levels of skill. It is this variety that makes the voiceover business a potentially profitable career for anyone willing to invest the time and effort.

For years, voiceover was one of the best-kept secrets around. The job can be loads of fun and very profitable, but it is not an easy business to break into. Today, there are roughly five times as many people who claim to be voiceover talent as there are actors trying to break into TV and movies. Add to that the major film stars who have discovered that voiceover work is more fun than spending many hours in makeup each day. The simple truth is that competition is tough, and it is easy to become frustrated when just beginning.

Voiceover work is part of "show business." As such it has all the potential excitement, celebrity status, and opportunities as the other areas of show business, as well as the long periods of waiting, frustrations in getting "booked," and problems dealing with agents and producers.

The Realities of Voice Acting

You have probably heard most of the pros of voiceover work: big money, short hours, celebrity status (fame and fortune without anyone actually knowing who you are), and more. For some voiceover performers, these things are true but it takes a long time, and constantly being in the right place, to get there. In other words, they had to work at it. Most overnight successes are the result of many years of hard work, constant study, dedication to the craft, and a mastery of business skills. One voiceover coach I know suggests that it takes 15 years to become successful in voiceover. I disagree with that! Everyone defines "success" differently. Sure, if you define success as being in high demand and making the "big bucks," it might take 15 years or longer to get there. But if you are doing voiceover because you really love it, and you wonder why you're not paying them to let you get in front of the mic, then success can be as soon as next week.

Like most of the performing arts, voice acting is a hurry-up-and-wait kind of business. By that I mean you will spend a lot of time waiting: waiting at auditions, waiting for a callback, waiting in the lobby of a recording studio, waiting for the email with your script, and waiting to get paid. Once a voiceover recording session begins, things tend to happen very fast. But you may still find yourself waiting as the producer works on copy changes, or while the studio engineer deals with a technical problem.

If you are recording in your home studio, which has become a standard practice for voiceover work at all levels, you will be expected to deliver studio-quality recordings. You'll also be expected to know how to do some limited production and editing—even though you are not a recording engineer. That means you need to be computer-literate and you'll need to invest in the training, equipment, software, and acoustic improvements necessary to build a functional recording facility in your home.

From a performance standpoint, producers assume that you know what you are doing and expect you to deliver your lines professionally. You are expected to be able to deliver a masterful interpretation of a script after only a short read-through—usually within the first two or three takes. Direction (coaching) from the producer or director often comes very fast, so you must listen closely and pay attention. Sometimes, the producer or director completely changes the concept or makes major copy changes in the middle of an audition or session—and you need to be able to adapt quickly. And more often than not, you won't get any direction at all. If you're recording in your home studio, the session may be director-less and producer-less, meaning you are on your own! You need to develop excellent interpretive skills and be a versatile performer with the ability to self-direct and provide what your client is asking for, even when you're not certain exactly what that is.

Your job as a voice actor is to perform to the best of your abilities. When you are hired, either from your demo or after an audition, your voice has been chosen over many others as the one most desirable for the job. Unless there is a serious technical problem that requires your being called back, or if there are revisions that are made after the session has ended, you will not get a second chance after leaving the studio or sending your files.

Full-Time or Part-Time

If you think voiceover work is for you, you may have some decisions to make. Not right this minute, but soon. Do you want to do voiceover work as a full-time career, or as a part-time avocation? What niche area of voiceover do you want to focus on? Should you move to a different city in search of work in your niche area? The choices may be many and may not be easy!

Doing voiceover work on a full-time basis is unlike just about any other job you can imagine. You must be available on a moment's notice when you are called for an audition. In addition, you must constantly market yourself, even if you have an agent.

Full-time voiceover work may also mean joining a union, and possibly even moving to a larger city—if that's where your destiny leads you. Los Angeles, Chicago, New York, and many major cities are strong union towns for voiceover work, and you must be in the union to get well-paying jobs in these cities. Although the possibility for nonunion work does exist in larger cities, it may require some additional effort to find it.

In smaller cities, the union for voiceover, AFTRA (the American Federation of Television and Radio Artists), is not as powerful, and there is a much greater opportunity for freelance voiceover work than in bigger cities. You'll find more about unions in Chapter 21.

OK—you've decided eating is still a pleasurable pastime, and you've made the wise decision that you would rather not quit your day job just yet. So, how about doing voiceover work on a part-time basis? Good question!

Doing voiceover work part-time is quite possible, although you probably won't be doing the same kind of work as you would if you devoted more time to it. You will most likely do some corporate/industrial work, telephone messages, and smaller projects for clients who have a minimal or nonexistent budget. Some of your work may be voluntary, barter, or you will do it just because you want the experience. The pay for nonunion freelance work is usually not terrific—but freelance work is a very good way of getting experience doing voiceover. You can gradually build up a client list and get copies of your work that you can use to market yourself later on when, or if, you decide to go full-time.

The biggest problem with doing voiceover work part-time is that you may find it difficult to deal with last-minute auditions or session calls. If you have a regular full-time job, you usually will need to arrange your voiceover work around it, unless you have a very understanding employer. Part-time voiceover work can be an ideal opportunity for the homemaker or self-employed individual with a flexible schedule.

With the advent of Internet audition services and advanced computer technology, it has become very convenient to record auditions and paid projects in a home studio and submit them as MP3 files via the Internet.

Doing voiceover work can be very satisfying, even if you only do an occasional session. Yet, the day may come when you decide to go for the big money in L.A., Chicago, or New York. In the meantime, don't be in a hurry. Make the best of every opportunity that comes along and create your own opportunities whenever possible. Networking is extremely important! You never know when you might be in just the right place to land that important national spot that changes your entire life!

7 Things You Must Know about Voiceover Work

On the surface, voiceover appears "easy," but in reality there is a LOT to learn! Here's a list, inspired by VO pro Michael Minetree, of some essential things you need to know about voiceover before you take the leap:

1. You can't learn how to perform for voiceover on your own. You need the guidance of a qualified coach who knows the business.

2. You can't learn how to perform for voiceover by reading a book. Any VO book (yes, even this one!) is only as good as the information it contains. The purpose of a book is to provide you with the information you need so you can more effectively learn the skills. You need talent, dedication, passion, and training that goes beyond the information contained in a book.

3. You can't learn how to perform for voiceover from a tele-course. A tele-course will give you lots of information, but by its very nature, will be limited in the effectiveness of any performance coaching. You may get the general idea of how to use a technique, but it won't qualify you to compete in this business. Personal coaching and experience are your best training.

4. You can't learn this craft from a single workshop. Some workshops are excellent—and some are, well... not. Any workshop (yes, even ours!) will only be good enough to get you started on the path. You need to take the next steps with additional training. Professional film, stage, and television actors are constantly taking classes between projects. Continued training is essential in the voiceover business. Throughout this book, I'll encourage you to do the same.

5. If you produce your demo immediately following a workshop, you will be wasting your money. Your money will be better spent on additional training and personalized coaching. Do not even think about spending money on producing your demo until **you** *know* you are ready. See Chapter 18, "Your Voiceover Demo," to learn more about how to prepare for your demo.

6. If a demo is included as part of a course... find a different course! No one is ready for a demo after a single workshop. Your demo must be great—it cannot be merely "good." Even more than that, your performance must be comparable to the best voice talent out there. That level of skill only comes with time and proper training.

7. Be wary of workshops and coaches who promise success and a substantial income from taking their course. No one can promise you success, and no one can promise your demo will even be heard. Your degree of success in voiceover will be directly related to your dedication to running your business in a professional manner.

3

Where to Start:
Voiceover Basics

The Voiceover Performer as Actor and Salesperson

When you stand in front of a microphone as voice talent, your job is to effectively communicate the message contained within the words written on the paper in front of you. You are a storyteller. You are an actor! The words, by themselves, are nothing but ink on a page. As a voice actor, your job is to interpret the words in such a way as to effectively tell the story, bring the character to life, and meet the perceived needs of the producer or director. I use the words "perceived needs" because many producers or writers only have an idea in their heads. The producer may think he knows what he wants, when, in reality, he hasn't got a clue as to the best way to deliver the message. This is where your acting skills and performance choices come in. You may find yourself in the enviable position of solving many of your producer's problems simply by performing the copy in a way that you feel effectively communicates the message. In other words, your acting abilities are the vital link between the writer and the audience.

YOUR ROLE AS A VOICE ACTOR

You are the actor playing the role of the character written in the script. On the surface, that may sound like a fairly simple task. However, mastering the skills to create interesting and compelling characters on a consistent basis can be very challenging. Unlike stage performers, who may have several days, weeks, or months to define, internalize, and develop their characters, you may have only a few minutes. You must use your best acting skills to deliver your best interpretation of the copy—and you must do it quickly. Your job is to breathe life into the script, making the thoughts of the writer become real through the character you create. You need to be able to quickly grasp the important elements of the script, figure out who you are talking to, understand your character in great detail, find the key elements of the copy, and choose what you believe to be the most effective delivery for your lines. Every script is written for a purpose and you must be able to find and give meaning to that purpose, regardless of how or where the voice track will be recorded. In many cases, especially in studio sessions, the producer or director will be coaching you into the read that gets you as close as possible to his or her vision. However, with the increasing prevalence of high quality home studios, more and more voice talent are being asked to provide self-directed, unsupervised, sessions.

One mistake made by many beginning voiceover performers is that they get nervous when they approach the microphone. They are focused on their voice, not their performance. They fidget, stand stiff as a board, cross their arms, or put their hands behind their backs or in their pockets. It is impossible to perform effectively under those conditions.

What is needed is to get into the flow of the copy, breathe naturally, relax, have fun, and let the performance take you where it needs to go. Discover your character and let that character come into you so that you can create a sense of truth and reality for the character. If you think too much about what you are doing, your performance will usually be forced and sound like you are acting. When you allow yourself to "become the character" you will be able to "live the voice."

UNDERSTAND YOUR AUDIENCE

Every message (script) has an intended (or target) audience. Once you understand who the audience is and your role in the copy, you will be on your way to knowing how to perform the copy for the most effective delivery. Figure out who you are talking to. Narrow it down to a single individual and relate to that person on an emotional level. This is the first step to creating an effective performance and a believable character.

Chapter 10, "The Character in the Copy," goes into greater detail about analyzing the various kinds of copy and creating characters.

WHAT TO LOOK FOR IN A SCRIPT—CD/11

Now that you know some of the basics for creating an effective voiceover delivery, here's a script for you to work with. Read it through once to get a feel for the copy. Notice that you instinctively make some choices as to how you will deliver the copy. Deliver the script using the choices you make. Then listen to track 11 on the CD.

It happens everyday . . . in hotels, restaurants and other public buildings . . . without warning. It's responsible for 20,000 fatalities – and it's the second leading cause of death and disability. Slip and fall accidents – learn how to protect your rights if it happens to you. Tonight at 11 on Eye Witness News.

Do you think your delivery achieved the objective of communicating the message effectively? An effective voiceover delivery requires looking beyond the words of a script to dig out the details and subtlety hidden in the message. Now, read it a second and third time, looking for the following points. Finally, read it out loud for time, to see how close you can come to 15 seconds.

- Who is the audience this copy is trying to reach?
- How can you create interest within the first few words?
- How can you create an emotional response to keep the audience listening?
- What is the single primary message in the copy?
- What are the supporting statements for the primary message?
- What is your role (your character) in the story?
- Why is your character telling this story?
- What does your character want or need from telling this story?
- What is the primary emotion, if any?
- What sort of delivery do you think would be the most effective to create the strongest memory of the message—strong, hard-sell, happy, smiling, mellow, soft-sell, fast, slow?
- What is your attitude as the character in this spot—serious, comfortable, happy, sad, and so on?
- In what way can you make the audience feel safe, comfortable, and in control of their decision to keep listening?
- What visual images come into your mind as you read the copy?

OK, how did you answer the questions? By the way, there are no wrong answers! Each of your answers represents a choice that ultimately results in your personal interpretation of the story. This spot is a TV promo, so there are visuals that go with the copy. You might know that from the format of the script, from a notation at the title, from written directorial notes, or from the producer telling you. Sometimes, however, you will not have anything more than the words on the page. Here's an interpretation of this copy:

- The target audience is men and women who spend time in public places, and who are concerned about safety issues. The focus is primarily on adults who travel or work in large buildings. To effectively reach this audience, you need to speak to one person.
- The message does not answer any questions, but instead, creates awareness of a potential problem. Your delivery of the first line of copy should instantly grab the listener's attention and you should deliver the remaining copy in a way that hints at solutions, but reveals that solutions to the problem can only be resolved by the viewer watching the program.

- There are several very visual and emotional references that can be used to help create a mood or tone for the message: The words "hotels," "restaurants," and "other public places," all conjure up powerful images in both your mind as a performer, and in the mind of the listener. By the same token, "it happens every day," and "without warning" are phrases the listener can identify with. The words "death" and "disability" are emotionally charged words that are intended to illicit a response. Each viewer's response to these words will be unique, but the intent is to create an impact in the viewer's mind as to the potential for serious injury.

- The person speaking here is telling a story—a sort of mystery story about a serious problem. The telling of this story is done with a sense of drama and suspense, which leads up to the revelation of exactly what the problem is and what the listener can do.

- The overall delivery is sincere, compassionate, and concerned, with a serious, almost foreboding tone. This attitude is retained throughout the delivery. Each element—almost every word—of the message is given value and importance. The end of the copy is delivered in a more matter-of-fact manner, but still keeping a tone of compassion about the story.

All these answers combine to provide the basic information you need to effectively deliver the copy. The visual image is important because it sets the scene, a solid framework for your character, and helps establish the attitude of the spot. As an actor, you need to know these things. Otherwise, you are just reading words on a page—and that's boring!

With experience, you can analyze a script in a matter of seconds, just from a single read-through. You'll instantly know what you need to do in your performance to make the message both interesting and compelling. Trust your instincts and use what you have learned from your interpretation to give depth to your character and life to the copy. Above all, bring your unique personality into the copy and everything else will come naturally.

THE VOICE ACTOR AS A SALESPERSON

It can be argued that virtually all voiceover is "selling" something. Commercials sell products or services, or try to get an emotional response to motivate action; instructional products sell procedures; audio books sell entertainment; and so on. Acting is the means by which any of these messages can be effectively communicated, the story told, and the listener motivated to take action. So, you are not only a performer, but you are also a salesperson. For the time you are in the recording studio, you are an employee of your client's business. In fact, you, as an actor, are the advertiser's top salesperson and must present yourself as a qualified expert. And you, as the character, must be perceived as real and honest.

Your acting job may only last a few minutes in the studio, but that performance may be repeated thousands of times on radio or TV. Your voice may be heard by more people in a single minute than might walk through the front door of a business in an entire year. But even though you may be a salesperson, you must never sound like you are selling. The credibility of the product or advertiser—and the success of an advertising campaign—may be directly related to the authenticity, effectiveness, and believability of your performance. Tell the story... Never sell it!

Are you beginning to see there's more to this thing called voiceover than merely reading words on a page? And we're just getting started!

Getting the Skills You Need

The bottom line here is to get experience—as much as you can, wherever you can, any way you can! Take classes in acting, voiceover, improvisation, business, and marketing. Get as much experience as you can reading stories out loud. Read to your children. Read to your spouse. Practice telling stories with lots of variety in your voice.

Analyze the characters in the stories you read. Take more classes. Read the same copy in different ways, at different speeds, and with different feelings or emotional attitudes—loud, soft, slow, fast, happy, sad, compassionate, angry. If possible, record yourself and listen to what you did to see where you might improve. Take some more classes. Become a master of performing on a microphone. You can't take too many classes!

One of the best ways to acquire skills as a voice actor is to constantly be listening to what other voiceover performers are doing. Mimicking other performers can be a good start to learning some basic performing techniques, but your ultimate goal is to develop your own, unique interpretive skills and your own, unique delivery style. To really get an understanding of communicating on an emotional level, listen to how other professional voice actors deliver their lines and tell their story:

- How do they interpret the message?
- How do they reach you emotionally?
- How do they use inflection, intonation, pacing, and express feelings?
- Is their delivery conversational or screaming?
- How do you respond to their interpretation?

In short, do they sound as if they are reading or do they sound natural and believable? Use what you learn from studying others and adapt that information to your own voice and style. Learn how to "make the copy your own." This simply means that you bring to the performance something of yourself to give the character and copy truth and believability. That's good acting! Chapters 5 through 10 will show you how to do it!

A TWIST OF A WORD

You will notice that the better commercials and voiceover work do not sound like someone "doing" voiceover work. They sound like your best friend talking to you—comfortable, friendly, and most of all, not "announcery." A good performer can make even bad copy sound reasonably good—and what they can do with good copy is truly amazing.

Create an emotional, visual image in the mind of the audience with a twist of a word. A slight change in the delivery of a word—a shift of the nuance—can change the entire meaning of a sentence. Speaking a word softly or with more intensity, or perhaps sustaining a vowel, making the delivery crisp, or taking the inflection up or down can all affect the meaning of a sentence and its emotional impact in the mind of the listener. These are skills that are acquired over time and are all basic acting techniques that help to create an emotional connection with the audience.

To be an effective voice performer you need to discover the qualities and characteristics of your voice that will make you different from all those other voices out there. Keep developing new techniques. Keep practicing and studying the work of others in the business. Find your unique qualities and perfect them. Learn how to make any piece of copy your own, and you will be in demand. Remember, it's not about your voice, but what you can do with it.

CLASSES

One frequent observation that has appeared in discussions of this book over the past several editions is my repeated recommendation for continued training. The necessity to keep up with business trends and constantly hone performance techniques cannot be over-emphasized! It is impossible to take too many classes! There is always something new to be learned. Even if you leave a class with only one small piece of useful information, that small gem may someday pay big dividends. The same is true of books and articles. You will be amazed at where you can find a tip or trick that will help you create a believable performance.

There are four types of classes that are most valuable for the voiceover performer: acting, voiceover, improvisation, and business. Acting classes will give you opportunities to learn about directing, dramatic structure, comedic timing, stage presence, emotional delivery, and innumerable other fine points of performing. Voiceover classes will give you opportunities to practice your skills on-mic and study new techniques with personalized coaching. Improvisation in voice work is common with dialogue or multiple voice copy and is an essential skill for commercials, animation, video game and other niche areas of the business. This type of training helps improve your spontaneity and ability to adapt quickly. You will also learn skills that can be applied to character development and copy interpretation. And because the nature of voiceover work today is largely entrepreneurial, it is imperative that you have at least a basic understanding of fundamental business skills. I truly encourage you to take some classes, attend a workshop, or even spend a few days learning from the pros at a voiceover convention. Continued training is an incredibly worthwhile investment in your performing career. I promise you will learn a lot, and you might actually have lots of fun. Here are some of the places you can find classes:

- Community theater groups are constantly in need of volunteers. Even if you are working on a crew, you will be able to study what goes on in the theater. Watch what the director does, and learn how the actors become their characters. Don't forget that voice acting is theater of the mind—without props, scenery, or lighting.

- Most community colleges offer continuing education classes, often in the evenings or on weekends. Tuition is usually reasonable and the skills you can learn will pay off later on. Suitable courses can also be found in most college theater arts curriculums.

- Many cities have adult education classes in voiceover, acting, comedy, improvisation, and other subjects that can give you opportunities to acquire the skills you need. Check your local adult or continuing education office, or local colleges and universities for classes offered in your area.

- Many cities have private acting and voiceover courses. They are usually not advertised in the phone book, so they may be somewhat difficult to locate. An Internet search for "voiceover (or voice acting) Your City" may bring up some interesting results. Talent agents in most cities may be aware of local training and may be able to refer you to a class or coach. Check the classifieds of the local subscription and free newspapers in your area. You can also call the drama department at high schools and colleges for any referrals they might be able to make. Your local professional and community theater groups may also be able to give you some guidance. You'll find a comprehensive listing of voiceover coaches in the Resources area at **www.voiceacting.com**.

- For voiceover classes, try calling some of the recording studios in your area. Many recording studios work with voiceover performers every day and can offer some valuable insights or give you some good leads. Some studios offer classes or do the production work for a class offered by someone else. Or they might be able to simply point you in the right direction by suggesting local workshops or refer you to a local talent agent who might be able to give you some direction.

A WORD OF CAUTION

Larger cities, such as Los Angeles and New York, have many voiceover workshops and classes available. Most are reputable and valuable resources. Be careful, though, because some classes are little more than scams designed to take your money. Usually the scam classes will begin with a short "teaser" class or workshop where they provide you with information that you can often find elsewhere for free or from a book. They tell you just enough to get you excited—usually conveniently underplaying the true realities of the business. Then they tell you they will produce and market your demo for a fee—anything from $500 to $5,000. You may even be required to take their class if you want them to produce your demo. Demo fees are usually in addition to the fees you pay for the class, although some will include a demo as part of their overpriced tuition. You may get a demo from these classes, but the quality will likely be poor, and their promises of marketing your demo or sending it out to agents are usually worthless.

Many legitimate classes will also offer their services to assist with your demo. The difference is that you will not be pressured into buying their services and the demo will not be a condition of taking the class. An honest and reputable voiceover instructor will not encourage you to do a demo until you are ready. When they do assist with your demo, the production quality is generally very high. Regardless of who you hire to produce your demo, be sure to check them out. Get copies of some demos they have done and get a list of former clients who you can call to ask about their experience with the producer. If they are legitimate, the demo producer will be happy to help you. Some will even give you a free consultation.

Be aware that no workshop coach or demo producer can guarantee your demo will be heard by an agent or talent buyer, or even that you will be accepted for voiceover work. No matter what they tell you, you are the only person who will determine your success in this business. Do not rely on someone else to do it for you.

4

The Business of Voiceover: Getting Paid to Play

It's Show-biz, Folks!

One thing many people seem to either not realize—or simply forget—is that voiceover is part of show business—and the larger part of show business is business! Before making the investment in time, energy, and money for workshops, training, and equipment to become a voice actor, it is important to have an understanding of what this business entails, how it works, and what is expected of you as an independent business owner.

This chapter will introduce you to the business of voiceover so you will be able to make an educated decision as to whether or not this type of work is right for you. Demos, marketing, auditions, and many other aspects of this business are discussed in detail later in this book.

Acting for voiceover may be one of the best-kept secrets around. You get to be serious, funny, and sometimes downright silly and your voice may be heard by thousands. Voiceover can be an incredible outlet for your creativity and it can often seem like you get paid to play!

To be perfectly honest, voice acting can be very challenging at times. The reality is that you are an entrepreneur running your own business and you can expect all the ups and downs that go along with that. Depending on the type of voiceover work you choose to do and the clients you work with, you may be on call 24/7/365. Vacations may be difficult to schedule and there will be moments when you wish you were somewhere else. You will encounter producers and/or directors who do not seem to know what they are doing and who will test your patience. You will be faced with cramming :40 of copy into :30—and the producer will expect it to sound natural and believable. All of this—and more—is just part of working in the world of voiceover. That's show-biz!

Fortunately, the uncomfortable moments are relatively rare, and the majority of voiceover work is enjoyable and often downright fun. If you really enjoy what you do, and become good at it, even challenging sessions can seem like play, although it may appear to be hard work to everyone else. If you approach voiceover work with a positive attitude, a mindset of teamwork, and an eagerness to help your clients achieve their objectives— rather than as just a way to make money, your likelihood of success will be much greater. To a large extent, your level of success as a voice actor will depend on your mental attitude and how you approach your work.

Many successful voice actors do much more than just perform as voice talent. It is not uncommon to find voice actors wearing many hats—ad-agency rep, copywriter, producer, studio engineer, and of course, performer. Many voice actors also work as on-camera talent or in theatrical productions. After all, acting is acting, and the more versatile you are as an actor, the greater your likelihood of success in voiceover. As you master voice-acting skills, you may find yourself developing other talents as well. This diversification can provide income from several sources.

Making Money Doing Voiceover Work

There are only two ways to get paid for voiceover performing: union jobs and nonunion freelance jobs. If you are just starting out, it is a good idea to do as much nonunion work as possible before joining a performing union. It's sort of like "on-the-job training." Nonunion voiceover work will provide the opportunities to get the experience you need and accumulate some recordings of your work.

If you pursue voiceover work as a career, you may eventually join a union, especially if you live in a large market. However, it is not necessary to join a union to become successful. There are many independent voiceover performers in major markets who are earning substantial incomes, even though they are not members of any union. The choice of whether or not to join a union is one that only you can make—and you don't need to make that decision now.

THE UNIONS

Nothing in this book is intended to either promote or discourage union membership. However, joining a performing union is an important decision for anyone pursuing the art of voice acting. If you are just beginning to venture into the world of voice acting, a basic knowledge of the unions is all you need. As you gain experience and do more session work, at some point you may want to consider union membership. Much of the information in this section can be found in the information packet available from your local AFTRA or SAG office[1] or online at **www.aftra.com** or **www.sag.org**.

There are two unions that handle voiceover performers in the United States: AFTRA (American Federation of Television and Radio Artists) and SAG (Screen Actors Guild). In Canada, voiceover work is handled by ACTRA (the Alliance of Canadian Cinema, Television, and Radio Artists) —**www.actra.ca**. British Columbia has UBCP (the Union of BC Performers), **www.ubcp.com**, which is the BC branch of ACTRA. In the United Kingdom, the voiceover talent union is Equity, **www.equity.org.uk**. Other countries with collective bargaining unions will also have one or more unions that work with voice talent. It may be necessary to contact a local talent agent to learn which union applies in your country.

The job of all unions is to ensure proper working conditions, to make sure you are paid a reasonable fee for your work, to help you get paid in a timely manner, and to provide health and retirement benefits. The degree to which these are accomplished may vary. Since the focus of this book is on the general craft and business of voiceover, I'll limit the discussion of performing unions to AFTRA and SAG. If you will be doing voiceover work outside of the United States, you should contact the performance union in the country where you will be working. Many performance unions have agreements or affiliations with unions in other countries, so your original union will be the best place to start.

In the U.S., the two major performing unions came into being in the early days of film, radio, and later, television. Unscrupulous producers were notorious for taking advantage of the actors and not paying performers a decent wage—some not even paying them at all. So, the unions were set up to make sure performers got paid and were treated fairly.

As the unions grew, it was decided that it was unfair for a person just working once or twice a year to have to join the union and pay dues every six months. The result was the Taft-Hartley Act, which resulted in major changes in U.S. labor-management relations. In regards to voiceover, this law gives you (the actor) an opportunity to work under the jurisdiction of the union for 30 consecutive days without having to join AFTRA or SAG. You then become "Taft-Hartley'd" or "vouchered" and must join the union if you do another union job. What this means is that if you do a lot of freelance work, you can still do a union job without having to join the union or pay union dues. The trick is that the next union job you do, you must join the union, whether it is three days or three years after your first union job. Immediately after the 30-day grace period you have the option to join or not join the union. At the time of this writing, with AFTRA, you can immediately join after your first union job. With SAG, you can join after working one job as a principal performer. As a background player, you must work three union jobs (three vouchers) before you can join. All membership details are explained at **www.aftra.com** and **www.sag.org**.

One of the advantages of being in the union is that you are more likely to be paid a higher fee, or scale, than if you did the same job as a freelancer—although, in some situations, you can actually negotiate a higher fee as a freelancer. Union *scale* is the fee set by the union for a specific type of work. By the time you reach the level of skill to have been hired for a union job, you will most likely be ready to join the union.

AFTRA is an *open union*. Anyone can join by simply paying the initiation fee and current dues. SAG works a little differently in that you must be hired for a union job in order to join the union, and after your third voucher, you must join the union when you are hired for a union job. It used to be that the only way to join SAG was to be hired for a SAG job. However, today you can join SAG if you are a paid-up member of AFTRA or another affiliated union for one year, and have worked at least one job as a principal performer during that time in that union's jurisdiction.

AFTRA and SAG cover different types of performing artists and do not duplicate the types of performances covered. Voiceover work for radio, television, and sound recordings are covered by AFTRA, while SAG covers performances that are released on film and multimedia. For example, if you were hired to work voiceover for a CD-ROM interactive program, you probably would be working a SAG job (although some interactive work is covered by AFTRA). A radio commercial or corporate video would be covered by AFTRA.

Sound confusing? Well, it can be, and there are some gray areas. But if you are a member, the union office will help sort out the details, and if you're not a union member yet, you don't need to worry about it. Although separate unions, AFTRA and SAG work closely together and even share office space in many cities.

Joining AFTRA and/or SAG requires payment in full of a one-time initiation fee and current dues. Dues are set at a base fee until a minimum income from union work is reached. Above the minimum, dues are a calculation of the base plus a percentage of union income. Visit the websites or call the AFTRA or SAG office in your area for current fees. New member information packets, which will answer most of your questions about the unions, can be purchased for a nominal fee. You can also ask the union what the current scale is for the type of work you are doing (commercials, industrial, etc.) and you can find current talent rates for most types of voiceover work online at **www.aftra.com**. The staff at the AFTRA and SAG offices are union members and will be happy to answer your questions.

One function of the unions is to protect your rights as a performer. A recording of your performance can be used for many different projects, and unless you are a union member, there is little you can do to protect yourself. A voiceover performance for a radio commercial can also be used in a TV spot or for an industrial video. There are some 400 different AFTRA and SAG agreements for different types of projects, each of which has a different pay scale. Radio and TV commercials are paid based on the market in which they air and how long they will be aired. Industrial videos and CD-ROMs are handled in other ways. Without the union you are potentially at the mercy of the person hiring you, and your voice may end up being used for projects you never agreed to.

AFTRA and SAG work under the principal of *Rule 1*, which simply means that union members agree not to work without a guild or union contract. A union member working in a nonunion production cannot be protected if the producer refuses to pay, pays late, makes unauthorized use of the performance, or in any other way takes advantage of the performer. Any legal action taken by a performer working outside of Rule 1 is at the performer's expense, and the union may actually discipline the member with fines, censure, suspension, or even expulsion.

As a member of AFTRA, you are free to audition for any job, including nonunion jobs. If you are hired for a nonunion job and the employer is not a signatory, the union may contact the producer and have him or her sign a signatory agreement before hiring you. If you are a union member, and are not sure about your employer's status with the union, call the union office in your area.

One way for a union member in the U.S. to work a nonunion job is a waiver called a *One Production Only* (O.P.O.) *Limited Letter of Adherence*. This waiver is good for one job only, and the work you do on that job is considered union work. The advantage is that the nonunion producer agrees to the terms of the union agreement, but does not have to become a union signatory. The O.P.O. contract must be signed before any session work.

There are producers who, for one reason or another, will not work with union performers. Money is usually not the reason. It may be unrealistic demands from an agent, company policy to work only with nonunion talent, or simply a dislike of the paperwork. To get around the paperwork and other issues, some agents and production companies will work as a union signatory effectively separating a nonunion producer from the union. This is a win-win situation—because the producer does not have to deal directly with the union, the quality of the talent remains high, and union performers have the opportunity to work for a greater variety of clients at a fair level of compensation. Some voiceover performers operate their own independent production companies as signatories and essentially hire themselves. It is also possible for you, as a union member, to handle the paperwork, thus making it more attractive for a producer to hire you.

FINANCIAL CORE

Financial core, or *fi-core* is an aspect of union membership in the U.S. that has been and remains very controversial. Fi-core is a level of union membership at which an actor can be a member of AFTRA or SAG and still be able to work nonunion jobs.

Since the beginning of labor unions, states would make their own laws about whether they would be a "union shop," or a "right-to-work" state. In a Union Shop state, laws were passed that required a person to be a union member and pay dues in order to do union work. Right to Work states allowed unions to exist, but membership was (and is) voluntary, and the union cannot require a person to pay anything as a condition of employment.[2]

Financial core came about as a result of union members who disagreed with the way their union was using a portion of their dues for political activities. They also disagreed with their union's control over work they could and could not accept. A series of U.S. Supreme Court legal battles beginning in 1963[3] eventually culminated in a 1988[4] landmark decision that changed the way all unions work (not just AFTRA and SAG). In 2001, President George Bush signed an executive order that requires all unions to inform prospective members of their "financial core rights," or "Beck Rights," before they join the union.

The resulting legal decisions for Financial Core require that an individual must first be a union member, and then formally request a change to Financial Core membership status. Upon declaration of Fi-Core status, the union member loses specific membership rights: the right to vote, hold union office, receive the union newsletter, declare their union status, and participate in union-sponsored events, among others. Payment of semi-annual dues is still required, however at a slightly reduced rate. The union determines the portion of dues that are spent for political and other activities that do not directly apply to the union's collective bargaining efforts, and for those at Fi-Core status, that percentage of dues is deducted from the dues payment.

At its essence, Financial Core creates a nonmember, dues-paying status that allows a performer to work both union and nonunion jobs. Those who favor Fi-Core will mention that the performer regains control over the kind of work they do and their compensation. Those against Fi-Core claim that this membership status seriously disables the effectiveness of collective bargaining. Ultimately, as a voiceover talent, it is up to you to fully research Fi-Core so you completely understand it's ramifications when the time comes for you to join AFTRA or SAG.

An Internet search for "financial core" will bring up dozens of websites that discuss both sides of this controversial aspect of union membership.

WHEN SHOULD I JOIN THE UNION?

It is generally a good idea to put off joining AFTRA or SAG until you have mastered the skills necessary to compete with seasoned union talent. Producers expect a high level of performance quality and versatility from union performers and it takes time and experience to master the skills necessary to perform at that level. Joining AFTRA too soon not only may be an unwise financial expense, but could have the potential for adversely affecting your voice-acting career. Most voice talent need the seasoning of working lots of nonunion jobs before they will be at a level of skill that can be considered competitive with union talent.

Here are some reasons to consider union membership when you feel you are ready, or when you begin getting audition calls for union work:

- Union membership is considered an indicator of professionalism and quality. Producers know they will get what they want in 2 or 3 takes instead of 20.
- Your performance is protected. Union signatories pay residual fees for use of your work beyond the originally contracted period of time.
- You will also be paid for any time over one hour on first and second auditions, and paid a fee for any additional callbacks.

WORKING FREELANCE

Nonunion, freelance work is an excellent way to get started in the business, and there are lots of advertisers and producers who use nonunion performers. As a nonunion performer, you negotiate your own fee. The fee will be a one-time-only *buyout* payment. There are no residuals for nonunion work, including work done at *financial core*. The going rate for freelance voice work can be anywhere from $50 to $250 or more depending on the project, the market, your skill level, and what you can negotiate. For nonunion work, or work booked without representation, the negotiated terms are between you and the producer, as are the terms of payment.

If a nonunion producer should ask your fee, and you are not sure what to say, the safest thing to do is to quote the current minimum union scale for the type of project you are being asked to do. You can always negotiate a lower fee. If you have an agent, the correct thing to do is to ask the client to contact your agent. A complete discussion of setting rates, negotiating fees, and getting paid is in Chapter 20, "How to Work in the Business of Voiceover."

Talent Agencies, Casting Agencies, Personal Managers, and Advertising Agencies

The jobs of talent agents, casting agents, and personal managers are often misunderstood by people not in the business or just starting out. They all have different functions in the world of voiceover.

THE TALENT AGENCY

Talent agents represent performers. Talent agencies are licensed by the state and must include the words "Talent Agent" or "Talent Agency" in any print advertising, along with their address and license number. The talent agent works with advertising agencies, producers, and casting directors to obtain work for the performers they represent.

A talent agent receives a commission of 10% to 25% based on the scale they negotiate for their performer and whether their performer is union or nonunion. For AFTRA/SAG work the commission is above and beyond the performer's fee (scale plus 10%). In some cases, the commission may be taken out of the talent fee, especially for nonunion freelance work obtained by an agent. For talent agencies to book union talent, they must be franchised by the local AFTRA and SAG unions. Contact the union office in your area for a list of franchised talent agents.

Unfortunately, this is not a perfect world, and there are many unscrupulous agents who will attempt to relieve you of your money. If anyone asks you for money up front to represent you or get you an audition, he or she is operating a scam. Period! The same is true for 1-900 numbers that charge a fee for information on auditions and casting. Most of the information is available elsewhere, either for free or a minimal charge. The best thing to do is find a reputable agent and stay in touch with him or her. Even if you are freelance and must pay your agent a 25% commission, the advantages of representation may well be worth it.

Do you need a talent agent to do voiceover work? No. Will a talent agent benefit you in your voiceover career? In most cases, yes. Chapter 20 includes a complete discussion on how to find and work with a talent agent.

THE CASTING AGENCY OR CASTING DIRECTOR

A casting agency is hired by an advertiser or production company to cast the talent for a particular project. They may also provide scriptwriting and some producing services, such as directing talent. They may even have a small studio where some of the production is done. Casting agent fees are normally charged directly to the client and are in addition to any fees paid for the talent they cast.

Most voice casting agencies work with talent agents and have a pool of talent that covers all the various character styles they use. Talent from this pool are used for all projects they work on and they will rarely add a new voice to their pool unless there is an opening or special need. The talent in their pool may be represented by several talent agents. Casting agencies may occasionally hold open auditions to cast for their projects but they are generally not a good resource for nonunion voice talent.

THE PERSONAL MANAGER

A personal manager is hired to manage a performer's career. The personal manager attempts to get the talent agent to send the performer out on auditions, and encourages the agent to go for a higher talent fee. Managers usually work on a commission of up to 20% of the performer's fee, which is taken out before payment to the performer and in addition to the agent's commission. Some managers may work on a retainer. Either way, a manager can be expensive, especially if you are not getting work.

Personal managers are fairly rare in the world of voiceover, and the voice actors they work for are generally well-established on-camera performers who will only occasionally do voiceover work.

HOW ADVERTISING AGENCIES WORK

Advertising agencies work for the companies doing the advertising, coordinating every aspect of an advertising or marketing campaign. They write the scripts, arrange for auditions, arrange for the production, supervise the sessions, handle distribution of completed spots (spot announcement, or commercial) to radio and TV stations, purchase air time, and pay all the fees involved in a project.

Ad agencies are reimbursed by their clients (advertisers) for production costs and talent fees. They book airtime at the station's posted rate and receive an agency discount (usually about 15%). They bill their client the station rate and get their commission from the station as a discount. If the advertising agency is an AFTRA or SAG signatory, they will also handle the union fees according to their signatory agreement. Since the ad agency books all airtime, they also handle residual payments, passing these fees on to their clients.

Most advertising agencies work through production companies that subcontract everything needed for the production of a project. Sometimes the production company is actually a radio or TV station that handles the production. In some cases a casting agent might be brought in to handle casting, writing, and production. Some larger ad agencies, with in-house facilities, may work directly with talent agents for casting performers.

Ad agencies can be a good source of work. Your agent should know which agencies use voiceover and will send out your demo accordingly. You can also contact ad agencies directly, especially if you are nonunion. As part of your marketing, you can telephone ad agencies and let them know who you are and what you do. You will find many ad agencies work only in print or use only union talent. When you call, ask to speak to the person who books voiceover talent.

The ad agency assigns an account executive (AE) or on-staff agency producer (AP) to handle the account. Sometimes both an AE and AP are involved, but it is usually the AP who knows more about the production than the AE. The AE is more involved with arranging the schedules for airtime purchases. The AP is the person who is generally in charge of selecting talent. The AE is less involved, but often approves the AP's talent choices.

Either the AE or AP may be present during auditions and one or both is almost always present at the session. If the ad agency is producing the spot, they will want to make sure everything goes as planned. If the spot is being produced by a casting agency, someone from that company may also be at the session. Casting agencies are more common for television on-camera productions than for voiceover, but a casting agency rep may be present at an audition or session if their agency is handling the production. And, of course, advertisers are very likely to be at the audition and session to provide their input.

HOW PRODUCTION COMPANIES WORK

As their name implies, production companies are where the work of creating the radio commercial, TV spot, industrial video, video game, or other production is done. They come in all shapes and sizes, from the one-man shop to the large studio with hundreds of people on staff. Most production companies have a small staff of 2 to 10 people, many of whom may be freelancers.

Production companies generally work directly for a client, or as a production resource for an ad agency or a corporation's on-staff producer. Many large corporations have their own in-house production facility.

Although some production companies can be a good source of freelance voiceover work, most work primarily with talent booked through a talent agent by the producer or with talent hired by an ad agency. Learn which production companies do the kind of voiceover work you want to do, and get to know the producers and directors. You can find production companies in your area by checking your phone book under "Recording Services—sound and video," through an Internet search for "production company your city," or by contacting your city's Chamber of Commerce. Many cities have a film bureau that maintains a list of local production companies.

Copyright 2010 - Voice-overload.com - JeffreyKafer.com

[1] Source: *http://www.AFTRA.com* and *http://www.SAG.org*

[2] Source: *http://www.nrtw.org/d/rtwempl.htm*

[3] U.S. Supreme Court, *NLRB v. General Motors*, 373 U.S. 734, 1963.

[4] U.S. Supreme Court, *Communications Workers vs. Beck*, 487 U.S. 735, 1988.

5

Using Your Instrument

As a voice actor, the tool of your trade—the instrument for your performance—is your voice. Just as any other craftsperson must know how to care for the tools of his or her trade, before you can begin to learn the craft of performing for voiceover, it is vital that you first learn how to properly use, and care for, the most important tool you have... your voice! So, with that in mind, this chapter includes some essential information about how your voice works, how to deal with common vocal problems, simple warm-up exercises, and tips for keeping your voice healthy. You'll also find some resources for further research if you feel that necessary. If you've never thought much about your voice, you'll probably find most of the exercises and tips helpful, some merely interesting, and a few perhaps totally weird.

Where Do You Sit in the Voiceover Orchestra?

All voice actors are not created equal! Sorry, but that's just the way it is. The world of voiceover is one of diverse talent, abilities, and sounds. Some people seem to master voiceover quickly and easily while others struggle for years to "break in." You must begin with some basic talent. You simply have to have it—talent cannot be taught. If you didn't have at least some level of talent, chances are you wouldn't be reading this book, so I'll assume that isn't an issue. Once you've discovered your basic talent, the next step is to build upon it and nurture it as you develop performing and business skills.

Learning the craft and business of voiceover is much like learning how to play a musical instrument. Some people are more adept at learning piano, while others choose to study flute, some will play string instruments, and still others have the ability to play a variety of instruments. Some dedicated musicians become virtuosos while others never advance their level of skill beyond the beginner stage. The simple truth is that some people simply have more talent for learning what it takes to play their chosen instrument. If you've ever taken lessons to learn how to play guitar, piano, violin, oboe, or some other musical instrument, you have a good idea of what to expect as you begin your study of voiceover. If you have the basic talent, and you're willing to dedicate yourself to mastering the necessary skills, there's a very good chance that you'll find your place in the voiceover orchestra.

In the context of an orchestra, each voice actor has a seat—and not everyone can be section leader. Each section of an orchestra consists of several musicians seated according to their skill level and expertise. A musical composition is broken down into several parts for each section. For example the first violin part may be played by several musicians and will usually carry the melody and be technically demanding. The second violin part, also played by several musicians, will be less demanding, but still critical to the overall composition. The third and successive parts are progressively less demanding, but all are essential parts of the whole. The violinist seated to the conductor's left is also known as the Concertmaster and is second in command after the orchestra conductor. This individual has earned their position through constant study and a demonstration of a high level of expertise with their instrument.

Do you know where you sit in the voiceover orchestra? Do you know which part you play? Do you know your level of expertise at playing your instrument? Do you even know what instrument you play?

If you're reading this book to learn what voiceover is all about and how to get started, your answers to the above questions are most likely all "no." And that's OK. By the time you finish this book you should be in a much better position to answer these questions with a resounding "yes."

At this point you should be aware that your instrument is your voice. But there's more to it! Just as every musician in an orchestra plays an instrument, every voice actor uses their voice. In an orchestra some instruments have a deep, resonant, low tone (string bass, cello, bassoon, and tuba), while other instruments have a high, clear tone (piccolo, flute, trumpet, and percussion bells). Other instruments have a raspy, edgy tone (violin, clarinet, and saxophone), or a percussive, harsh attack (piano, harpsichord, percussion). Even within sections of the orchestra there are a variety of instruments that are of the same basic design, yet exhibit a uniquely different tone:

- Brass: trombone, trumpet, French horn, tuba
- String: violin, viola, cello, upright bass, guitar, piano
- Wind: flute, piccolo, clarinet, oboe, bassoon, saxophone
- Percussion: xylophone, tympani, drums, cymbals, bells

Voice actors are no different than the instruments in an orchestra. As you study this craft you'll begin to discover things about your instrument. Do you have a smooth, mellow, clear tone like a trumpet? Do you have a voice that is high pitched, or as a friend of ours says: "baritone challenged," like a piccolo? Or perhaps you will discover that your voice is deep and resonant with limited range, like a bassoon.

Your determination of the tonality and texture of your voice is a very important discovery because it will ultimately guide you through your study of this craft. If you have a voice with deep "golden tones" you'll find it a challenge to perform a script written for a high pitched, fast-talking character voice. By the same token, if your vocal tone resides in the mid-range, you may find it difficult to work at either extreme without sounding artificial and unreal. All music uses the same written notes, just like all voiceover copy uses the same words. Although you might hear a tuba solo, you'll never hear a tuba trying to sound like a flute.

There is still one instrument I haven't mentioned yet, and you may discover that this is where you fit in the voiceover orchestra. That instrument is the digital MIDI keyboard. Press a button on this keyboard and you have a string section. Press a different button and you're playing a piano. Press yet another button and it's now a trumpet. The possibilities are endless.

Many voice actors specialize in mastering the skills for performing within the primary range and tone of their voice. They become the best violin, trumpet, or bassoon they can be, with an ability to convey the subtlest nuance through their performance.

Most voice actors who work in animation or video games fall in the digital keyboard category. Through their years of study, they have mastered the ability to create a wide range of very real and believable voices on demand.

What instrument do you play? Are you a highly proficient first violinist capable of playing complex melodies at ease? Or are you a third trombone, able to get all the notes right, but still learning how to master the nuance of your instrument? Maybe, just maybe, you're a digital keyboard with the ability to create radically diverse voices with different tonalities and textures, all of which sound completely authentic. The only way you'll know is to discover your unique talent, study performing techniques, and experiment to learn what works best for you.

The beauty of both music and voiceover is that the performance is not dependent on what's on the paper. The performance is the end result of how the performer plays their instrument.

All about Breathing

Your voice is a wind instrument. To do any voiceover work that reveals subtlety and nuance through a performance, it is essential that you know how to play your instrument properly. In other words, you need to know how to breathe. Proper breathing provides support for your voice and allows for emotional expression. It allows you to speak softly or with power, and to switch between the two styles instantly. Proper breathing is what makes possible the subtleties of communicating a broad range of information and emotion through the spoken word.

Breathing comes naturally, and it is something you should not be thinking about while performing. From the moment we are born, we are breathing. However, during our formative years, many of us were either taught to breathe incorrectly, or experienced something in our environment that left us with an improper breathing pattern. It may be that we learned to breathe from our chest, using only our lungs. Or perhaps, we adapted to our insecurities and created a mental block that inhibits our ability to breathe properly.

YOUR VOCAL PRESENTATION

Arthur Joseph, a voice specialist and creator of Vocal Awareness, describes vocal presentation as the way in which others hear and respond to you. The way you are perceived by others is directly related to your perception of yourself. If you perceive yourself to be outgoing, strong, forceful, and intelligent, your voice reflects these attitudes and perceptions with a certain loudness and assertiveness. By the same token, if you perceive yourself to be weak, helpless, and always making mistakes, your voice reflects your internal beliefs with qualities of softness and insecurity. How you breathe is an important factor in your individual vocal presentation because breath control is directly related to the loudness, tonality, and power behind your voice.

Your perception is your reality. So, if you want to change how you are perceived by others, you must first change how you perceive yourself—and that requires awareness. In most cases, a problem with vocal presentation is a habit directly related to a lack of vocal awareness—and habits can be changed. Changing a habit requires an extreme technique, discipline, conscious diligence, and constant awareness. A number of vocal presentation problems, and exercises for correcting them, are discussed later in this chapter.

Many of the exercises in this book will help you discover things about yourself and your voice, of which you might not have been aware. They will also help you improve or change your breathing technique and vocal presentation, and maintain the new qualities you acquire. The lessons you learn about your voice from this and other books will help give you awareness of your voice and will be of tremendous value as you proceed on your voice-acting journey. From this new awareness, you will be able to adapt and modify your vocal presentation to create believable, compelling characters.

Joni Wilson has written an excellent series of books for improving and maintaining the sound of your voice. The first book of the series, *The 3-Dimensional Voice* is the much-needed owner's manual for the human voice and introduces her ideas and techniques. You can learn more about Joni and her books by visiting her web site at **www.joniwilsonvoice.com**. On Track 2 of the CD, Joni describes 20 facts you should know about your voice (CD/2).

BREATH CONTROL FOR THE VOICE ACTOR

The first lesson you must learn before you can begin mastering the skills of voice acting is how to breathe properly. Take a moment to observe yourself breathing. Is your breathing rapid and shallow? Or do you inhale with long, slow, deep breaths? Observe how you breathe when you are under stress or in a hurry, and listen to your voice under these conditions. Does the pitch of your voice rise? When you are comfortable and relaxed, is the pitch of your voice lower and softer? Feel what your body is doing as you breathe. Do your shoulders rise when you take a deep breath? Does your chest expand? Do you feel tension in your shoulders, body, or face? Your observations will give you an idea of how you handle the physical process of breathing that we all take for granted.

Of course, the lungs are the organ we use for breathing, but in and of themselves, they cannot provide adequate support for the column of air that passes across your vocal cords. Your lungs are really nothing more than a container for air. It is the diaphragm, a muscle situated below the rib cage and lungs, that is the real source of support for proper breathing.

Allowing your diaphragm to expand when inhaling allows your lungs to expand more completely and fill with a larger quantity of air than if a breath is taken by simply expanding your chest. When you relax your mind and body, and allow a slow, deep, cleansing breath, your diaphragm expands automatically. Contracting your diaphragm, by pulling your lower abdominal muscles up and through your voice as you speak, gives a constant means of support for a column of air across your vocal cords. For a performer, correct breathing is from the diaphragm, not from the chest.

Good breath control begins with a relaxed body. Tense muscles in the neck, tongue, jaw and throat, usually caused by stress, constrict your vocal cords and cause the pitch of your voice to rise. Tension in other parts of your body also has an effect on the quality of your voice and your ability to perform. Relaxation exercises reduce tension throughout your body and have the additional benefit of improving your mental focus and acuity by providing increased oxygen to your brain. Later in this chapter, you'll find several exercises for relaxing your body and improving your breathing.

Good breath control and support can make the difference between a voice actor successfully transcending an especially unruly piece of copy or ending up exhausted on the studio floor. A voice actor must be able to deal with complex copy and sentences that seem to never end, and to make it all sound natural and comfortable. The only way to do it is with good breath control and support.

The following piece of copy must be read in a single breath in order to come in at :10, or "on-time." Even though the words will go by quickly, it should not sound rushed. It should sound effortless and comfortable, not strained or forced. It should be delivered in a conversational manner, as though you are speaking to a good friend. Allow a good supporting breath and read the following copy out loud (CD/3).

> Come in today for special savings on all patio furniture, lighting fixtures, door bells, and buzzers, including big discounts on hammers, saws, shovels, rakes, and power tools, plus super savings on everything you need to keep your garden green and beautiful.

How did you do? If you made it all the way through without running out of air, congratulations! If you had to take a breath, or ran out of air near the end, you need to increase your lung capacity and breath support. Long lists and wordy copy are commonplace and performing them requires a relaxed body, focus, concentration, and breath support. You need to start with a good breath that fills the lungs with fresh air.

Check your breathing technique by standing in front of a mirror. Place your fingers just below your rib cage, with thumbs toward the back and watch as you take a slow, deep breath. You should see and feel your stomach expand and your shoulders should not move. If your hands don't move and your shoulders rise, you are breathing from your chest.

As the diaphragm expands, it opens the body cavity, allowing the rib cage to open and the lungs to expand downward as they fill with air. If you breathe with your chest, you will only partially fill your lungs. It is not necessary for the shoulders to rise in order to obtain a good breath. In fact, rising shoulders is a sign of shallow breathing, indicating that the breath is getting caught in the chest or throat. Tension, fear, stress, and anxiety can all result in shallow breathing, causing the voice to appear weak and shaky and words to sound unnatural.

Breathing from your diaphragm gives you greater power behind your voice and can allow you to read longer before taking another breath. This is important when you have to read a lot of copy in a short period of time, or when the copy is written in long, complicated sentences.

Do the following exercise and then go back and read the copy again. You should find it easier to get through the entire piece in one breath.

- Begin by inhaling a very slow, deep, cleansing breath. Allow your diaphragm to expand and your lungs to completely fill with air. Now exhale completely, making sure not to let your breath get caught in your chest or throat. Rid your body of any remaining stale air by tightening your abdominal muscles after you have exhaled. You may be surprised at how much air is left in your lungs.

- Place your hands below your rib cage, lower your jaw, and allow two very slow preparatory breaths, exhaling completely after each one. Feel your diaphragm and rib cage expand as you breathe in and contract as you exhale. Your shoulders should not move. If they do, you are breathing from your chest for only a "shallow" breath.

- Allow a third deep breath and hold it for just a second or two before beginning to read. Holding your breath before starting gives stability to your performance by allowing you to lock your diaphragm so you can get a solid start with the first word of your copy.

A slow, deep, cleansing breath is a terrific way to relax and prepare for a voice-acting performance (see Exercise 1 on page 46). It will help center you and give you focus and balance. However, working from a script requires a somewhat different sort of breathing. You will need to find places in the copy where you can take a breath. For some scripts you may need to take a silent catch breath. At other times you might choose to vocalize a breath for dramatic impact, or take a completely silent breath so as not to not create an audible distraction.

If you breathe primarily from your chest, you will find that breathing from your diaphragm makes a difference in the sound of your voice. Your diaphragm is a muscle and, just as you tone other muscles in your body, you may need to tone your diaphragm.

Here's a quick exercise from Joni Wilson that will help you develop strong diaphragmatic breathing. You'll find other exercises in her book *The 3-Dimensional Voice*[1]:

- Put the fingers of both hands on the abdominal diaphragm and open the mouth in a yawn position. Inhale the air, then say as you exhale the air, "haaaaaaaaaaaaaa," manually pushing the diaphragm with your fingers in toward the spine for as long as air comes out of the mouth.

- When there is no more air, and what comes out begins to resemble a "death rattle," slowly relax the pushing and allow the diaphragm to drop back down and suck the air back into the lungs. You may experience some dizziness. Stop for a moment, and let it pass before you do the exercise again. You can do this throughout the day to strengthen the diaphragm.

BREATHE CONVERSATIONALLY

One of the secrets for proper breathing with a voiceover performance is to only take in enough air for what you need to say. We do this instinctively in normal conversation. If you only need to say a few words, there's no point in taking a deep breath. Inhaling too much air may result in a sudden and unnatural exhale at the end of your line.

Listen to how you and others speak in conversation. You'll notice that no one takes a deep breath before they speak. You'll also notice that no one waits until someone else finishes talking before they take a breath. In conversation, we breath in a natural and comfortable manner—even when others are speaking. When we speak, we only take in enough air for the words we say, and we breathe at natural breaks in our delivery without thinking about what we are doing. When you understand how to properly use your diaphragm to provide breath support you will eliminate the need for frequent deep breaths and rapid catch breaths.

You need to breathe, and you will sometimes be working a script with extremely long, complicated sentences. Breath points in most copy usually occur after a portion of a thought has been stated. Listings provide natural break points between each item. You probably won't want to breathe between each item, but there is usually an opportunity if you need it. To make lists more effective, try to make each item in a list unique by slightly altering your delivery or inflection.

Of course, when we are performing from a script, the words aren't ours, but as voice actors we must make the words sound conversational and believable as if they are ours. A common problem for many people just starting in voiceover is that they become too focused on the words or feel like they may not be able to "get through" the script, especially if the sentences are long. The result is that they tend to take a deep breath and read as much as they can until they begin to run out of air, somehow thinking that by reading the copy without breathing will help. Although some may be able to deliver the words with a reasonable interpretation for short bursts, most will sound rushed and detached from the meaning of the words. In this case, there is no acting or performance taking place and no connection with the intended audience. The voice talent is merely being themselves while struggling through a highly stressful situation.

The remedy for this common ailment is to realize that the words are just words on a piece of paper, and that our job is to simply speak those words in an appropriate manner. The stress of the moment is completely self-imposed and need not exist. Nowhere is it written that a voice actor must read, or "get through," a script without breathing. The truth is that breathing is an essential part of communicating the meaning of those words.

The challenge is in learning how to breath naturally while reading from a script, allowing the breath to happen at appropriate places, and truly telling the story. In order to find the natural breath points in a script, you need to understand the story, your character, and the myriad other details in the script. When you play the role of a character you create, the stress of working with a script can be completely eliminated because the character already knows how—and where—to breathe.

Be Easy on Yourself

My first recommendation as you begin studying the craft of voiceover is for you to record yourself reading copy every chance you get. I guarantee that you will likely not care for the way you sound, and what you hear as you listen to your recorded voice may surprise you—and for good reason. When you speak, you are not actually hearing your own voice in the same way others do. Much of what you hear is actually resonance of vibrations from your vocal cords traveling through your body and bones to your inner ear. When other people hear you, they don't get the advantage of that nice resonance. The way your voice sounds to other people is what you hear when your voice is played back from a recording.

I suggest you find a way to record your voice. In this age of digital audio, there are new devices coming out every day for recording audio. Some offer extremely high-quality recordings, while others are marginal. What you need, at least to start, is a way to record a reasonably high quality voice recording so you can play it back to study what you are doing. For the purpose of rehearsal and mastering your technique, you don't need to spend a lot of money on building a home studio. The time for that will come when you start marketing yourself as a professional voice talent. For now, an old tape recorder, handheld digital recorder, or some simple recording software and a microphone for your computer will do the job.

Practice reading out loud—the newspaper, magazine ads, pages from a novel—anything that tells a story. Record yourself reading a few short paragraphs with different styles of delivery and different emotions. Create different characters and read the copy with their attitudes. Allow the characters you create to discover how they breathe. Change the pitch of your voice—make your voice louder or softer—and vary the dynamics of pacing, rhythm, and emotion. Practice looking for, and giving value to, the key elements of the copy. Now, go back and read the same copy again—this time, read with an entirely different attitude, emotion, and character. All of these techniques are explained in detail later in this book.

One of the best ways to learn this craft is to listen to voiceover work at every opportunity. How do you compare? Adapt your style to imitate the delivery of someone you have heard on a national radio or TV commercial. Don't try to be that other performer, but rather imitate the techniques and adapt them to your style. If you are still looking for your style, the exercises in this chapter will help you find it.

Listen to your recordings to evaluate what you are doing, but don't be too hard on yourself. Don't be concerned about what your voice sounds like. Focus on what it feels like as you work on your reading. Listen to where you are breathing and if your delivery conveys an understanding of the story. Listen for your pace, rhythm, and overall believability. Be as objective as you can and make notes about the things you hear that you would like to correct. Practice the exercises and techniques in this book that apply. Recording and listening to yourself can be an enjoyable process and a great learning experience that helps give you an awareness of what you are doing with your voice. Remember, it's not about your voice, it's about what you can do with your voice—and it might take some time before you really discover what you can do, especially if you're doing it on your own.

One other tip: You might want to videotape yourself as you work on your vocal delivery. It may sound odd, but studying your physical movement will make a big difference in the way you sound.

Exercising Your Voice

Two things are essential when exercising your voice: (1) a deep breath with good breath control and (2) making a sound. Your vocal cords are muscles, and as with all other muscles in your body, proper exercise and maintenance will provide greater endurance and stronger performance. The vocal cord muscles are little more than flaps that vibrate as air passes over them. Sound is created by a conscious thought that tightens the vocal folds, enabling them to resonate as air passes by. Overexertion and stress can cause the vocal cords to tighten too much, resulting in hoarseness and an impaired speaking ability. A sore throat, cold, flu, or other illness can also injure these muscles. If injured, your vocal cords will heal more rapidly if they are allowed to stay relaxed. However, if you don't correct the source of the vocal injury, the problem will reoccur.

The manner in which we speak, breathe, and use our vocal and facial muscles, can often be traced to our childhood. Cultural and regional speech patterns influence the way we speak, as do family attitudes and speaking habits. From the time we first began to talk, we developed speaking habits and attitudes that remain with us today. We became comfortable with these habits because they worked for us as we learned to communicate with others. Some of these habits might include a regional accent, rapid speech, slurred speech, not thinking thoughts through before speaking, a lack of confidence in our ability to communicate, and poor breathing. These and many other speech habits can be corrected through exercise and technique.

Changing a habit will take approximately 21 days and at least 200 or more repetitions. For most people, it takes about seven days of repetition of a new behavior pattern before the subconscious mind begins to accept the change. It takes another 14 days, more or less, for a new habit pattern to become established in the mind. This time frame is true for changing just about any habit and will vary from person to person. As much as we might wish otherwise, achieving the desired results of a changed habit will take a concentrated effort and constant awareness.

Discover which of the exercises in this chapter are most helpful and do them on a regular basis, setting aside a specific time each day for your voice exercises. A daily workout is especially important if you are correcting breath control or a specific speaking habit.

Correcting Speech Problems and Habits

As you exercise your voice, awareness of what is happening physically is vital to improving your ability to experience yourself as you work on changing a habit. Observe what is happening with your voice, diaphragm, body, and facial muscles. Self-awareness helps you discover and correct problems with your speech. Without it, you will not be able to recognize the characteristics you need to work on. As you develop self-awareness skills, you will also be developing instincts for delivery and interpretation that will be of tremendous benefit during a performance.

It is often helpful to have another set of ears listening to you as you work on correcting a problem or speaking habit. A speech therapist, voice coach, or a local voiceover professional can be invaluable to improving your speaking voice. You can also get constructive criticism designed to improve your communication skills from acting classes and workshops.

There are many common speech problems that can be corrected by simple exercise and technique. However, all these problems have an underlying cause that requires self-awareness to correct them. In her book *Voice and the Actor* (1973),[2] Cicely Berry discusses the human voice and methods to improve a vocal performance in great detail. She also explains some of the following common speech problems and how to correct them.

UNCLEAR DICTION OR LACK OF SPEECH CLARITY

Usually, unclear diction or lack of speech clarity is the result of not carrying a thought through into words. A lack of focus on the part of the performer or an incomplete character development can affect diction. This problem can be heard in the voice as a lack of clarity or understanding, often communicated through inappropriate inflection or attitude.

To correct this, you'll need a clear understanding of each thought before you speak. Then, speak more slowly than what might feel comfortable to you. Speaking slowly forces you to focus on what you are saying.

Stuttering can be classified in this problem area. Although the actual cause of stuttering is still not known, research has shown that it may have different causes in different people and is generally a developmental disorder. Even though research has found three genes that appear to cause stuttering, there is no evidence that all stutterers have these genes or that stuttering is an inherited trait.

There are two traditional therapies to correct stuttering. The first is *stuttering modification therapy,* focusing on reducing fears and anxieties about talking. It can be done with a self-therapy book or with a speech pathologist. The second is *fluency shaping.* This therapy teaches the stutterer to talk all over again by beginning with extremely slow, fluent speech and gradually increasing the speaking rate until speech sounds normal. This therapy is normally done at a speech clinic.[3]

OVEREMPHASIS, EXPLOSIVE CONSONANTS, AND OVERENUNCIATION

The source of overemphasis or overenunciation usually derives from the actor's insecurity or lack of trust in his or her ability to communicate. As a result, the tendency is to push too hard to make sense and start to explain. The moment you begin to overemphasize, you lose the sense.

To correct this problem, don't worry about the listener understanding what you are saying. Stay focused on your thought and just tell the story. Don't explain it, just tell it. It may help to soften the tone of your voice, lower your volume, slow down, or simply to focus on talking to a single person. If you find yourself overemphasizing, you may be trying too hard to achieve good articulation.

Sibilance, the overemphasis of the "s" sound, is often caused by not differentiating between the "s," "sh," and "z" sounds. It can also be the result of a clenched or tight jaw, dental problems, loose dentures, or missing teeth. Minor sibilance problems can be corrected in the studio with a "de-esser," but serious problems can only be corrected with the help of a speech therapist or perhaps a good dentist.

LOSING, OR DROPPING, THE ENDS OF WORDS

A habit common to many people who are just starting in voiceover and acting is to simply not pronounce the ends of words. Words ending in "b," "d," "g," "p," "t," and "ing" are especially vulnerable.

One cause of this problem is simply not thinking through to the end of a thought. The brain is rushing from one thought to another without giving any thought an opportunity to be completed. This is usually due to a lack of trust in one's abilities, but can also be the result of a lack of focus or concentration. Another cause of this problem is a condition known as "lazy mouth," which is simply another way of saying poor articulation.

This problem can be corrected by forcing yourself to slow down—speaking each word clearly and concisely as you talk. Think each thought through completely before speaking, then speak slowly and clearly, making sure that the end of each word is spoken clearly. You may find this difficult at first, but stick with it and results will come. Awareness of this problem is critical to being able to correct it. Exercise #9, *The Cork*, on page 48 addresses this problem.

LACK OF MOBILITY IN THE FACE, JAW, AND LIPS

A person speaking with lack of mobility is one who speaks with only minimal movement of the mouth and face. This can be useful for certain types of characterizations, but is generally viewed as a performance problem. Lack of mobility can be due in part to insecurity or a reluctance to communicate; however, it can also be a habit.

To correct this problem, work on the facial stretching exercises described later. Practice reading out loud in front of a mirror. Watch your face as you speak and notice how much movement there is in your jaw, lips, forehead, and face. It may help to incorporate other body movement into your exercises. Body movement and gestures can help you discover the emotions associated with facial expressions, which will in turn, help you to be more expressive. Work on exaggerating facial expressions as you speak.

Raise your eyebrows, furrow your brow, put a smile on your face, or frown. Stretch your facial muscles. Go beyond what feels comfortable.

CLIPPED VOWELS

Many people think in a very logical sequence. Logical thinking can result in a speech pattern in which all parts of a word are treated equally. This often results in a monotone delivery with vowels being dropped or clipped. There is little emotion attached to the words being spoken even though an emotional concept may be the subject.

Vowels add character, emotion, and life to words. To correct the problem of monotony, search for the emotion in the words being spoken and commit to the feeling you get. Find the place in your body where you feel that emotion and speak from that center. Listen to your voice as you speak and strive to include emotional content and a variety of inflections in every sentence. For someone who is in the habit of speaking rapidly or in a monotone, this problem can be a challenge to overcome, but the rewards are well worth the effort. Once again, slowing down as you speak can help you overcome this problem.

BREATHINESS AND DEVOICING CONSONANTS

Breathiness is the result of exhaling too quickly while speaking, or exhaling before starting to speak. Improper breath control, resulting from nervousness or an anxiety to please, is the ultimate cause. Consonants and ends of words are often dropped, or unspoken, and breaths are taken at awkward or inappropriate places within a sentence.

To correct this problem, work on breathing from your diaphragm. Take a good breath before speaking and maintain a supporting column of air as you speak. Also, be careful not to rush, and think each thought through completely.

EXCESSIVE RESONANCE OR AN OVEREMOTIONAL QUALITY

This problem arises from an internal involvement with an emotion. It is usually the result of becoming more wrapped up in the emotion than understanding the reason for the emotion.

To correct this, you may need to learn how to look at things a bit more objectively. People who exhibit this problem are generally reactive and live life from an emotional center. For them life is drama. Work on looking at situations from a different angle. Try to be more objective and less reactive. When you feel yourself beginning to react, acknowledge the feeling and remind yourself to step back a bit from your emotional response.

ACCENT REDUCTION OR MINIMIZATION

Many people feel their natural accent or dialect is a problem when doing voiceover. This can certainly be true if you are unable to adapt your style of vocal delivery. In some cases, an accent or dialect can be used to your advantage to create a distinctive style for your performance, when you create a character, or when you are working in only a certain region. However, if you want to be well-received on a broad geographic level, you will need to develop the skill to modify your delivery style to one that is expected, and accepted, by the general population. In the United States, most people have come to expect a certain "sound" for a voiceover performance, commonly referred to as "nonaccented American English." But even though there may be a generally accepted "standard," different regions of a country may respond better when hearing a message in their regional accent. If you want to do voiceover, and have a foreign, or thick regional accent, you have two choices: 1) develop your acting skills to a high degree and create a niche for the sound of your voice, or 2) learn how to adapt your voice to create characters with an accent different from yours, and that includes the "expected" generic accent. This may require some time spent taking some training for accent modification or accent reduction.

Many famous actors have learned how to either use their accent to enhance their performance image, or have learned how to adapt their voice to create uniquely believable characters: Sean Connery, Mel Gibson, Patrick Stewart, Nicole Kidman, Meryl Streep, and Tracy Ullman to mention only a few. Mel Gibson has a thick native Australian accent, yet he can play a very believable American. Tracy Ullman has a native British accent, yet she creates dozens of characters from around the world. And Meryl Streep has developed a reputation for creating incredibly authentic and believable foreign accents, even though she is American.

When we first learn to speak, we imitate and mimic those around us as we develop our speaking skills. By the time we are two or three years old, the mannerisms and vocal styling that we adopt become the habit pattern for our speaking. Over the years, we become very comfortable with our speaking patterns to the point where it can be difficult to modify them.

Accent reduction, modification, or minimization is, in essence, a process of learning new habit patterns for speaking. For most adults, it is impossible to eliminate completely their native accent. However, reducing the accent or modifying the way words are formed is certainly possible. There are many good books and audio programs designed to help people speak with a more "natural" American, regional, or foreign accent. An Internet search for "accent reduction" will result in a wealth of resources.

The process of retraining your speaking habits can be lengthy, and may involve working with a dialect coach or speech pathologist. Contact your local University's speech department for recommendations of a licensed speech pathologist, or look into an English as a Second Language (ESL) program in your area. The time and energy required can be more than most people are willing to invest. But a basic level of accent reduction or modification can be achieved if you simply listen to someone with the desired accent, study the sound of their speech, mimic the sound of their words, and practice the speaking pattern until it feels comfortable. This is essentially how actors do it.

In the United States, most voiceover talent perform with the standard nonaccented American English. Regional inflections, dialects, and other tonalities are, for the most part, absent unless required for a character in the script, or unless the production is intended for a regional audience. Although this has become the generally accepted sound for American voiceover, it does not mean that someone who speaks with an accent or dialect cannot be successful. The most successful voice actors are those who are versatile with their speaking voice and who possess the ability to create a variety of believable characters. If you have an accent (foreign or domestic) there are several things you can do to make yourself more marketable as a voice actor:

1. Refine your accent and learn how to use it to your advantage. Although you may be able to create a unique performing style, you may find that you are limited in the types of projects you can do if you focus only on improving your native accent.

2. Learn how to adapt your speaking voice to mimic other accents for the purpose of creating believable characters. Learn to do this well and you can develop the ability to create any character on demand.

3. Work with a diction coach or study methods of modifying your speech patterns. All of these will require some time and effort on your part, but the results will be well worth it.

Voice and Body Exercises—CD/4

A variety of methods to use to care for your voice are covered later in this chapter. But first, let's begin with some ways to create a relaxed body and mind. That will be followed by a variety of exercises designed to tune your voice and exercise the muscles that comprise your vocal instrument. When doing breathing or relaxation exercises, it is important for you to breathe correctly. Most of us were never taught how to breathe as children—we just did it. As a result, many of us have developed poor breathing habits. See the *All about Breathing* section starting on page 33 for breathing techniques and exercises to help you become comfortable breathing from your diaphragm.

You will find it much easier to get into the flow of a script and concentrate on your performance if you are in a relaxed and alert state of mind. The exercises that follow will help you relax and serve to redirect your nervous energy to productive energy that you can use effectively as you perform. Breathe slowly and deeply, and take your time as you allow yourself to feel and experience the changes that take place within your body. Try to spend at least a few minutes a day with each of these exercises. It's best if you can do these in a quite place where you won't be disturbed.

EXERCISE 1: RELAX YOUR MIND

This exercise is a basic meditation technique best done while sitting in a quiet place. Begin by allowing a very slow, deep breath through your nose. Expand your diaphragm to bring in as much air as you can, then expand your chest to completely fill your lungs. Hold your breath for a few seconds, then slowly exhale through your mouth—breathe out all the air. As you do this, think calm thoughts, or simply repeat the word "relax" silently to yourself. Take your time. Do this about 10 times and you will find that your body becomes quite relaxed, and your mind will be much sharper and focused. You may even find yourself becoming slightly dizzy. This is normal and is a result of the increased oxygen going to your brain.

This exercise is an excellent way to convert nervous energy into productive energy. Do this in your car before an audition or session—but not while driving.

EXERCISE 2: RELAX YOUR BODY

Deep breathing to relax your mind will also help to relax your body. Even after some basic relaxation, you may still experience some tension in certain parts of your body. An excellent way to release tension is to combine breathing with stretching. There are several steps to this stretching exercise, so take it slow and if you feel any pain, stop immediately.

Stand with your feet about shoulder width. Close your eyes and breathe deeply from your diaphragm, inhaling and exhaling through your nose. Extend your arms over your head, stretching to reach the ceiling. Stretch all the way through the fingers. Now, slowly bend forward at the waist, lowering your arms as you stretch your back. Try to touch the floor if you can. If you need to bend you knees, go ahead. The idea here is to stretch the muscles in your arms, shoulders, back, and legs. When you feel a good stretch, begin to slowly straighten your body, allowing each vertebra to straighten one at a time as you go. Don't forget to keep breathing.

Now that you are once again standing, with your arms still over your head, slowly bend at the waist, leaning to the left, reaching for a distant object with both arms. You should feel a stretch along the right side of your body. Slowly straighten and repeat with a lean to the right, then straighten.

Next, lower your arms so they are directly in front of you. Rotate your body to the left, turning at the waist and keeping your feet pointing forward. Allow your hips to follow. Slowly bend at the waist as you stretch your arms out in front of you. Keep your head up and your back as straight as you can. Now, rotate forward and repeat the stretch as you reach in front of you. Finally, repeat to the other side before returning to an upright position.

EXERCISE 3: RELAX YOUR NECK

A relaxed neck helps keep the vocal cords and throat relaxed. Begin by relaxing your mind and body with the techniques described in Exercises 1 and 2. If you want to close your eyes for this one, feel free.

This exercise should be done very slowly and it can be done sitting or standing. If, at any time, you feel any pain in your neck, stop immediately. There may be a neck injury present that your doctor should know about. Begin by sitting or standing up straight. Slowly tilt your head forward until your chin is almost resting on your chest. Allow your head to fall forward, slightly stretching your neck muscles. Slowly rotate your head to the left until your left ear is over your left shoulder; then move your head back and to the right. Continue to breathe slowly as you move your head around until your chin returns to its starting point. Now rotate your head in the opposite direction. This exercise will help release tension in your neck and throat.

EXERCISE 4: RELAX YOUR ARMS

This exercise helps remind you to keep your body moving and converts locked-up nervous energy into productive energy you can use. When you are in a session, it often can be helpful to simply loosen up your body, especially if you have been standing in front of the mic for a long time. Remember that moving your body is a very important part of getting into the flow of the script. Loosen your arms and upper body by letting your arms hang loosely at your side and gently shake them out. This relaxation technique works quickly and can be done inconspicuously. You can also expand your shake out to include your entire upper body.

EXERCISE 5: RELAX YOUR FACE

A relaxed face allows you to be more flexible in creating a character and can help improve articulation. You can use your facial muscles to add sparkle and depth to your delivery. Your face is one of the best tools you have as a voice actor.

Begin by relaxing your body. Then, scrunch up your face as tight as you can and hold it that way for a count of 10. Relax and stretch your face by opening your eyes as wide as you can. Open your mouth wide and stretch your cheeks and lips by moving them while opening and closing your jaw. The process of stretching increases blood flow to your face and gives a feeling of invigoration.

EXERCISE 6: HORSE LIPS

Take a long deep breath and slowly release air through your lips to relax them. Let your lips "flutter" as your breath passes over them. This is a good exercise to do alone in your car on your way to a session. By forcing the air out of one side of your mouth or the other, you can also include your cheeks as part of this exercise. As with the face stretch, this exercise will help you in creating character voices and aid in improving articulation.

EXERCISE 7: RELAX YOUR TONGUE

This may sound odd, but your tongue can get tense too. A simple stretching exercise can relax your tongue, and also helps relax the muscles at the back of your mouth. You may want to do this exercise in private.

Begin by sticking out your tongue as far as you can, stretching it toward your chin. Hold for a count of five, then stretch toward your right cheek. Do the same toward your left cheek and finally up toward your nose.

Another tongue stretch that also helps open up the throat is to gently grasp your extended tongue with your fingers. You might want to use a tissue or towel to keep your fingers dry. Begin with a deep breath and gently stretch your tongue forward as you slowly exhale and vocalize a "HAAA" sound, much like the sigh you make when yawning. In fact, this exercise may very well make you feel like yawning. If so, good. Yawning helps open your throat.

EXERCISE 8: YAWNING

As you do these exercises, you may feel like yawning. If that happens, enjoy it. Yawning is a good thing. It stretches your throat, relaxing it and opening it up. More important, yawning helps you take in more air, increasing the flow of oxygen to your brain, improving your mental acuity. It also helps lower the pitch of your voice and improves resonance.

To increase the feeling of relaxation, vocalize your yawn with a low pitch "HAAA" sound, concentrating on opening the back of your throat. It is also important that you allow yourself to experience what happens to your body as you yawn.

EXERCISE 9: THE CORK EXERCISE—CD/5

You may find this exercise a little odd at first, but the results will most likely amaze you. Although a pencil is a suitable substitute, using a cork will give you quicker results simply because it forces you to work your muscles harder.

Get a wine bottle cork—save the wine for later, or have it first (your choice). Now, find a few good paragraphs in a book or newspaper. Before doing anything with the cork, begin by recording yourself reading the copy

out loud. Stop the recorder. Now place the cork in your mouth horizontally so that it is about one-quarter inch behind your front teeth—as though biting on a stubby cigar. If you use a pencil, place it lengthwise between your teeth so you are gently biting it in two places. Don't bite hard enough to break the pencil, and don't place the pencil too far back—it should be positioned near the front of your mouth. Now read the same paragraphs out loud several times. Speak very slowly and distinctly, emphasizing every vowel, consonant, and syllable of each word. Don't cheat and be careful not to drop the ends of words. In a very short time your jaw and tongue will begin to get tired.

After you have spent a few minutes exercising your mouth, remove the cork, turn the recorder back on, and read the copy one more time. Now, play back both recordings. You will notice a remarkable difference in the sound of your voice. The *after* version will be much clearer and easier to listen to.

The cork is an excellent warm-up exercise for any time you feel the need to work on your articulation or enunciation. You can even do this in your car, singing to the radio, or reading street signs aloud as you drive to an audition or session.

EXERCISE 10: THE SWEEP

Vocal range is important for achieving emotional attitudes and dynamics in your performance. By vocal range, I am referring to the range from your lowest note to your highest note. Start this exercise by taking a deep breath, holding it in, and releasing slowly with a vocalized yawn. This will help to relax you. Now fill your lungs with another deep breath and release it slowly, this time making the lowest note you can with a "HAAAA" sound. Gradually increase the pitch of your voice, sweeping from low to high. It may help to start by holding your hands near your stomach and gradually raise your hands as you raise the pitch of your voice.

You will quite likely find one or two spots where your voice breaks or "cracks." This is normal and simply reveals those parts of your voice range that are not often used. Over time, as you practice this exercise, your vocal range will improve and as your vocal cords strengthen, the "voice cracking" will become less or may even go away entirely. This is also a good breathing exercise to help you with breath control. If your recordings reveal that you take breaths in midsentence or that the volume (overall loudness) of your voice fluctuates, this exercise will help. Practicing this regularly will improve your lung capacity and speaking power, as well as vocal range.

EXERCISE 11: ENUNCIATION EXERCISES

The following phrases are from a small but excellent book titled *Broadcast Voice Exercises* by Jon Beaupré (1994).[4]

To improve diction and enunciation, repeat the phrases that follow. Do this exercise slowly and deliberately making sure that each consonant and vowel is spoken clearly and distinctly, stretching your lips and cheeks as you read. Don't cheat on the ends of words. Watch yourself in a mirror, listen to yourself carefully, and be aware of what you are feeling physically and emotionally. Remember that consistent repetition is necessary to achieve any lasting change. For an extra challenge, try these with the cork.

Specific Letter Sounds—do each four times, then reverse for four more. Make a clear distinction between the sounds of each letter.

Gudda-Budda (Budda-Gudda)
   [Emphasize the "B" and "G" sounds.]
Peachy-Weachy (Weachy-Peachy)
   [Emphasize the "P" and "W" sounds.]
Peachy-Neachy (Neachy-Peachy)
   [Emphasize the "P" and "N" sounds.]
Peachy-Leachy (Leachy-Peachy)
   [Emphasize the "P" and "L" sounds.]
Fea-Sma (Sma-Fea) [pronounce as FEH-SMA]
   [Emphasize the difference between the "EH" and "AH" sounds.]
Lip-Sips (Sip-Lips)
   [Make the "P" sound clear and don't drop the "S" after lips or sips.]
TTT-DDD (Tee Tee Tee, Dee Dee Dee)
   [Emphasize the difference between the "T" sound and the "D" sound.]
PPP-BBB (Puh Puh Puh, Buh Buh Buh)
   [The "PUH" sound should be more breathy and have less vocalizing than the "BUH" sound.]
KKK-GGG (Kuh Kuh Kuh, Guh Guh Guh)
   [Emphasize the difference between the "K" and "G." Notice where the sounds originate in your mouth and throat.]

Short Phrases—make sure every syllable is spoken clearly and that the ends of words are crisp and clear.

Flippantly simpering statistics, the specifically Spartan
strategic spatial statistics of incalculable value

   [This one works on "SP" and "ST" combinations. Make sure each letter is clear.]

She stood on the steps
Of Burgess's Fish Sauce Shop
Inexplicably mimicking him hiccuping
And amicably welcoming him in.

[Make each word clear—"Fish Sauce Shop" should be three distinctly different words and should not be run together. Once you've mastered this, try speeding up your pace.]

TONGUE TWISTERS—CD/4

Tongue twisters are a great way to loosen up the muscles in your face and mouth. Go for proper enunciation first, making sure all letters are heard and each word is clear. Begin slowly at first, then pick up speed. Don't cheat on the end of words. For an extra challenge, practice these using your cork. With repeated practice, they will be a bit easier to do.

I slit a sheet; a sheet I slit, upon the slitted sheet I sit.

A proper cup of coffee in a copper coffee pot.

A big black bug bit a big black bear, and the big black bear bled blood.

The sixth sick sheik's sixth sheep's sick.

Better buy the bigger rubber baby buggy bumpers.

Licorice Swiss wrist watch.

Tom told Ted today to take two tablets tomorrow.

The bloke's back brake block broke.

Most Dr. Seuss books can provide additional tongue twisters, and can be lots of fun to read out loud in a variety of styles. Some excellent tongue twisters can be found in *Fox in Sox* and *Oh, Say Can You Say* (1979). Another good book of tongue twisters is *You Said a Mouthful* by Roger Karshner (1993). Most retail and online booksellers can help you find a variety of other tongue twister books.

In 1984, while at a dinner party with people from 12 countries representing more than 15 languages, Michael Reck, of Germany, began collecting tongue twisters. Since then, he has compiled the largest collection of tongue twisters to be found anywhere—"The 1st International Collection of Tongue Twisters" at **www.uebersetzung.at/twister/en.htm** and at **www.voiceacting.com** (linked in the Free Stuff menu area). You'll find more than 2,000 tongue twisters in 87 different languages. If you think the English tongue twisters are challenging, try some of the other languages (assuming, of course, you can read them!).

Tips for Maintaining Your Voice and Improving Your Performance

Keeping your voice in good condition is vital to maintaining peak performing abilities. Some of the following tips may seem obvious, and you may already be aware of others. Some of the tips here were taken from the private files of some top professional voice actors. None of them is intended to be a recommendation or endorsement of any product, and as with any remedy, if you are unsure please consult your doctor.

TIP 1: SEEK GOOD TRAINING

A good performer never stops learning. Continued training in acting, improvisation, voiceover, singing, and even classes in marketing and business management can be helpful. Learn the skills you need to become the best performer and business person you can be. Study other voiceover artists. Watch, listen, and learn from television and radio commercials. Observe the trends. Practice what you learn to become an expert on the techniques. Rehearse regularly to polish your performing skills. Take more classes. Learn everything you can about your home studio equipment so you can provide the best possible recordings of your work. Master marketing techniques, develop strong negotiating skills, and learn how to run your business. You can get a lot of this information from books, but the best way to learn will be to study one-on-one with professionals who can teach you the skills you need to know.

TIP 2: NO COFFEE, SOFT DRINKS, SMOKING,
ALCOHOL, OR DRUGS

Coffee contains ingredients that tend to impair voice performance. Although the heat from the coffee might feel good, the caffeine can cause constriction of your sinuses or throat. Coffee is also a diuretic. The same is true for some soft drinks. Soft drinks also contain sugar that can cause your mouth to dry out.

Smoking is a sure-fire way to dry out your mouth quickly. Smoking over a long period of time will have the effect of lowering your voice by damaging your vocal cords, and presents potentially serious health risks.

Alcohol and drugs both can have a serious effect on your performance. You cannot present yourself as a professional if you are under their influence. Using alcohol and drugs can have a serious negative influence on your career as a voice actor. Word can spread quickly among talent agents, studios, and producers affecting your future bookings. I have seen sessions cancelled because the talent arrived at the studio "under the influence."

TIP 3: KEEP WATER NEARBY

Cold liquids can constrict your throat, so it's a good idea to keep a bottle of room temperature water nearby when doing voice work. Water is great for keeping the mouth moist and keeping you hydrated.

As your mouth dries out, tiny saliva bubbles begin to form, and as you speak, the bubbles are popping. Well-known voice coach Bettye Pierce

Zoller recommends keeping a bottle of water handy—the type with a squirt top. When *dry mouth* is noticed, squirt all areas of the mouth wetting the cheeks, teeth, and tongue—even underneath it. Then, do not swallow right away, but instead swish for about five seconds or more. The idea is to get all mouth tissues wet. Swishing water in your mouth will help reduce dry mouth temporarily, but only hydration will correct the cause.

Here are some interesting statistics about water, hydration, and the human body:

- It is estimated that up to 50% of the world population is chronically dehydrated.
- It is estimated that in 37% of Americans, the thirst mechanism is so weak that it is often mistaken for hunger.
- Mild dehydration can slow down the human metabolism up to 3%.
- One glass of water shuts down midnight hunger pangs for almost 100% of dieters studied in a University of Washington study.
- Lack of water is the #1 trigger of daytime fatigue.
- Research indicates that drinking 10 glasses of water a day could significantly ease back and joint pain for up to 80% of sufferers.
- A drop of 2% in body water can trigger fuzzy short-term memory, trouble with basic math, and difficulty focusing on a computer screen or a printed page.
- Drinking five or more glasses of water daily may decrease the risk of colon cancer by 45%, plus it may slash the risk of breast cancer by 79% and reduce the likelihood of bladder cancer by up to 50%.
- It takes about 45 minutes for a drink of water to achieve proper hydration. You may want to start drinking water well before you leave for a session.

TIP 4: DEALING WITH MOUTH NOISE (a.k.a. DRY MOUTH)

Every voice actor dreads the inevitable *dry mouth*. There are many causes of mouth noise including stress, illness, smoking, antihistamines, decongestants, pain relievers, and other medications. But it is most often simply the result of saliva bubbles popping because the mouth is not hydrated. The next several tips in this section can help minimize mouth noise or reduce the possibility of it happening—but eventually it will.

Here are a few of the many solutions voice actors have come up with to deal with the symptoms of dry mouth: distilled water with Emergen-C (one packet per quart); no dairy for two days prior to a VO session; or a swish of carbonated water (flavored or not).

Allowing a throat lozenge or cough drop to slowly dissolve in your mouth can help keep your throat and mouth moist. However, most lozenges are like hard candy and contain sugar that can actually dry your mouth.

Exceptions to this are Ricola Pearls natural mountain herbal sugar-free throat lozenges and breath mints (**www.ricola.com**), Fisherman's Friend (**www.fishermansfriend.com),** and Grether's Redcurrant or Blackcurrant Pastilles (**www.grethers.com**). Some throat sprays such as Entertainer's Secret (**www.entertainers-secret.com**), Singer's Saving Grace (**www.herbsetc.com**) and Thayer's Dry Mouth Spray (**www.thayers.com**), can help keep your throat lubricated. The best time to use a lozenge is about 30 minutes before a session. Lubricating sprays can be used at any time.

An alternative to dissolving a lozenge in your mouth is to drop one or two lozenges into a bottle of water. The lozenge-treated water will not only give you the benefits of the lozenge, but will also help keep you hydrated.

There are also a few over-the-counter remedies that work nicely to control dry mouth. One of the most effective is Oasis Moisturizing Mouthwash (with Tri-Hydra® Technology) (**www.oasisdrymouth.com**). Oasis was developed as a therapy for hospital patients who suffered from dry mouth as a result of medication. But it is so effective at temporarily hydrating the mouth that it has become very popular as a general mouth rinse and is also available as a mouth spray. Oasis is PH balanced, contains no alcohol, and can give a feeling of hydration for up to two hours.

TIP 5: SWISH VIRGIN OLIVE OIL

Swish a small amount of virgin olive oil to reduce or kill mouth noise and clicks. About a capful will do nicely. Work the olive oil into every corner of your mouth. The olive oil has a mild taste and leaves a coating on the inside of the mouth that holds moisture in. This clever trick came from one of our students who is also an opera singer—and it really does work!

TIP 6: EAT GREASY POTATO CHIPS

I warned you that some of these tips might sound a bit weird. Well this insider secret and the next are two that fit that category. During a session a singer asked if I had any potato chips handy. This, of course, raised my curiosity. She then explained that a trick singers will use is to eat greasy food, like potato chips, before a session to lubricate their mouth and throat. Odd as it may sound, it does seem to work.

TIP 7: HAVE SOME JUICE

Some juices can be helpful in keeping your mouth moist and your throat clear. Any of the Ocean Spray brand juices do a good job of cleansing your mouth. A slice of lemon in a glass of water can also help. Grapefruit juice, without pulp, can help strip away mucus and cleanse the mouth. Any juice you use to help clear your mouth and throat should be a clear juice that contains no pulp. Be careful of fruit juices that leave your throat "cloudy" or that leave a residue in your mouth. Orange juice, grape juice, carrot juice, and others can be a problem for many people.

TIP 8: THE GREEN APPLE THEORY

This is a good trick for helping reduce "dry mouth." Taking a bite of a Granny Smith or Pippin green apple tends to help cut through mucous buildup in the mouth and clear the throat. Lip smacks and mouth noise are the nemesis of the voice actor, and a green apple can help with this problem. This only works with green apples. Red apples may taste good, but they don't produce the same effect.

TIP 9: AVOID DAIRY PRODUCTS

Dairy products, such as milk and cheese, can cause the sinuses to congest. Milk will also coat the inside of the mouth, affecting your ability to speak clearly. Stay away from milk and cheese products when you know you are going to be doing voiceover work.

TIP 10: CLEARING YOUR THROAT

When you need to clear your throat, do it gently with a mild cough rather than a hard, raspy throat clearing, which can actually hurt your vocal cords. Try humming from your throat, gradually progressing into a cough. The vibration from humming often helps break up phlegm in your throat. Always be sure to vocalize and put air across your vocal cords whenever you cough. Building up saliva in your mouth and swallowing before a mild cough is also beneficial. Be careful of loud yelling or screaming and even speaking in a harsh, throaty whisper. These can also hurt your vocal cords.

TIP 11: AVOID EATING BEFORE A SESSION

Eating a full meal before a session can leave you feeling sluggish and may leave your mouth in a less-than-ideal condition for performing. If you do need to eat, have something light and rinse your mouth with water before performing. Avoid foods that you know will cause digestive problems or that might result in your saliva glands working overtime.

TIP 12: BE AWARE OF YOURSELF AND YOUR ENVIRONMENT

Get plenty of rest and stay in good physical condition. If you are on medication (especially antihistamines), be sure to increase your intake of fluids. If you suspect any problems with your voice, see your doctor immediately. Be aware of dust, smoke, fumes, pollen, and anything in your environment that may affect your voice. You can also have reactions to food that will affect your voice. If you have allergies, you need to know how they might affect your performance, and what you can do about them. An Internet search for "allergies" will reveal resources with lots of information you can use.

TIP 13: AVOID ANYTHING THAT CAN DRY OUT YOUR THROAT

Air conditioning can be very drying for your throat. Be careful not to let cold, dry air be drawn directly over your vocal cords. Smoke and dust can also dry out your throat.

TIP 14: DON'T COVER UP THROAT PAIN

Covering up throat pain will not improve your performance and may result in serious damage to your vocal cords. If you feel you cannot perform effectively, the proper thing to do would be to advise your agent or client as soon as possible so that alternative plans can be made. The worst thing you can do is to go to a session when you are ill. If you must attend a session when your voice is not in top form, be careful not to overexert or do anything that might injure your vocal cords.

TIP 15: KEEP YOUR SINUSES CLEAR

Clogged or stuffy sinuses can seriously affect your performance. The resulting denasal sound (no, or limited, air moving through your nose) may be appropriate if it is consistent with a character, or if it is part of a style that becomes something identified with you. Usually, however, stuffy sinuses are a problem.

Many performers use a decongestant to clear their sinuses. Nasal sprays tend to work more quickly than tablets or capsules. Be careful when using medications to clear your sinuses. Although they will do the job, they can also dry your mouth and can have other side effects. Even over-the-counter decongestants are drugs and should be used in moderation.

When used over a period of time, the body can build up an immunity to the active ingredient in decongestants, making it necessary to use more to achieve the desired results. Once the medication is stopped, sinus congestion can return and may actually be worse than before. Some decongestants can make you drowsy, which can create other problems.

An alternative to decongestants is a saline nasal rinse, technically known as Buffered Hypertonic Saline Nasal Irrigation. That's a technical phrase that simply refers to rinsing the nasal passage with a mixture of warm saline solution. This is a proven method for treating sinus problems, colds, allergies, post-nasal drip, and for counteracting the effects of environmental pollution.

There are a variety of ways to administer the nasal wash, including a syringe, bulb, and water pik. However, one of the easiest to use, and most effective, is a Neti™ Pot. This is a small pot with a spout on one end. Although the nasal wash can be done using only a saline solution, some studies have shown that the addition of baking soda (bicarbonate) helps move mucus out of the nose faster and helps the nose membrane work better. An Internet search for *nasal rinse* will bring up numerous resources and recipes. You'll find an assortment of Neti Pots at **www.netipot.com**.

TIP 16: IF YOU HAVE A COLD

You know what a cold can do to your voice! If you feel a cold coming on, you should do whatever you can to minimize its effects. Different precautions work for different people. For some, Alka Seltzer changes the blood chemistry and helps to minimize the effects of a cold. For others, decongestants and nasal sprays at the first signs of a cold help ease its onset. Lozenges and cough drops can ease cold symptoms or a sore throat, but be aware that covering up the soreness may give you a false sense of security and your vocal cords may actually be more easily injured in this condition.

The common cold is a viral infection characterized by inflammation of the mucous membranes lining the upper respiratory passages. Coughing, sneezing, headache, and a general feeling of "being drained" are often symptoms of the common cold. In theory, there are more than 200 strains of rhinovirus that can enter the nasal cavity through the nose, mouth, or eyes. Once in the nasal cavity, the virus replicates and attacks the body. Most cold remedies rely on treating the symptoms of a cold to help you "feel better" while your body's immune system attempts to repair the damage.

Zicam® is a homeopathic cold remedy that has been shown in clinical studies to reduce the duration and severity of the common cold. According to the manufacturer, Gel Tech, LLC (**www.zicam.com**), Zicam's active ingredients are Zincum Aceticum and Zincum Gluconicum. I'm not quite sure what they are, but I do know it works for me, and many people I've recommended Zicam® to.

Other OTC remedies that claim to reduce a cold's severity and duration include Airborne® (**www.airbornehealth.com**), developed by second-grade teacher Victoria Knight-McDowell; Cold-Eeze®, manufactured by Quigley Pharma, Inc. (**www.coldeeze.com**); and Emergen-C®, manufactured by Alacer Corp. (**www.emergenc.com**).

Health food and online specialty stores are a good source for herbal remedies. Many voice actors recommend special teas from online stores like **www.traditionalmedicinals.com** (Throat Coat products and Breathe Easy Tea), **www.yogiproducts.com** (Breathe Deep Tea), and Chinese cold remedies from **www.yinchiao.com**, drinking at least 1/2 gallon of white grapefruit juice a day, Echinacea, and others. As with OTC remedies, some herbal remedies may work better for some people than others.

Many people swear by Grapefruit Seed Extract (GSE) as a means of boosting the immune system to either head-off or minimize the effects of a cold and other ailments. GSE is available at health food stores.

TIP 17: SOME REMEDY RECIPIES

There are literally dozens of herbal remedies that reportedly reduce the symptoms of a cold. If you have a favorite remedy recipe you find especially effective, I'd love to know about it. You can send it to me at **bookmail@voiceacting.com**. If not, you might give some of these a try.

One cold and sore throat remedy that seems to do the job for many people is this rather tasty recipe: 1 can of regular Dr. Pepper (not diet), 1 fresh lemon, 1 cinnamon stick. Pour Dr. Pepper into a mug and add 1 slice (circle) of lemon. Heat in the microwave to your preferred temperature. Remove and add one cinnamon stick. Relax and sip slowly.

For the more adventurous, here's a recipe for Cold Killer Tea given to us by one of our workshop students. To one cup of tea (Green Tea is an excellent choice) add 1 tsp. lemon juice, 1 tsp. honey, 1 tsp. apple cider vinegar, and a dash of cayenne pepper. The key ingredients are the vinegar and the cayenne pepper. Ingredients can be adjusted for taste.

Many performers find that they can temporarily offset the effects of a cold by drinking hot tea with honey and lemon. The heat soothes the throat and helps loosen things up. Honey is a natural sweetener and does not tend to dry the mouth as sugar does. Lemon juice cuts through the mucus, thus helping clear the throat. The only problem with this is that tea contains caffeine, which may constrict or dry the throat.

Bill Smith of The Acting Studio in Denver, CO recommends this mixture of Tabasco sauce and water: Mix 8 drops of Tabasco sauce into an 8 ounce glass of luke-warm water. Stir. Gargle and spit. Gargle and spit. Then drink and swallow regular water. According to Bill, at this solution level, you may taste the Tabasco sauce but you won't "experience" its hotness. You'll notice that most of the phlegm has been cleared from the back of the soft palate and all the way down past the vocal cords to the esophagus. One of the key benefits of this mixture is that the key ingredient in Tabasco sauce is Capsaicin. Although this ingredient is what gives Tabasco sauce its hotness, Capsaicin is also used in pain neuropathy to reduce inflammation of damaged tissues and nerve endings. As a result of gargling this mixture, the tendency to cough or clear the throat is reduced.

You may not be able to prevent a cold, but if you can find a way to minimize its affects, you will be able to perform better when you do have a cold. If you have a cold and need to perform, it will be up to you to decide if you are fit for the job.

TIP 18: LARYNGITIS

There can be many causes of laryngitis, but the end result is that you temporarily lose your voice. This may be the result of a cold or flu infection that has moved into the throat and settled in your larynx, or another cause.

When this happens to a voice actor, it usually means a few days out of work. The best thing to do with laryngitis is nothing. That is, *don't* talk and get lots of sleep! Your vocal cords have become inflamed and need to heal. They will heal faster if they are not used. Also, the remedy of drinking hot tea with honey and lemon juice will often make you feel better.

A classic remedy is a mix of hot water, Collins mix, and fine bar sugar. This is similar to hot tea, lemon juice, and honey with the benefit of no caffeine. The idea is to create a hot lemonade that can be sipped slowly. Many performers claim that this mixture has helped to restore their voice.

Another remedy that is said to be effective is to create a mixture of honey, ground garlic cloves, and fresh lemon juice. This doesn't taste very good, but many have reported a quicker recovery from laryngitis after taking this remedy. Garlic is known to strengthen the immune system, which may be a factor in its effectiveness.

Similar to hot tea with honey and lemon is a remedy popular in the eastern United States. This was given to me by one of my voice students and seems to work quite well. Boil some water and pour the boiling water into a coffee cup. Add 1 teaspoon of honey and 1 teaspoon of apple cider vinegar. The mixture tastes like lemon tea, but with the benefit of having no caffeine. Slowly sip the drink allowing it to warm and soothe your throat.

TIP 19: ILLNESS

The best thing you can do if you have a cold, laryngitis, or just feel ill is to rest and take care of yourself. If you become ill, you should let your agent, or whoever cast you, know immediately and try to reschedule. Talent agents and producers are generally very understanding in cases of illness. However, there are times when you must perform to the best of your abilities, even when ill. These can be difficult sessions, and the sound of your voice may not be up to your usual standards. In situations such as this, be careful not to force yourself to the point of causing pain or undue stress on your voice. Use your good judgment to decide if you are capable of performing. You may cause permanent damage to your vocal cords.

TIP 20: DEALING WITH GERD

Most people will experience **G**astroesophageal **R**eflux **D**isease at one time or another. For some it is an ongoing condition that must be dealt with on a daily basis. It is a condition in which stomach acid backs up from the stomach into the swallowing tube or esophagus. For a voiceover talent, this condition can present a serious problem. It's not that GERD will directly affect the sound of the voice, but the physical discomfort of the condition, and some of it's symptoms, can get in the way of an effective performance.

Many of the tips in this section will have a direct effect on GERD, and there are several over-the-counter and prescription remedies that address the problem. Basically, all digestive processes will produce stomach acid which can result in GERD. Some recommendations for dealing with the condition are: 1) avoid acidic foods, 2) avoid eating *anything* within two hours of bed, and 3) avoid alcohol for at least two hours before bed. Any digestive disorder lasting more than a few days should be checked by consulting a physician.

TIP 21: BE PREPARED

Sooner or later you may find yourself at a session where you are recording in a very strange environment, or the studio may be out of pencils or not have a pencil sharpener, the water may be turned off, or any number of other situations might occur. It's a good plan to arrive prepared.

Enter the Voiceover Survival Kit! You can purchase a small bag or pouch to hold the essential items and keep it with you whenever you go to an audition or session. You'll find the complete list of recommended items for your "survival kit" at the end of Chapter 24, "Auditions."

TIP 22: PRACTICE CREATING VISUAL MENTAL PICTURES

Visual images will help you express different emotional attitudes through your voice. Close your eyes and visualize the scene taking place in the copy or visualize what your character might look like. Lock the image in your imagination and use it as a tool to help feel and experience whatever it is that you need to express in the copy. Visualization will also help create a sense of believability as you read your lines. Don't worry if you can't visualize in "pictures." However you use your imagination is how you visualize: colors, sounds, or images. Use whatever works for you.

TIP 23: HAVE FUN

Voiceover work is like getting paid to play. Whether you're working from your home studio or at a recording studio, your auditions and sessions will go more smoothly when you are relaxed, prepared, and ready to perform. Choose to not worry about mistakes you might make. Use these as opportunities to learn more about your craft and to hone your skills.

---

[1] Wilson, J. (2000). *The 3-Dimensional Voice*. San Diego: Blue Loon Press.

[2] Berry, C. (1973). *Voice and the Actor*. New York: Macmillan.

[3] Kehoe, TD. (1997). *Stuttering: Science, Therapy and Practice*. Boulder: Casa Futura Technologies.

[4] Beaupré, J. (1994). *Broadcast Voice Exercises*. Los Angeles: Broadcast Voice Books.

6

The Seven Core Elements of an Effective Performance
(AKA: The A-B-C's of Voice Acting—CD/6)

Acting is an art! As with any art form, acting has some very specific processes, techniques, and component parts—or elements—that must be understood and properly applied in order to achieve the desired result of creating a sense of believability. If any element is overlooked or omitted from a performance, the audience will sense that something is missing.

During the course of teaching The Art of Voice Acting workshops, we've boiled down the essence of acting for voiceover to Seven Core Elements that we refer to as *The A-B-C's of Voice Acting*. Traditional acting classes for stage, film, and television teach many of these concepts, but not quite the way you'll learn here.

Take a look at the title of this chapter again. I'll wait.

You'll notice that the title includes the words *effective performance*. Voiceover work is not about what most people think of as "acting" or performing. To be effective, an actor must create a sense of drama. Miriam-Webster defines *drama* as "a situation or series of events in which there is an interesting or intense conflict of forces." In other words, drama is what gives a performance the appearance of believable reality. Any actor can deliver words from a script, but to be effective, an actor must be believable. All drama contains elements of conflict, humor, mystery, emotion, and feelings. Drama also creates tension, suspense, and anticipation for what will happen next.

As you learn how to apply the concepts in this chapter, you will be able to create compelling, believable, and real characters in the mind of your audience. That's what an effective performance is all about.

So now the question you should be asking is: "How do I create drama?" The answer is simple in concept but complex in execution. The answer is: "You must make choices and you must commit to those choices." The Seven Core Elements of a performance are all about making choices.

It Starts with Pretending

A voiceover performer is an actor—period. It doesn't matter what the copy or script is for. It doesn't matter if the copy is well written or poorly written. It doesn't matter if you are delivering the copy alone or with others. You are an actor when you stand in front of the microphone.

It is truly a rare individual who is born with natural acting ability. For most people, acting skills take time to learn and master. Acting is not difficult; it's just that as we've grown, we've simply forgotten how to play. As a child, you were acting whenever you pretended to be someone you were not. Pretending is where it starts. But there's a lot more to it than that.

Voiceover performing—or, more accurately, voice acting—is an opportunity to bring out your inner child. Regardless of the copy you are reading, there will always be some sort of character in the words. To be believable, that character must be brought to life. To do that effectively, you must start by becoming a master of pretending.

By definition, the word *pretend* means "to give a false appearance of being." So, if you are strictly pretending, you are not being real, but the objective of all acting is to create the illusion of reality. Learning how to pretend believably (or act) allows you overcome this apparent contradiction so you can step outside of yourself, using what you know as you move down the path of creating that illusion of reality.

The major problem most people have in performing for voiceover is in creating a believable illusion of reality while reading from a script. Reading is a left-brain, linear process, while performing is a right-brain, non-linear, process. The tendency is to "read" the words, rather than allowing the words to become real by pretending to be the person speaking the words.

This is where the Seven Core Elements of an Effective Performance come in. By applying these seven elements, you will be able to take your acting from simple pretending to creating a completely believable reality.

If you remember nothing else from this book, the following concepts will take you further in voice acting, or any other performing craft, than just about anything else. You can also apply these ideas in any area of your personal or professional life to achieve a high level of communication skill.

These techniques do not have to be done in sequence. In fact, most of the time one element will help define another. As you work on your performance, begin by making choices in whichever element seems to be a good place to start, but be sure that you include them all.

A = AUDIENCE: Core Element #1

Who are you (or, more accurately, who is your character) talking to? Decide on who will be hearing the message—the ideal person who needs to hear what you have to say. Different styles of delivery are appropriate for different audiences. In most cases, the copy will give you a good idea of who the ideal audience is. It may be helpful to ask the producer who he or she is trying to reach, or you may need to make a choice based on your gut instincts. By knowing your audience, you will be able to figure out the most appropriate and effective way to speak to them.

The most important thing to remember about your audience is that no matter what the script or project may be, you are *always* talking to *only one person*. Attempting to *shotgun* your performance, by trying to connect with many people at once, will generally result in the listening audience losing interest and becoming uneasy with you as a performer. There is a very subtle difference between focusing attention on an individual versus focusing on a mass of people. You've no doubt experienced seminars where the speaker just doesn't seem to reach the audience, and yet there are others where everyone is hanging on the speaker's every word. In the first instance, the speaker is most likely "shotgunning" their message in an attempt to reach everyone in the audience. In the second, the speaker is getting eye contact with individuals in the audience—one at a time, and has a crystal clear idea of the ideal person who needs to hear the message. When you focus your attention on one person, and speak with honesty and sincerity, everyone listening will feel drawn in, as though you are speaking only to them. This is an incredibly powerful technique that many voice talent simply don't understand or apply.

For the following line of copy, make some choices as to who the one, ideal person who needs to hear the message might be:

Some people think they're a mistake! But most people
think they're delicious! OK... so they've got a big seed and
they're green... Avocado's are still my favorite fruit.
Great in salads... or all by themselves. Get some today.

Here are some possible choices:

- A shopper in a grocery story also looking at avocados
- Someone who has never seen an avocado before
- A grocery clerk who is carefully stacking avocados
- A customer in a restaurant ordering a meal with avocados

The choice you make for your audience will help determine your tone of voice, your attitude, and the overall approach to your performance. Focus your attention on speaking to just one person as though you were having a conversation with them. Describe the person you are speaking to in as much detail as possible and give him or her a name. Use a photograph to get the feeling of having eye contact with a real person. Doing this may help make your delivery more conversational and believable. The RISC AmeriScan process discussed in Chapter 10, "The Character in the Copy," will give you additional tools you can use to define your one-person audience.

Here's a tip when choosing your one-person audience: Don't choose to be speaking to someone you know. The reason for this is that when you select someone you know as your audience, the speaker of the words becomes *you* and it will be considerably more challenging to create a believable performance when reading from a script intended to be spoken by a character who is not you.

It's entirely possible that the original choice you make for your audience may not be the best choice and you will need to change it. There may be many reasons for this, but regardless of how it happens, you will need to make an *adjustment* and make a new choice.

B = BACK STORY: Core Element #2

In voiceover, a *back story* is the specific event that takes place immediately before the first word of copy. It is what the character in the script is responding to. The back story is the reason why your character is saying the words in the script. If the back story is not clearly defined in the script—your job, as an actor, is to make one up! This is a very important aspect of performing from a written script because the back story sets your character's motivation, attitude, and purpose for speaking.

Acting coaches will often refer to a back story as "the moment before." Technically, a back story consists of the character's entire life experience that has brought them to the moment in time for the story in the script. For voiceover work, that's too much information, and we don't have the time to deal with a long, involved story leading up to the first word of the script. So, I suggest that you define a back story in specific terms that can be described in a single sentence. It must be something very immediate and powerful that has caused your character to speak. It can't be a vague description of a scene—it must elicit a specific response.

In some scripts, the back story is pretty obvious. In others, you'll have to make up something. Either way, the back story is essential to the development of your character. By understanding what brought your character to this moment, you will know how your character should respond. This, in turn will make it much easier for you to sustain your character and effectively communicate your character's feelings, attitudes, and emotions as he or she interacts with the audience and other characters.

For the following line of copy, make some choices as to the specific event that occurred, or words spoken immediately before this statement, and to which this statement is in response:

Some people think they're a mistake! But most people think they're delicious! OK... so they've got a big seed and they're green... Avocado's are still my favorite fruit. Great in salads... or all by themselves. Get some today.

To discover the back story, look for clues in the script that reveal specific details about what is taking place. Use these clues to create your own idea of what took place *before* the story in the script. This is the essence of your back story, and this is what brought your character to this moment in time.

Here are some possible choices for a back story for the first line:

- The person you are speaking to has asked you what this big green thing is with all the bumps. You respond with the first line of the script.
- The person you are speaking to has mentioned that they absolutely love avocados. You respond with a silent lead-in "I love them too, but…" followed by a short pause, then the first line of the script.
- The person you are speaking to is ordering a meal and is uncertain about whether or not to add avocado by saying "… would you recommend avocado?" You respond with the first line of the script.

Any given script may have several opportunities for a back story—possibly for every line. For each of those back story opportunities a very short phrase or one-word lead-in may help to *bridge* lines of copy to help add reality to the delivery. When you bridge lines of a script in this way, what you are actually doing is adding a thought process to your performance, which is a direct reflection of how our minds really work when we are having a conversation. For example, a bridging back story for the line "OK… so they've got a big seed and they're green…" might be this: The person you are speaking to says "I've heard avocados have a huge seed!" Obviously, we don't have time to actually verbalize or deal with these bridge lines in real time. Simply writing the word or phrase on your script or holding it in your imagination will usually be enough to trigger the thought process and thus create a sense of reality.

One way to use a back story to your advantage is to create a *lead-in line*, or *pre-sentence*. This is simply a verbalization of the back story to assist you in creating a believable response. For example, if you are speaking to someone who has never seen an avocado before you might create a lead-in line like: "So… you've never seen an avocado before? Well…" and then begin the script. A lead-in line is not intended to be spoken out loud, but, rather, should be said silently to set up the intonation and attitude of the words that will be spoken.

Each of these choices will have a different affect on your approach to the performance, including intonation, rate of delivery, attitude, dynamics, and underlying meaning (or subtext). As with the other choices you'll be making, one of these may be more suitable than the others. The only way you'll know which choice works best is to test them. When you make a choice, commit to it until you determine (or are given direction) that the choice needs to be adjusted or changed.

C = CHARACTER: Core Element #3

Who are you as the speaker of those words on the paper? Define your character in as much detail as you like. How does your character dress? What does the character's voice sound like? Does the character speak with an accent, dialect, or have any speaking quirks? Does the character exhibit any sort of attitude or personality quirks? How does the character move? How does the character think? What is the character's lifestyle? How does the character interact with other characters in the story, known or unknown? In what ways does the character respond to events that take place during the telling of the story? How does the character feel about the product, service, or subject of the script? The more details you can come up with, the more believable your character will be to you and to your audience. Every script has a character, regardless of how poorly the script may be written or what the content of the script may be. Your job is to find that character and give it life.

Just as in life, scripted characters have feelings and experience emotions about the stories they tell. And, just as in life, characters respond, evolve, and express emotions during the course of their stories. Learn how to reveal those emotions and feelings through your voice and you will create believable characters. Chapter 10, "The Character in the Copy," will explain many ways for you to do this, and you will find additional tools for creating and documenting characters in Chapter 13, "Character Copy."

For the following line of copy, make some choices that will clearly define and describe the person speaking:

Some people think they're a mistake! But most people
think they're delicious! OK... so they've got a big seed and
they're green... Avocado's are still my favorite fruit.
Great in salads... or all by themselves. Get some today.

Here are some possibilities for the character speaking these words:

- A grocery clerk stocking the shelves
- A shopper (talking to another shopper)
- An avocado grower or farmer
- A waiter or waitress in a restaurant
- A person speaking to a friend about fruits and vegetables

As with the other choices you make, your choice and definition of the character you are playing will have an impact on every aspect of your performance. Your other choices may affect your choice of character, and, of course, your choice of the character may require that you adjust some or all of your other choices.

D = DESIRES: Core Element #4

All characters have wants and needs! Theatrical actors will refer to this aspect of character development as the character's *objectives* or *intentions*. *Desires*, *objectives*, and *intentions* all refer to what your character ultimately wants as a result of his or her words and actions. Use whichever term works best for you, but for the purpose of this alphabetical mnemonic, "D" for *desires* works best. A-B-C-O, just doesn't seem right!

The character always wants something very specific from speaking the words. It may be simply to enlighten the listener with a valuable piece of information, it may be to entertain, or it may be to instruct the listener in the fine points of operating a complex piece of machinery. Whatever it may be, your character wants, needs, and desires to accomplish something from speaking those words. If that desire is not clearly explained in the context of the script—use whatever information is available to make it up.

Here's a quick test: What does the character in the following script want and need (desire) as a result of speaking these words? Come up with some choices of your own before reading further. No fair cheating.

> Some people think they're a mistake! But most people
> think they're delicious! OK… so they've got a big seed and
> they're green… Avocado's are still my favorite fruit.
> Great in salads… or all by themselves. Get some today.

Here are some possibilities for the character's desires and the words that might be clues to the ultimate desire:

- Establish curiosity (Some people think they're a mistake!)
- Tease to create interest (…they're delicious…)
- Add a touch of humor (…so they've got a big seed…)
- Intrigue the listener (…they're green…)
- Provide important information (they're a fruit and good in salad)
- Create urgency (Get some today.)

As you can see, there are many possibilities. There is really no single, correct way to interpret or deliver any piece of copy. As an actor, you need to make a choice as to what might be the most appropriate message that your character wants to communicate. And there may be more than one. As with your other choices, the only way you will know what works best is to test them when rehearsing the copy.

There are no wrong choices. But there are choices that may be more effective than others in terms of communicating the message. One key to choosing your character's desires is to consider the interaction between your character and the one-person audience. Also, be wary of choosing a desire of "selling." People love to buy, but they hate to be "sold." Choosing a desire of helping by providing important information that allows the listener to make an educated decision to buy will almost always be best.

E = ENERGY: Core Element #5

Voice acting comes from your entire body.
If only your mouth is moving, that's all anyone will hear.[1]
Cory Burton

There are three levels of energy in every performance: psychological energy, physical energy, and emotional energy. All three must be present. Leave one of these out and your character will lack a sense of truth and honesty.

PSYCHOLOGICAL ENERGY

Think back to a time when you said one thing, but what you really meant was something else entirely—and the person you were speaking to somehow knew exactly what you meant. We've all done this at one time or another. In fact, this is the basis of all sarcasm. The thought you hold in your head can directly affect the way the words come out of your mouth.

Try this: Say the phrase "That's a really nice hat." You most likely just spoke the words without any objective, intention, or desire, so it probably sounded pretty flat and uninteresting. Now hold the thought in your head that the hat you're looking at is the most incredible hat you've ever seen, and on the person you're talking to, it looks amazing! You want them to know how excited and happy you are that they have found a "look" that works for them. Say the phrase again and notice how different it sounds.

"That's a really nice hat."

Now, change the thought in your imagination to be that you are very jealous to see the other person wearing a hat that is exactly like your favorite hat. Your desire is to outwardly compliment them on their hat, but on the inside you really don't think it looks very good (even if it does). You're not happy, and you want them to know it without really saying it.

"That's a really nice hat."

The words are exactly the same in both situations, but the thoughts you held in your mind were different. The result is that the perceived meaning of the words is different.

In theater, the term *subtext* is used to refer to the underlying personality, and unspoken thoughts of a character that define the character's behavior and reveal what they really believe. *Psychological energy* is simply another way to understand *subtext*.

Psychological energy is a powerful concept when applied to voice acting. In voiceover, the sound of our voice is all we have to communicate the message in a script, and we need to use every tool available to create a believable reality. Applying psychological energy to a performance allows you to emulate the thought process of your character, which in turn allows the words to sound honest, real, and authentic.

The trick to using psychological energy properly is to keep the true belief just under the surface and to not reveal it during the performance, except through subtle intonation and behavior. By keeping the true belief hidden behind the words, it allows other characters to respond more appropriately, and it keeps the audience curious. This is especially important if the true meaning is in direct opposition to the textual meaning.

Although the concept of psychological energy may sound relatively simple, putting it to work as part of a voiceover performance may take some practice. Once you've mastered this aspect of a performance you'll be well on your way to creating consistently believable characters and highly effective performances.

PHYSICAL ENERGY

Physicalize the moment… and your voice will follow.
Bob Bergen (based on teachings of Daws Butler)

I think it's pretty safe to say that when you are in conversation with someone, you are not standing or sitting perfectly still, without moving. OK, maybe some of you reading this don't move, but most people use much more than just their mouth when talking. Facial expressions, body language, physical movement, and gestures are all part of the way we communicate when speaking to others. I'll bet you move your body even when you're talking on the phone.

Have you ever noticed that your physical movements are a big part of the way you speak? You use *physical energy* to give power to the thoughts and emotions that lay just under the surface of the words you speak.

Physical energy is absolutely essential in any voiceover performance. When you move your body with appropriate energy to support the emotions and thoughts of the words you speak, the result can be a totally believable performance.

A mistake many beginning voice actors make is that they will stand perfectly still and stiff-as-a-board when they are in front of a microphone. Their hands will hang at their sides and their faces will show no expression. Their performance will be flat and uninteresting, with often an almost monotone delivery. Once they start moving, everything changes. Words come to life, we can hear how the character feels, and we are actually drawn in to the drama of the story.

Unfortunately, for some, the idea of putting physical movement to words while reading from a script is much like walking and chewing gum at the same time—it can be a challenge to learn how to do it. Fortunately, it is an easily acquired skill. Usually, lack of movement is the result of nervousness or comes from a feeling of discomfort from being in an unfamiliar environment (often called "mic fright"). But the simple truth in voice acting is that you <u>must</u> move. It is one element of a performance that is essential to creating compelling and believable characters.

EMOTIONAL ENERGY

> *Life will give you what you need.*
> *Situations are your tools.*[2]
> Christina Fasano

Understanding how your character feels about an event, situation, thing, product, or person is an aspect of *subtext*. Your character's emotional energy is different from psychological energy in that psychological energy deals with the thoughts behind the words, whereas *emotional energy* is the expression of the feelings and emotions that underscore the thoughts. The two go hand in hand.

Using the hat example, consider how your character feels emotionally about the discovery that someone else has the exact same hat they have—and that they look great wearing it. Your character might feel devastated, frustrated, angry, happy, proud, or even excited. A full range of emotions is possible, but the most appropriate emotion will be determined by looking at the overall context of the story—understanding the big picture. Based on your choices as to how your character behaves and speaks within the context of the whole story, you will better understand the how and why of the character's feelings and emotional responses.

Keep in mind, that as actors, our job is to create a sense of reality, so any expression of emotion that is *over-the-top* might destroy any chance of believability. The best way to use emotional energy is to keep the emotions just under the surface. Start by allowing yourself to remember how you felt in a similar situation, and then base your performance from that feeling. By using a personal experience the emotional response will have truth and honesty, which will support the thoughts held in your imagination, which will result in an authentic and believable performance.

The essence of how the three levels of energy affect your performance can be summed up as:

- Change your thoughts—it will change the way you move
- Change your physical movement—it will change the way you feel
- Change your emotions—it will change the way you sound

For the following line of copy, make some choices as to how your character might think (psychological energy), how he/she might move and where tension is held in the body while talking (physical energy), and how he/she feels about the subject (emotional energy).

Some people think they're a mistake! But most people think they're delicious! OK... so they've got a big seed and they're green... Avocado's are still my favorite fruit. Great in salads... or all by themselves. Get some today.

F = FORGET WHO YOU ARE AND FOCUS:
Core Element #6

Acting is all about listening and forgetting who you are.[4]
Shirley MacLaine

A key principal of acting is to "get out of your own way" so the character or role you are playing can emerge and appear real to your audience. It sounds simple on the surface, but this idea may be confusing to some people. After all, isn't it an actor's job to figure out how a particular role should be played? Doesn't the actor need to be present during a performance? Aren't there a whole bunch of techniques that an actor can use to make a role believable? And doesn't all this mean that an actor needs to put a lot of thought into their performance?

Although all of these things are true to some degree, they are all just part of the process of creating a performance. They are not the performance. The reality of all acting is that the role you are playing is *not* you. The secret to excellent acting is to do everything that needs to be done to understand the story, character, relationships, responses, moods, attitudes, dynamics, and energy; apply the appropriate acting techniques to give meaning to the story, breathe life into the character, and "take the words off the page"; then put all of that behind you as the real you steps aside to let the character come to life. And to do all of this invisibly without giving the appearance that you are "acting." If there is any part of the real you that is apparent in a performance, it is you "doing" the character—not the character being authentic. You're thinking too much about what you need to do, or you're giving too much importance to the techniques you are using. In other words, when you put too much effort into the process of creating a performance, your acting becomes apparent and the performance will suffer, often by sounding as though you are "reading" the script.

One of the most difficult things for any actor to learn is how to forget who they are so the character can become real. The reason this is often a difficult task is because, as human beings, we have an ego that can cause us to second guess ourselves or stand in the way of what we know needs to be done. We can be a master of performing techniques and still be in our own way on an unconscious level. Often the only way we know it's happening is when our director asks us to make an adjustment in our performance.

Learning how to get out of our own way is, for most of us, an acquired skill that can take many years to master—or it can be achieved in an instant. This is one of the reasons acting is a craft and not a skill. A skill is a specific talent or ability, while a craft is the application of multiple skills to achieve a specific end result. Mastering any skill or craft takes time, patience, and dedication.

Listen to your director, listen to your instincts, listen to the unspoken words to which your character is responding, listen to the other actors in the studio, listen to everything. It is only through listening that you will be able to *focus* on doing what needs to be done to create the reality of the moment. When you are fully focused, you will discover that you no longer need to think about what you are doing. The characters you create will almost magically come to life. The second you allow yourself to drift off focus, or start to think about what you are doing, you will fall out of character.

If you don't fully grasp the idea of *forget who you are and focus*, don't be concerned. Many very successful actors and performers don't fully understand this concept and may never experience what it is like to truly forget who they are and get out of their way. For most actors, the experience is erratic at best, happening only occasionally. Achieving this state of performance on a consistent basis usually comes only with consistent work and study. The best I can say is that when you achieve this state of performance, you'll know it! It will feel as though you are outside of yourself observing your performance. Sort of like an "out-of-body" experience, except that you have complete control. This is the state of performance we strive for.

G = GAMBLE: Core Element #7

Be willing to gamble. Be willing to take a chance.

You must be willing to risk. Every performance requires that the performer be willing to step outside of their comfort zone to do or be something that most people would feel uncomfortable doing or being. It could be as simple as making an announcement at a party, standing on stage in front of an audience of thousands, or standing all by yourself in front of a microphone in a voiceover booth.

All performing is about risk. You risk the chance of not being liked, you risk making inappropriate choices, you risk the chance of not being believable, you risk the chance of not being hired again, you risk many things on many levels.

All performing is about taking a chance on an uncertain outcome. You may never know exactly what the producer or director is looking for in your performance, if your performance truly meets their needs, or how your performance will ultimately be used. Even though you may not know, you must be willing to take a chance, based on experience and observation, that what you do will be best bet for a successful outcome.

All performing is a *gamble*. You are gambling that the choices you make for creating your character and delivering your lines will bring the character and the story to life.

Building your business as a voice actor requires a willingness to risk. You'll risk your money as you build your home studio, invest in your marketing, produce your demo and study to master your craft. You'll risk rejection when you audition, when you call prospective clients, and when you think you've come up with exactly what a script needs—and the producer doesn't think so.

If you are not willing to take a risk, performing as a voice actor is probably not something you should pursue any further. Just stop reading right now and give this book to someone who is willing to take the risk of doing something they have never done before. A simple truth of this business is that, with relatively few exceptions, you cannot achieve any level of success if you insist on being only you as you read a script.

Voiceover is a craft based on creating compelling characters in interesting relationships. The only way you can create a character that is not you is to be willing to *gamble* that you can do what needs to be done for a believable performance.

It's completely natural to be reluctant, or fearful, of taking a risk, but there is a huge difference between taking an educated risk versus one that is a random shot in the dark. Gambling on your performance is *not* about winning or losing. It *is* about using the tools of your trade, your experience, your training, and your many performing and business skills to create more certainty for an otherwise uncertain outcome. In other words, you can stack the deck to improve the odds for a masterful performance and successful voiceover career each time you stand in front of a microphone.

This book is about giving you the basic knowledge that will enable you to take an educated risk if voiceover work is something you truly want to pursue.

[1] Burton, C. (2003). *Scenes for Actors and Voices by Daws Butler,* Bear Manor Media.
[2] Fasano, C. (1999). Lyrics from the song "Welcome to the Workshop," *Spiritually Wet,* published by FWG Music.
[3] Bergen, B. Warner Bros. voice of Porky Pig and other characters—www.bobbergen.com.
[4] MacLaine, S. in an interview by James Lipton, *Inside the Actor's Studio,* Bravo Television Network.

The A-B-C's of Voice Acting
the complete alphabet

A	**Audience**	Authentic in attitude	Articulate (cork exercise)
B	**Back story**	Be real	Believe in yourself
C	**Character**	Commit to choices	Critical thinking
D	**Desires**	Different approach	Dynamics for variety
E	**Energy**	Emotion	Environment
F	**Forget who you are**	**Focus**	Feelings
G	**Gamble**	Gestures	Go for it!
H	Honesty in character	How does your character… ?	Have alternatives ready
I	Intentions	Improvise	Imagination
J	Juxtapose (change words to find emotion)	Jargonize (when appropriate)	Journey (explore options)
K	Key words & phrases	Keep it real	Kid (let yours come out)
L	Listen carefully	Less is more	Lose yourself
M	Mouth work	M.O.V.E.	Moment before
N	No guessing	Never touch the mic	Nuance
O	Objectives	Out of the black	Off the page
P	Pitch	Pitch Characteristics	Physicalize
Q	Quality (always do your best)	Question everything	Quickly find your character
R	Rhythm	Respond	Relax
S	Sense memory (use your past experience)	Script analysis (woodshed)	Suspension of disbelief
T	Tempo	Think out of the box	Teamwork
U	Understand the whole story	Underplay	Use tools & techniques
V	Voice act (not "voiceover")	Visualize the scene	Vision (the big picture)
W	Warm-up	Water (to stay hydrated)	Woodshed copy (the 6 Ws)
X	X-periment	X-plore	X-citement
Y	Yawn to open throat	Yourself (don't be)	Yell (only if appropriate to character)
Z	Zicam (homeopathic cold remedy)	Zeppo (a famous Marx brother)	Z end of Z list

7

Developing Style & Technique

Think of *technique* as the tools of your trade, and there are always new techniques to study and learn. The application or use of any technique is something that becomes very personal over a period of time as the process of the technique evolves into a *style* that is uniquely yours.

A voiceover technique is really nothing more than a skill you develop or a process that you use that allows you to become a better performer. Sure, you can do voiceover without mastering any skills, or you may already have an innate ability with many of them. However, having an understanding of basic acting and voiceover techniques gives you the knowledge necessary to work efficiently under the pressure of a recording session—and to make your performance more real and believable.

As a voice actor, your job is to give life to the words in a script. The writer had a sound in mind when writing the script and you must find that sound by making the words compelling, interesting, real and believable. Technique is the foundation for your performance. It is the structure on which your character, attitude, and delivery are built. Technique must be completely unconscious. The moment you begin thinking about technique, the illusion is broken and the moment is lost.

As you study and learn the techniques in this book, you will find yourself at first thinking a lot about what you are doing. Be careful not to get too analytical as you work with these concepts. Just know that these techniques work. They are much like a frog... you can dissect the thing, but it will die in the process. As you gain experience and become more comfortable, your technique will become automatic, and you will be able to adapt quickly to changes without having to think. Acting techniques are much like riding a bicycle. Once you've mastered the process, it becomes automatic.

Voice exercises can help you develop and perfect your acting techniques. Chapter 5, "Using Your Instrument," includes many exercises, tips, and suggestions for improving your voice and developing your skills.

Style

It is interesting to note that using the voice is the only art form in which an individual style may be developed out of an inability to do something. It may be an inability to form certain sounds, or it may be a cultural affectation (an accent or dialect) that results in a quality uniquely your own. This is especially true with singers, but the same idea can apply to voice artists.

One person's vocal style might emphasize lower frequencies, creating an image of strength and power. Someone else may not be able to reach those low tones, and his or her style might be based on a somewhat warped sense of humor expressed through attitude as he or she speaks. Each of us has developed a unique vocal style for speaking in our everyday lives, and for most, it is possible to build upon this natural style to create a "sound" or performing style that can be a marketable commodity.

Your natural speaking style is a reflection of how you perceive yourself, and it may change from moment to moment as you move from one situation to another. When you are confident of what you are doing, you might speak with determination and solidarity. But when your insecurities take over, your voice might become weak, breathy, and filled with emotion.

Your style as a voice actor comes first from knowing who you are, and then expands on that by adding what you know about human nature, personality, character development, and acting. Developing your vocal style is an ongoing process. You start with your voice as it is now, and as you master new acting and performing skills your style will begin to develop. Your vocal range will expand, as will your ability to express attitude, emotion, subtlety, and nuance in your delivery.

You may believe that you have certain limitations with your vocal range, perhaps due to the way you vocal instrument is constructed, and that these limitations may prevent you from developing a marketable style. The truth is that there are many very successful voice artists who have taken what might be viewed as a vocal limitation, and developed it into a highly successful performance style. With proper training, your perceived "limitations" can often be polished and honed into a style that is uniquely yours. The challenge is to first discover your potential, and second, have the dedication and persistence to discover where that potential may lead. The style you ultimately discover may be that of a single "signature voice" or a style that covers a broad range of characterizations and attitudes.

The Road to Proficiency

Acquiring a skill, and becoming good at that skill, is called *competency*. Becoming an expert with the skill is called *proficiency*. You must first be competent before you can become proficient. Sorry, but it just doesn't work the other way around.

BECOMING COMPETENT

Your degree of competency with any skill actually falls into the following four distinct levels. Each person works through these levels at his or her own pace and with varying degrees of success.

LEVEL #1: *Unconscious incompetence.* At this level you are not even aware that you don't know how to do something. You have absolutely no skill for the task at hand.

LEVEL #2: *Conscious incompetence.* You become aware that there is something you don't know or understand, and you begin to take steps to learn what you need to know.

LEVEL #3: *Conscious competence.* You have acquired the basic skills necessary to accomplish the task. However, you must consciously think about what you are doing at each step of the process.

LEVEL #4: *Unconscious competence.* When you reach this level, you have mastered the skills necessary to accomplish the task without thinking about what you are doing.

THREE STAGES TO ACHIEVING PROFICIENCY

There are three stages to acquiring a proficient level of skill to become an expert. These must be worked through regardless of the skill that is being learned. Playing the piano, building a table, or performing in a recording studio all require the same three stages of learning and perfecting the skills needed to achieve the end result.

STAGE #1: *Understand the underlying mechanics.* Every skill requires an understanding of certain basic mechanical techniques that must be learned before any level of expertise is possible. In the craft of voice acting, some of these mechanics include: breath control, pacing, timing, rhythm, inflection, acting, and effective use of the microphone, computer and recording software.

STAGE #2: *Understand the theory and principles that are the foundation for using the skill effectively.* In voice acting, these principles include script analysis, character development, audience psychology, and marketing.

STAGE #3: *Apply the knowledge learned in the first two stages and continually improve on the level of skill being achieved (practice and rehearsal).* For the voice actor, this means constantly studying acting techniques, taking classes and workshops, studying performances by other voice actors (listening to commercials, etc.), following the trends of the business, and working with what you learn to find the techniques that work best for you.

Three Steps to Creating an Effective Performance

In all areas of performing, there are three steps to creating an effective performance; the end result of any task can be considered a performance. For example, when building a table, you are performing a series of tasks required to result in a finished table. Your degree of proficiency (expertise) at performing the various tasks will determine how sturdy your table is and what it looks like when you are finished.

The following three basic steps to performing any task are necessary in the business of voice acting as well:

1. Practice—learning the skills and techniques
2. Rehearsal—perfecting and improving techniques and skills
3. Performance—the end result of learning and perfecting

The steps must be done in that order. You, no doubt, have heard the phrase "practice makes perfect." Well, guess what! It's a misnomer. Even *perfect* practice may not make perfect, because it is possible to practice mistakes without realizing it—only to discover too late that the end result is ineffective—and you may not understand why.

A voiceover performance will rarely be "perfect." So what we need to do as a voice actor is to practice with a mind-set of knowing that there may be dozens of ways to apply a certain technique or deliver a line of copy. Our mastery of a technique will be achieved through testing and experimentation as we discover how it works when combined with other techniques. This is one of the reasons why continued training from qualified professionals is so important.

PRACTICE

Practice is the process of learning what is needed to achieve the desired result—acquiring the skills and applying the underlying mechanics and techniques to achieve proficiency. In voiceover work, the practice phase begins with the initial read-through; having any questions answered by the producer; doing a character analysis; doing a script analysis; working on timing, pacing, and delivery; locking in the correct pronunciation of complicated words; and possibly even recording a few takes to determine how the performance is developing. To discover problems in the copy or character, and correct them, practice is an essential step in voiceover.

If problems are not corrected quickly, they will need to be addressed later during the rehearsal phase. In the real world of voiceover, there are two aspects to this phase. The first is when you are practicing on your own or with a coach to learn basic skills and techniques, and the second is the initial practice read-through at a session while woodshedding. Personal practice should be a life long quest to learn new skills and techniques. The practice phase at a recording session generally lasts only a few minutes.

REHEARSAL AND PERFORMANCE

Rehearsal begins once all the details of the performance are worked out. The choices for character, attitude, voice placement, vocal texture, delivery, and timing are set and committed to during practice. The process of perfecting the performance progresses through a series of takes as the choices are tested and modified. Each take is subject to refinement by direction from the producer, director, or engineer. Every rehearsal, or recorded take, has the potential of being used as the final performance, either in whole or in part.

Once an aspect of the performance is set, it should be rehearsed in the same manner, as much as possible, until adjusted or modified by the director. When the delivery on a line is set, don't vary it too much in the takes that follow. Set the tone of the delivery in your mind so that you can duplicate it as you polish the rest of the copy.

Eventually, every line of copy will be set to the liking of the producer. In some cases, a producer may actually have the voice actor work line-by-line, getting just the right timing and delivery on one line before moving on to the next line. Later, the engineer will assemble each line's best take to create the final track. This process is considerably different from acting for stage or film

Theatrical and film actors practice their lines as they work on their blocking and staging. The director gives them some instruction, but for the most part, actors are in the practice phase as long as they are working with a script. By the time they are ready to put down their scripts, they are at a point where they know what they are doing on stage—and rehearsal begins.

As they rehearse, the director makes adjustments and polishes the performance, most often in terms of blocking and staging. Finally, there is a dress rehearsal where all the ingredients of the show—music, scenery, props, lighting, special effects, actors, and so on—are brought together. The dress rehearsal is normally the final rehearsal before opening night and is usually considered to be the first complete performance. There is no such thing as a dress practice! Some theatrical directors even consider the entire run of a show as a series of rehearsals with an audience present.

As voice talent, we're fortunate if we receive the script a day or so prior to recording. Quite often the time we have for practice and rehearsal is very limited. The fact that all three elements may occur simultaneously means that it is essential that our use of technique be instinctive.

Never assume you have perfected a technique. There will always be something new, more, or different that you can learn to expand your knowledge. There will always be new techniques for you to try and use. There will always be a different way you can approach a character or piece of copy. There will always be new trends in performance style that require learning new techniques. To be an effective and versatile voice actor, you need to be aware of the trends and be willing to learn new techniques.

The Elements of a Voice Acting Performance
Techniques for Developing Style—CD/7

There are many aspects to voiceover performing, most of which must be learned over time. It is through the mastery and application of specific skills and techniques that a performer's unique style and business acumen is developed. And, as with any profession, the use of only a few, highly refined, skills and techniques may be the foundation of a performer's voiceover style. In today's world of voiceover, there are only two paths to success: either be the best you can be in a specific niche or be extremely good and as versatile as possible in multiple areas. Whichever path you choose, mastering technique is the name of the game. This is one of the reasons why I recommend continued training and development of both performing and business skills. The remainder of this chapter will cover a wide variety of skills and techniques that apply directly to voiceover.

LESS IS MORE

When understood and applied, this simple concept is one of the most powerful things you can do to create believability in your performing, and it works well in just about every aspect of the business from marketing to production.

Just because you love what you do does not necessarily mean you are good at what you do. In voice acting, accuracy with pronunciation or an obvious presentation does not necessarily create the highest level of believability. You will find that you can often create a greater level of truth and honesty in a character by simply holding back a little (or a lot). Some professionals refer to this as "letting go of your voice," "making it real," or "being conversational." It may be that speaking a bit slower, a bit softer, altering the phrasing, or being somewhat more relaxed might be just the thing to make that emotional connection with the listener. If your character has a specific regional sound or accent to his or her voice, you may find that softening the edge makes your performance more effective. If your character is intended to be an exaggeration, the *less is more* philosophy probably won't apply, and to be effective you may actually have to go overboard on the characterization.

Less is more is a technique often used by filmmakers to create tension and suspense or as a form of misdirection to set the audience up for a surprise. For example, in the Steven Spielberg film *Jurassic Park*, the initial appearance of the T. Rex was not accompanied by a huge roar. Instead, the tension of the moment was created by ripples in a simple cup of water, implying the approach of something huge and menacing.

The same technique of minimalizing in your voiceover performance can create a moment of dramatic tension, or wild laughter. It often has to do with the character's attitude, the twist of a word, the phrasing of a sentence, the pace of the delivery, or simply a carefully placed pause.

Understanding and applying *less is more* is an acquired skill, much like comedic timing. It requires a mastery of the craft of voice acting to a point where you are not thinking about what you are doing, and your delivery comes from someplace inside you. Although some people seem to have a natural instinct for interpretation and using the *less is more* concept to create a believable performance, most acquire this skill through experience.

MORE IS MORE

As powerful as the *less is more* concept is, there are times when a script simply calls for taking your performance a bit "over the top." To create a believable illusion of reality, you may occasionally need to present an attitude or emotion that feels slightly exaggerated. This *more is more* idea is common in dialogue commercials and character voice work for animation, cartoons, and video games, but will also be heard in single voice work.

More is more works in voiceover because the only thing the listener has for creating a scene in their imagination is the sound of the voice. They don't have the benefit of any visuals. Because of this, giving the performance a slight bit more that "real life" will often create a stronger, and more visual, sense of reality than if the words were spoken from a completely realistic perspective. The trick to using *more is more* effectively is to be careful that you don't take it too far over the top or you stand the chance of breaking the illusion of reality with a delivery that sounds forced.

PERFORMANCE DYNAMICS—PACING, VOLUME, RANGE

Performance dynamics are the fundamental elements of vocal variety and lay at the heart of any voiceover performance. It is the dynamics of your performance that makes *less is more* a powerful technique. When you understand and apply the dynamics of *pacing, volume,* and *range,* you will be able to make any vocal presentation interesting and captivating.

Pacing refers to the variations of speed in your delivery. It is closely related to the rhythm and timing of the copy and to the tempo of your delivery. *Pacing* is how fast or how slow you are speaking at any given moment. I'm sure you've heard commercials or other voiceover that is delivered at the same pace throughout. There is no phrasing, no pausing for impact, absolutely nothing that makes an emotional connection. Only intellectual information being delivered often at a rapid-fire pace. Or you've heard people who… seem… to… take… for… ever… to… say… what's… on… their… mind. Does either of these styles of delivery get and keep your interest? No! In most cases a steady pace is boring and uninteresting, if not downright hard to listen to. There are some exceptions in projects for which a steady or slow pace may be critical and necessary to the effective delivery of information, as in an educational or training program. However, in most cases, slowing down or speeding up your pacing to give importance to certain words, phrases, or ideas will make a big difference in your presentation. Create interesting phrasing by varying your pace or tempo. Within two or three read-throughs, you should be able to find the pace and phrasing that will allow you to read a script within the allotted time and in an interesting manner. Some directing cues that relate to pace are: "pick it up" (speed up), "stretch" (slow down), "fill" (you have extra time), and "tighten" (take out breaths or pauses between words).

Volume, or *dynamic range,* refers the variations in the loudness of your delivery, and is how soft or how loud you speak at any given moment. Just as volume changes in a piece of classical music keep things interesting, dynamic range in voiceover directly relates to the believability of a performance. Performing a script at the same volume throughout is much like speaking at the same pace throughout. Both result in loss of credibility in the mind of the listener, because real people change how fast and how loud they speak depending on how they feel about what they are saying. The dynamic range of a performance is directly related to attitude and tone—from soft and intimate to loud and aggressive.

Vocal range, or *vocal variety,* refers to a performer's ability to put variety into the performance by adjusting the pitch and placement of the voice to maintain interest. You've, no doubt, experienced a seminar or lecture at which the speaker spoke in a monotone, resulting in the audience tuning out and losing interest. Vocal *range* covers the spectrum from your lowest pitch to your highest pitch. Voice actors for animation have developed a wide range from which to create many characters. You have a normal vocal range for speaking in everyday conversation, and you can speak at a lower or higher pitch when necessary or when you are expressing an emotion. Practice speaking at a slightly lower or higher pitch and notice how a small change in vocal range can result in a big shift in interpretation.

Listen to the way people talk to each other and you will notice a wide range of speaking styles. Excitement, enthusiasm, awe, sarcasm, pity, wonder, sorrow, cynicism, and sadness are all expressed in different ways by different people. The variations in the way a person expresses herself or himself reflect that individual's *vocal range.*

Observe how you instinctively adjust your *pacing, volume,* and *range* in your everyday conversations. Practice altering your dynamics as you speak to your friends or at work, and notice how they pay more attention to what you have to say.

Be aware, however, that performance dynamics can be easily misused, forced, or overdone. The secret to understanding these dynamics is in the interpretation of a script. What is the writer's objective? Who is the intended audience? How should the words be spoken to achieve the maximum emotional and dramatic effect? How should the intellectual content be delivered so the listener can understand and use it?

When combined, the dynamics of voice acting serve to help create drama, humor, and tension in a performance. When effectively used, they go hand-in-hand to result in a performance that inspires, motivates, and is believable.

ARTICULATION

Complex sentences are an everyday occurrence that every voice actor must deal with. Words must be spoken clearly and concepts communicated in a way that can be understood. Voice acting, and effective communication in general, is a blend of intellectual and emotional information delivered in an interesting and understandable manner. Unless a specific speech affectation is called for in a script, it is generally unacceptable to stumble through words or slur through a piece of copy. *Articulation* refers to the clarity with which words are spoken. Most common problems with articulation are the result of *lazy mouth*, or the tendency to not fully use the muscles of the tongue, jaw, and mouth when speaking. Good articulation, or enunciation, can be especially tricky when copy must be read quickly.

The script we worked with earlier works well as an articulation warm-up exercise. Read the following copy, this time making sure that your articulation is crisp and clear. Don't worry about getting it in "on-time," just focus on making every word clear and distinct. For the purpose of the warm-up exercise, you'll want to force yourself to over-articulate —and don't forget to speak the ends of every word. (See "The Cork" exercise on page 48 and CD/5.) After doing this exercise, your conversational articulation will sound natural, but will actually be more clear because your vocal instrument is warmed-up.

Come in today for special savings on all patio furniture, lighting fixtures, door bells, and buzzers, including big discounts on hammers, shovels, and power tools, plus super savings on everything you need to keep your garden green and beautiful.

When the same letter is back-to-back in adjacent words such as the "s" in "hammers, shovels" and "plus super," it's easy to slide through the words sounding the letter only once. In a conversational delivery, it's fine to tie those letters together, but for this exercise speak the end of each word clearly. It is also easy to drop the letter "d" from words like "and" and "need," especially when the next word begins with a "t," "d," "g," or "b." The letter "g" on words, such as "big," can sometimes be swallowed resulting in the phrase "big discounts" sounding like "bih discounts." The suffix "ing" can often be modified when in a hurry, causing words, such as "lighting" and "everything," to sound like "lightin" and "everythin." With good articulation, the ends of words are clearly heard, but not overenunciated and suffixes are properly pronounced.

The "s" and "z" sounds should be clearly distinct. The "s" in "door bells" should have a different sound from the "z" in "buzzers." The consonant "s" should sound like the end of the word "yes," which is primarily a nonvocalized release of air over the tongue. To properly pronounce the more complex "z" sound, the tip of the tongue starts in the "es" position and a vocalization is added. Say the word "buzz" and hold the "z." You should feel a distinct vibration of your tongue and teeth.

Plosives are another articulation problem area. Plosives are caused by excessive air rushing out of the mouth when speaking letters such as "P," "B," "G," "K," and "T." When this sudden rush of air hits a microphone's diaphragm, the result is a loud "pop." Plosives can be corrected by turning slightly off-axis of the microphone, by using a foam windscreen, or placing a nylon "pop filter" in front of the mic. To feel the effect of plosives, place your hand directly in front of your mouth and say "Puh, Puh, Puh" several times. Turning your hand to the side will show you how the blast of air is reduced when turning off-mic.

To achieve a conversational and believable delivery, it is often necessary to violate some of the basic rules of crisp articulation. However, it is important to understand and to master the correct way to do something before you can effectively do that thing incorrectly and make it believable. In other words, you've got to be good before you can do bad, believably. When speaking in a conversational style, be careful NOT to over-articulate.

An important aspect of articulation is the ends of words. It is common in every-day conversation to drop the ends of words, and we instinctively fill in the missing sounds as we listen. But in voiceover, those ends of words are important and need to be heard. As you begin to work with copy, learn to listen to yourself to hear if you are dropping the ends of words. The technique of *Linking* on page 96 is a good way of correcting this problem.

DICTION

Diction is defined as the accent, inflection, intonation, and speaking style dependent on the choice of words. Diction is directly related to articulation, the clarity of your delivery, the correct pronunciation of words, and the sound of a character's voice. One of the best ways to improve your diction is simply to slow down as you speak and focus on your enunciation and clarity. Diction is important in all voiceover performances—you really do want to say the client's name correctly and clearly.

If you are creating a character voice, your diction becomes even more important. A character voice may be a dialect or specific speaking style, and it is vital that your words be understood. Listen to yourself closely to make sure you are speaking clearly and at the correct pace for the character. As with articulation, Exercise 9: "The Cork" on page 48, can help with diction.

TEMPO, RHYTHM, AND TIMING

All voiceover copy has an ideal tempo. *Tempo* refers directly to the speed at which the words are spoken. A performance may be delivered at a constant tempo or at a varying tempo. You speak at a comfortable tempo when you are in conversation. When performing, your delivery tempo may be slower, faster, or about the same as your normal, conversational tempo.

Voiceover copy also has a built-in rhythm. *Rhythm* is an aspect of phrasing and is closely related to tempo. Combined with tempo, rhythm gives a voiceover performance its sense of musicality. It is the flow of the words, the way the words are organized in sentences, and the placement of importance, or value, on certain words. Rhythm is also directly related to the emotional content of the copy. Poetic copy has an obvious rhythm (or meter). The rhythm of narrative copy is a bit more challenging to find, but it is there. Dialogue copy has a distinctive rhythm, which often includes a sort of verbal syncopation, gradually, or quickly, building to a punch line. Just as you speak with your own personal rhythm, the characters you create for a voiceover performance will each speak with their own rhythm. It may be choppy, staccato, smooth, or even vary throughout the delivery. Finding the proper rhythm is critical to an effective and compelling performance.

The combination of tempo and rhythm in a performance is known as timing. *Timing* refers to interaction between characters or the manner in which pauses between lines of copy, and general phrasing are handled. As a voice actor, where you place a pause or a beat can create tension, humor, or drama in a performance. How quickly does one character speak after another finishes a line? Do the characters step on each other's lines? Is there a long silence before a character speaks? These are all aspects of timing.

If you have a natural sense of timing, you are ahead of the game. If not, the producer will direct you into the timing, and you will get a sense of what is needed as the session progresses. As you become comfortable with your character's tempo and rhythm, timing becomes automatic.

Watch TV sitcoms to study tempo, rhythm, and timing. Study the interaction between characters and how they deliver their lines. Listen for the jokes, and how a joke is set up and delivered. Watch the physical characteristics of the actors as they work together. What are their gestures? What facial expressions do they use when they deliver a joke? What expressions do they have when they react to something? How do they express emotion and dramatic tension? Use what you learn to help develop tempo, rhythm, and timing for your performances.

The combination of tempo, rhythm, and timing works differently for different media. Theater has the slowest tempo and rhythm, then film, followed by television and finally, radio with the fastest tempo and rhythm. In some ways, radio can be performed at almost any rate, but generally a radio performance is faster than the same copy performed on-camera for television or film. Because radio uses only one of the senses, the rhythm, timing, and pace are set a bit faster to create a more real and believable interaction between characters. The faster tempo of radio gives the copywriter and talent an opportunity to quickly establish and develop an interesting story that will grab the listener's attention and hold it while the message is delivered.

PHRASING

One of the most common challenges when working with a script is to determine the proper delivery speed and variety. How quickly or slowly should you speak? And how will you adjust your phrasing or pacing to add variety to your delivery?

Phrasing in voiceover is very much like phrasing in music. It refers to the overall flow of your delivery; the variations in tempo, rhythm, and timing as you speak; and the subtle nuances of your tone of voice. More specifically, phrasing relates to the way you say certain words or sentences. For example, a short statement—"I would like some more, please"?—can be phrased in several different ways. The first word "I" can be emphasized to give personal emphasis. By the same token, changing the tempo, and emphasizing the word "would" can give an entirely different meaning. Breaking the phrase into two sentences by putting a period after the word "some" can result in a completely different delivery.

Try this exercise to discover different ways to express this simple phrase. Read each line at different tempos and rhythms, giving importance to the word in bold:

I would like some more, please!

I **would** like some more, please!

I would **like** some more, please!

I would like **some** more, please!

I would like some **more**, please

I would like some more, **please!**

SUSTAIN TO SLOW DOWN AND ADD INTEREST

An aspect of phrasing is sometimes referred to as *pulling words*. This technique focuses on *sustaining*, or stretching, specific sounds, words, or phrases. Sustaining an entire phrase can usually be achieved simply by slowing down the overall delivery of a sentence. But a phrase can be made more interesting by sustaining only the beginning, middle, or end of some words, rather than an entire sentence.

Experiment with this line of copy to get a sense of how you can elevate the interest level of a line by sustaining certain sounds. Start by delivering the line as written, at a steady pace without altering the tempo or rhythm:

So, you're thinking about buying a new car? Maybe you know something about cars, maybe not.

Here's how this phrase might be written to indicate sustaining sounds:

Ssssooooo... you're thinnnking about buying a new carrrr? Mmmmaybe, you knooow something about cars, mmmaybe not.

By sustaining the beginning, middle, or end of a word, or even an entire word, you can create anticipation for what will come next. Adding natural vocal sounds to the phrasing can add even more interest, realism, and believability to the character.

Although pulling lines can help to create a more compelling delivery, it takes up valuable time, and most voiceover projects don't have time to spare. So this technique is generally used in a shortened form, for a specific character's speaking style, or only when necessary.

Phrasing and *sustaining* are both elements of tempo, rhythm, timing, and pacing in that they refer to the way in which words are spoken within a sentence or paragraph. But, even more than that, phrasing allows you to make the words more real by adding compelling emotional content.

THEE AND THUH, AE AND UH

Few words in the English language are used improperly more often than the little words "the" and "a." When used correctly, these words can help add power and emotion to your delivery. Used improperly, your message may sound awkward, and might even create an impression of your being "uneducated." Here are a few quick rules to keep in mind when you see these words in a script. Keep in mind these rules are not set in stone, but are only guidelines. Ultimately, whatever sounds best in the context of your performance, or the way you are directed, is the way you should go:

Basic Rules for "the"
1. Pronounce stressed as "thee" (long ē):
   - When "the" precedes a vowel: *Thē English alphabet has 26 letters.* Exception: pronounce as "thuh" if the word starts with a long "U" as in "thuh university" or "thuh United States."
   - When "the" precedes a noun you wish to stress for emphasis (replacing "a" or "an"): *Yes, that is thē book you gave me.*
   - When "the" precedes a word you wish to indicate as unique or special, or is part of a title: *thē place to shop, thē King of France.*
2. Pronounce conversationally and unstressed as "thuh":
   - When "the" precedes a word that begins with a consonant: *The kitchen cabinet is empty. The car ran out of gas. The dog chased the cat.*
   - When "the" modifies an adjective or adverb in the comparative degree: *She's been exercising regularly and looks the better for it.*

Basic Rules for "a" and "an"
1. Use "a" before words that begin with a consonant, "an" before words that begin with a vowel: *a lifetime of choices, an extreme sense of duty.*
   - Words that begin with a vowel but are pronounced with the consonant sound "y" or "w" are preceded with "a" ("uh"): *a European farmer, a united front, a one-room school.*
   - Words that begin with a consonant but are pronounced with a vowel sound are preceded with "an": *an SST (es es tee), an F (ef) in English.*
2. Pronounce stressed as "ae" (as in "hay") (long ā):
   - When "a" is intended to emphasize the next word in a singular sense or is referring to the letter "A": *That is a singular opportunity. The letter A is the first letter of the alphabet.*

- The pronunciation of "a" in its stressed form (ae) will be relatively rare for most voiceover copy as it is not generally conversational. However some technical copy may require this pronunciation to properly convey the message or instructions for training purposes.
3. Pronounce unstressed as "ă" ("uh") when:
   - "a" precedes a consonant: a *horse, a new car, a cat, a personal debt.*
   - Your character is speaking conversationally or casually.
   - This unstressed form of "a" ("uh") is used in most situations.

ATTITUDE

What is it that you bring to the performance of voiceover copy? Are you happy? Sad? Angry? What is the mood of the copy? How do you visualize the scene? What is there—in your personal history—that you can tap into to help make the words real and your performance believable? Answer these questions and you will have your personal attitude. Answer these questions in terms of your script, and you will have your character's attitude.

Attitude is the mindset of the character in the copy. It gives a reason for the words, and motivation for the character's existence and behavior. When you read through copy for the first time, find something in the words that you can relate to. Find an emotional hook. Bring something of yourself to the copy as you perform and you will create more effective characters, a strong suspension of disbelief and a believable illusion of reality.

SENSE MEMORY

Every moment of your life is stored in your memory. And every emotional experience has a physical tension associated with it that might reside anywhere in your body. There is also a sensory experience associated with the emotional experience that is closely linked to the physical tension.

Your five senses are some of your most valuable tools as a voice actor. Constantin Stanislavski, founder of "method acting," developed this tool to help actors create believable characters, and most acting schools teach some variation of the technique. To truly master the technique of *sense memory* you may need to take some acting classes which involve creative exercises in which you tap into your senses of sight, touch, taste, sound, and smell.

It is said that all creativity originates in the sensory organs. So, to fully utilize your creative voice-acting abilities, you will need to develop skills for recalling and utilizing sensory memories. Once the basic concept of *sense memory* is understood, you can apply this technique to become a better communicator and achieve some amazing results. Here's how:

Close your eyes and think back through your life to a time, event, experience, sensation, or feeling that is similar to what your character is experiencing and hold that memory in your mind. Make the memory as visual as you possibly can. With that memory held in your mind, recall how your senses were affected by what took place. Was there a special smell? A certain sound? Did something taste odd, or especially good? Did you see something unusual? Do you recall touching something in your memory?

As your memory becomes more visual, observe where in your body the physical tension for that memory is being held: neck, shoulders, chest, stomach, legs, arms, and so on. Recall the physical tension, body posture, facial expression, and hold onto it. Keep that memory firmly fixed in your imagination. Now, open your eyes and allow your character to speak the words in the script, in a sense filtered through your experience.

Although it may take some time for you to master this technique, even doing just the basics will put you well on your way to becoming a successful voice actor. Many people who do voiceover either don't utilize this technique, don't understand how to use it, or simply are not aware of it.

The visualization exercise on page 160 (CD/9) takes the concept of *sense memory* to a higher level to help create a totally believable character.

SUBTEXT

All commercials have an attitude. In fact, all copy has an attitude. Your job is to find it and exploit it. One way to find the attitude is to uncover the thoughts or feelings behind the words. This is commonly known in theater as *subtext*. Subtext is what sets your character's attitude and establishes, or shades, the meaning of what you are saying. It is the inner motivation behind your words. Subtext allows you to breathe life into the words in a script and into the character you create.

Using your sense memory to unlock emotional hooks is a technique for setting attitude. Now take that process a step further and define the attitude in words to arrive at the subtext. For example, let's say you have this line: "What an interesting fragrance." If the thought behind your words is "What is that disgusting odor? You smell like something that's been dead for a week!" the perceived meaning will be quite different than if your thought and/or feeling is "Wow! You smell amazing! That perfume you're wearing makes me want to be close to you." Each of these subtexts results in a different mental and physical attitude that comes through in your voice.

What you are thinking and feeling as you deliver your lines makes a tremendous difference in the believability of your character. You have a subtext in your everyday conversations and interactions with others. The idea here is to include a subtext in your performance. Decide how you want the listener to feel or respond to your character—what emotional response do you want to produce? To get the desired response, all you have to do is internalize the appropriate thoughts and feelings as you perform.

For some copy, creating a believable character can be challenging, even with a well-understood subtext. The problem may lie in the subtext itself. If you have chosen a subtext that is weak or unclear, try changing the subtext to something completely different, using an entirely different set of emotional hooks. You may find that by shifting your subtext, your entire performance attitude will change.

TONE

Closely related to attitude and subtext is the tone. Occasionally referred to as "tone of voice," the *tone* of your performance is the sum total of *pacing, volume, range, articulation, diction, tempo, rhythm, phrasing, attitude,* and *subtext*. It is important to be consistent throughout your performance. Do not change your tone mid-copy. If you are doing a soft, intimate delivery with a friendly attitude, maintain that tone from beginning to end. If your copy is fast-paced, aggressive, and hard-sell, keep the attitude and tone throughout.

Tone can also refer to the quality of your performance. If you change tone as you read, you will fall out of character and your levels on the audio console will fluctuate, which will drive the engineer and producer crazy. To maintain a consistent tone, do not drift off-mic. Keep your head in the same position relative to the microphone from start to finish. Working close to the mic gives a warm, soft tone, while backing off as little as a few inches gives a cooler, more open, tone for straighter, more direct reads.

Occasionally a script is written that calls for a complete change of attitude and tone in mid-copy. If there is a logical motivation for your character to change attitude, then it would be out of character to maintain a consistent tone throughout the copy.

REMOVE OR CHANGE PUNCTUATION MARKS

Copywriters use *punctuation marks* because a script is originally written grammatically correct for the eye, to be read. However, we don't use punctuation marks when speaking in conversation. Part of our job as voice talent is to take the words "off the page" and make them real and believable. If you work the punctuation marks, your delivery will usually end up sounding like you're reading.

One of the best ways to create an illusion of reality in a performance is to remove or change the punctuation marks. Instead of instinctively pausing at a comma, or stopping at a period, try ignoring the punctuation to create a contiguous flow of words.

Removing the punctuation marks doesn't mean literally going through the script with white-out, although I do know of some voice actors who actually do that. What it does mean is performing the copy in a real, believable, and conversational manner. A real-life conversation is punctuated with pauses, changes of inflection, dynamics (soft, loud), emotional attitude (excitement, sadness, and so on), vocalized sounds (uh-huh, hmmm, etc.), and many other subtleties. To create a sense of reality, voiceover copy should be delivered the same way. Let your delivery dictate the punctuation.

Just because there is a comma in the script, it doesn't mean you have to pause or take a breath! Just because there is a period, doesn't mean you can't deliver the line as a question or as an exclamation. What would your delivery sound like if you changed a comma to a dash? What if you put a comma at a different place in the script? You have an almost infinite number of possibilities for delivering any line of copy.

Allow the scripted punctuation marks to guide you, but be careful not to take them too literally. Sometimes, a simple change of punctuation can make a big difference in the interpretation, thus improving the performance. Allow the lines of a script to flow into one another as they would if you were telling the story to another person—not reading it. Take the punctuation marks out of your performance and your performance will be on its way to being more believable.

Occasionally, you'll get a piece of copy that just doesn't make sense because the grammar or punctuation is wrong or grammatically incorrect. The writer may understand what she wants to say, and even how the words should be spoken, but because it isn't punctuated properly for the eye, the words are pretty much meaningless. It then becomes your job to figure out what the correct punctuation should be so you can give the words meaning. For example, punctuate the following phrase to give it meaning[1]:

that that is is that that is not is not is that it it is

There is only one correct way to punctuate this line of copy to give it meaning. Most copy will also have one punctuation that works best for the eye, but there may be multiple options from which to choose when those words are spoken. You'll find the correct punctuation for the above line of text at the end of this chapter.[2]

Changing and removing punctuation marks as you perform is a way of making the words your own to truly take them "off the page." This tool can help you find the inflection, energy, and dynamics you are looking for as you begin to make the critical choices for delivering your copy.

PAUSE FOR IMPORTANCE

A *pause* is much more than just a beat of silence between words or phrases. It is an aspect of phrasing, and a powerful tool you can use to take a voiceover performance to an entirely new level. A pause in your delivery can be any length from a fraction of a second to a few seconds, depending on the context of a script. You pause instinctively in normal conversation whenever you are thinking about what you'll say next. It's almost possible to hear the thought or the intention of importance that takes place during even the shortest pause. A pause implies that something big is coming and builds tension and suspense in the mind of the listener. When you pause, whatever follows is automatically perceived as being more important. And that's exactly what we want to achieve by using this tool.

Learning how to use a *pause* effectively can take some time, but once understood, the concept can be used to help create humor, drama, tension, suspense, and emotional response.

Another way to look at a pause is in terms of *timing*. Comedic timing requires just the right amount of time—or beat—between the set-up of a joke and its punch line. If the timing is off, the joke isn't funny. The same is true when using a pause in a voiceover performance. Timing is everything.

Improper use of a pause can result in an uneven or choppy delivery, or in a delivery that sounds as if the script is being read. If there is nothing happening in your mind during the pause, those beats of silence are little more than empty holes in the phrasing. To be effective, there must be something happening that fills those holes. There must be thoughts taking place that are in alignment with the *desires*, or wants and needs, of your character. Those thoughts won't be verbalized, of course, but their mere existence will be heard in your tone of voice, attitude, and overall delivery.

The following phrase will give you an idea of how you can use a pause to create value and importance. Begin by just reading the line once to get an understanding of its meaning and to come up with some initial delivery choices. Now deliver the line out loud as one continuous thought—no pauses.

Everything in our store is on sale this week only at Ponds.

Since there are no commas or other punctuation to give you hints as to the delivery, you're on your own to find the most effective way to say the phrase. Delivering the line as one continuous stream of words is certainly a valid choice, but it may not be the strongest. Now, deliver the same phrase, this time experimenting with placing a pause or two in your delivery. Use each hyphen in the lines below as a cue for a beat or brief pause in your delivery. Notice that no matter where you place a pause, you will instinctively give the words that immediately follow greater value (in **bold**).

Everything in our store—**is on sale** this week only at Ponds.
(the event receives natural emphasis)

Everything in our store—**is on sale—this week** only at Ponds.
(the event and time receive natural emphasis)

Everything in our store is on sale this week—**only at Ponds**.
(the location receives natural emphasis)

Everything in our store is on sale—**this week only—at Ponds**.
(the time and location receive natural emphasis)

The only way you'll find the most effective delivery when using the *pause* will be to experiment with the many possibilities in every script.

HOLD THAT THOUGHT—USING THE ELLIPSIS

Interruptions are a way of life. You experience them every day. You might be in the middle of saying something really interesting… and then someone breaks in or cuts you off before you finish what you are saying.

This also happens in voiceover, especially in dialogue, but it also often occurs in single voice copy. The challenge for the voice actor is to make the interruption sound real and believable.

In a voiceover script, an interruption is usually indicated by the ellipsis, or 3 dots (...). The ellipsis can also indicate a *pause* in the delivery, occasionally replacing a comma or other punctuation.

For example:

> Boss: Peterson... we seem to be having some problems in
>   your division. What do you have to say about that?
> Peterson: Well, sir, I...
> Boss: Now, listen up, Peterson. We need this taken care of
>   right away... Understand?

The trick to making an interruption sound real is to continue the thought beyond the last word to be spoken. Much like a pause, if the line is simply read as written, the performance can easily sound like the words are being read, or the interaction between characters may sound "off" or artificial. However, if the thought is carried beyond the last word, the interruption becomes real and natural.

To continue the thought, all you need to do is make up something your character might say that is appropriate to the context of the script. Write it on the script, if you like, but at the very least, keep the complete thought in your mind as you deliver the line, and be prepared to speak the words. Completing a thought will enable you to create a believable delivery of the words. This concept works well in a variety of situations.

In the following script, Peterson continues the thought until interrupted. By completing the thought "Well, sir..." you will set the tone, attitude, and pace for your delivery of the line.

> Boss: Peterson... we seem to be having some problems in
>   your division. What do you have to say about that?
> Peterson: Well, sir... (*I've taken steps to get things back on track.*)
> Boss: Now, listen up, Peterson. We need this taken care of
>   right away... Understand?

When the moment of the interruption occurs, simply hold the thought and let the interruption happen naturally. The continuation of the thought is often more realistic if verbalized, especially in a dialogue performance. If the other actor is a bit late with the interruption, no one will ever know, because you kept the thought going. If you are the actor who is interrupting, you need to make sure you deliver your line with the appropriate energy and attitude, and that you are cutting off the other person in a way that sounds like a real conversation.

THOUGHT PACING—ANOTHER USE OF THE ELLIPSIS

Thought pacing is a another tool that makes your character real! When you see ellipses in a script, you have an ideal opportunity to reveal the thoughts of your character. Not only can you keep the initial thought going until you are interrupted, but you can also make your character more real by vocalizing sounds during the ellipses. For example, in the above script, Peterson might interject unscripted responses during the ellipses, and the Boss might even put in some "umms," or "uhhs" to add believability.

Another aspect of *thought pacing* is to ad-lib natural, conversational, responses while delivering a script. This will most often occur in a dialogue script, but ad-libbed human sounds can also be quite effective in the delivery of a single voice script. The proper use of thought pacing can literally bring a script—and your character—to life.

REVERSE TEXT TO FIND INFLECTION

Occasionally, it can be challenging to find the best way to deliver a line of copy. Usually when this happens, it's because the copywriter wrote the script for the eye and not for the ear. Sentence structure for the written word is often quite different than for the spoken word. A trick I call *text reversal* can often help. The basic idea is to simply reverse the sentence structure to discover a different way of inflecting the words. Once found, put the sentence back as written, and deliver with the newly discovered inflection and energy. It works just about every time! Here's an example:

Created to bring you the ultimate home theater experience, our showrooms are stocked with the latest high-tech equipment.

By reversing the two parts of the sentence, you may discover a better way to inflect the words.

Our showrooms are stocked with the latest high-tech equipment, and are created to bring you the ultimate home theater experience.

Once you've found an inflection you like, deliver as written, but keep the new inflection.

RIDE THE ELEVATOR TO TWEAK YOUR TIMING

Commercial scripts are often written with too little, or too much copy. It's just a fact of life. Also, we may discover that the choices we make for our character result in a delivery that is too slow or too fast. We need to be able to adjust our delivery so that we complete the copy within the specified period of time. Sometimes this challenge can only be resolved through script revisions. But, more often than not, we can easily adjust the tempo of our delivery without affecting the meaning or intentions of our delivery.

The common way to think of this adjustment is to simply speed up or slow down the delivery. But thinking in these terms can have an adverse effect in that the words may sound rushed or unnaturally slow. A better, and much more practical, way to think of adjusting speed is to imagine that you are riding in an elevator. To speed up your delivery, simply imagine that you and the person you are speaking to get on at the same time, and that they are getting off at the next floor. You must tell them what you have to say by the time the elevator doors open. When you need to slow your pace, give yourself an extra floor or two for telling your story.

Changing the way you think about how you speak the words in a script can completely change the believability of your voiceover performance.

IMITATION

It has been said that *imitation* is the sincerest form of flattery. This may be true, but as a voice actor, you want to be unique. Mimicking the delivery style of an experienced professional can be an excellent way to learn the pacing, inflection, and nuance of a particular voiceover niche. But be careful that your imitation of other voiceover performers is for the purpose of developing your own unique style. If all you can do is mimic someone else's style, dynamics, or attitude, you are doing nothing unique.

Be yourself, and find the uniqueness of your voice. That's what will get you work! Only mimic what other voice talent do to learn their techniques. Then adapt what you learn to your personality and style. If you insist on imitating other performers, it could take a long time for you to find your unique voice-acting personality.

Shortcuts That Trick Your Brain—CD/8

Over the years, your brain has developed some very specific and predictable ways in which information is perceived and interpreted. It is because of this predictability that we can utilize some clever techniques that effectively "short-circuit" the normal processes so we can achieve our desired results in a performance. You'll be amazed at how effective some of these shortcuts are!

THE 2-4 SHORTCUT

When you speak conversationally with a fairly relaxed delivery, the result is that certain words are often pronounced in a manner that is not totally accurate. Regional accents and dialects will reveal a wide variety of how certain words are spoken. For example, the word "tomorrow" is often pronounced as "tahmarrow," or "tamarreh." "Forget" becomes "fergit," "our" becomes "are," and so on.

When you want to speak with the standard nonaccented American English to correctly pronounce words that have a "to" or "for" in them, simply replace the "to" or "for" with the numeral "2" or "4." Your brain is trained to say the numbers as "two" and "four," so as you are reading, your brain sees the number and you automatically speak the word more precisely.

RESPELL WITH SOUNDALIKES

The same basic idea as the *2-4 shortcut* can be used for other words as well. When you find you are mispronouncing a word, or need to speak with clearer diction, you can simply respell the word using a different word that has the sound you want. For a word like "our," change the spelling to "hour." One student of mine had difficulty speaking the word "cellular" when used in the context of a script discussing cellular telephones. By simply changing the spelling of the word on his script from "cellular" to "sell-ya-ler," he was almost immediately able to deliver the lines perfectly. This little trick fools the brain and works with most soundalike words. The possibilities are unlimited, and using this trick can truly be a life saver when working with technical or medical copy.

LINKING

A common problem is the dearticulation, or dropping, of the last letter or sound of a word. This condition is occasionally referred to as *lazy mouth*, and is simply the result of poor diction. Although it may be OK for general conversation, this can present a problem for recorded projects. When the last sound of a word is not spoken, or is spoken too softly, the word can get "lost in the mix" when combined with music or sound effects.

To correct for this, most people will mistakenly adjust their delivery to be overly articulated or over-enunciated. The result is an artificial sound that is not authentic. In some cases, where the character naturally speaks in a "lazy" style, this dearticulation of the ends of words can be completely appropriate. However, for most voiceover copy—especially copy that will eventually be mixed with music or sound effects—the delivery must be spoken with clear diction. Here's a way to do that without resorting to over-articulation.

The technique is called *linking*, and it's a trick that comes from the world of singing. The idea is to take the last letter of a word and attach that letter to the beginning of the next word. For example, the phrase "...and everyone was there" might sound like "an everyone was there," with the "d" not spoken on the word "and." To use the *linking* technique, the "d" on "and" is moved to become the first letter of the word "d-everyone." So the adjusted line will sound like "an deveryone was there."

Advanced Techniques

Many of the tools in this chapter are intended to help you discover the most effective, or appropriate, choices for delivering a line of copy. As with most things in life, voiceover work has many levels of skill and techniques that range from very simple to very difficult. The following techniques fall in the "Advanced" category, not because they are especially difficult, but, rather, because these concepts are most effectively applied after achieving a certain level of skill with other, fundamental processes and techniques.

SUBSTITUTION: CHANGING CONTEXT TO FIND ATTITUDE

Changing *context* is yet another way to look at your script from a different perspective. This simple trick, that I refer to as *substitution*, might make the difference between a flat delivery and one that lands on target. This trick is similar to reversing the text, except that instead of reversing the sentence structure to find alternative choices, only certain words are changed, while leaving the overall sentence structure intact.

When you have problems with a line, you can completely change the sentence to something that you understand and relate to. It's OK to change the words because this is only a process for you to discover choices. Once you've found a meaningful interpretation, go back to the script and use the same delivery style. For example, this probably won't mean much to you:

> The GMS 5502 and the H-27-R hybrid transducer were successfully tested during a trial period in October of last year.

So, let's change it to something like this:

> The red cherries and the yellow lemon were successfully eaten during a lunch break last week.

The new context doesn't need to make any more sense than the original script. But by using *substitution* to change the context to something you easily understand, you will be able to create a meaningful delivery. Now all you need to do is apply your chosen delivery to the original script.

MAKE THEM THE ONLY WORDS

Occasionally the way a paragraph is written can be troublesome, resulting in difficulty finding an effective interpretation. When you notice that you're throwing away the end of a sentence, or that your inflections are the same for every line, reduce the script to the one line of copy giving you trouble. Make that one line the entire script and deliver it out loud to hear how it sounds. Say it a few different ways and choose the best delivery. Then put the line back into the context of the script to hear how it works with the full text. This trick will usually make a big difference.

ADD A WORD OR TWO

As a general rule of thumb, you will want to deliver a script as written. However, the underlying job of a voice actor is to bring the words to life by creating a believable character who delivers the message in an interesting and compelling manner. Sometimes the way a script is written just doesn't lend itself to an effective rhythm. This is often true of lists in which every item of the list tends to have the identical inflection resulting in a monotonous rhythm.

To create greater interest and a more compelling rhythm try *adding conjunctions* between items in a list. This can give each item greater value and, depending on the intonation of your delivery, effectively convey a specific attitude or emotion such as excitement or frustration. For example, here's how a typical list might be written:

> Your burger comes with two patties, tomato, lettuce, cheese, onion, ketchup, mustard, and pickles.

Adding conjunctions will allow you to "play" with inflection, pacing, and attitude as you speak, making each item important in its own right. You will also be able to more effectively build interest throughout the list by creating the impression that you are thinking of each item just before you say the words:

> Your burger comes with two patties, and tomato, and lettuce, and cheese, and onion, and ketchup, and mustard, and pickles.

Adding words will almost always add length and time to your performance, which may require you to make timing adjustments in other parts of your delivery. A more advanced use of this technique is to add words silently as you deliver a line of copy. Not speaking the added words can be a bit tricky, but the effect can be very powerful as a means for creating an emotionally charged delivery.

This technique should be used judiciously as many producers will want you to deliver the script exactly as written.

SUBTRACT WORDS

Just as adding a word or two can enhance your delivery, the same can be true if you *subtract* a word or two. This technique can speed up your delivery by making statements more terse and abrupt. By removing the "and" that usually sits between the last two items of a list, the overall tone of the delivery can instantly take on a sense of authority.

> Your burger comes with two patties, tomato, lettuce, cheese, onion, ketchup, mustard, pickles.

BREAK THE RULES OF LOGIC

Logic dictates that we deliver a line of copy with proper sentence structure, articulation, grammar, and an interpretation based on our understanding of the text. Many of the most effective techniques for creating a believable character and compelling delivery require breaking the rules of logic. *Breaking the rules of logic* can have a powerful impact on your delivery.

Here's yet another nifty trick for keeping the listener's attention: Break a single word into two or more words. In most cases, we logically deliver a word the same way regardless of the context of the script. Breaking a single word into its component parts can result in an enhanced understanding of the underlying meaning. For example, speaking the simple phrase "Absolutely amazing!" will have a certain level of impact. However, the meaning will be completely different if you restate the phrase by treating each syllable as a separate word: "Ab so lute ly a maz ing!"

This technique won't work in every situation or with every script, but when you need to draw importance to a specific copy point this tool can be incredibly effective.

BREAK THE RULES OF GRAMMAR

Copy is written grammatically for the eye. Even when a script is written in the style of a specific character, the text will often tend to be written grammatically correct and may even include syntax that is not consistent with the character you create. *Break the rules of grammar* by dropping words, adding words or sounds, rearranging words, changing punctuation, and altering the rhythm or tempo to create more compelling characters.

A basic premise of all voiceover work is that, as a voice actor, you must do whatever it takes to bring the words to life. If the character you create speaks in a specific manner with an accent, attitude, or incorrect grammar, then you need to present that in your performance. In other words, *character has precedence over copy*. Of course, your producer may want you to deliver the copy in a certain way which may stifle your creative efforts, but that's the way this business works. This concept is covered in more detail in Chapter 10, "The Character in the Copy."

WORK BACKWARDS

This is a quick trick to quickly get a sense of the big picture of a story or script. The idea is to scan the script from the end first, working your way up to the first line of the script. Many times, simply reading the tag or the last few lines of a script will give you a very good idea of where you need to take your performance for the entire script. It doesn't work all the time or with all scripts, but when it does work, it can be a tremendous time saver.

MEMORIZE THE FIRST LINE

This is a neat trick given to me by fellow VO coach, Marc Cashman, to quickly achieve a conversational delivery. The general idea is to *memorize* the first sentence of a script and pick up reading the text from the second sentence. The process of memorizing the first line allows you to internalize the words, context, and interpretation so that when you begin reading you will already be in a conversational delivery. Memorizing the first line also makes it easier to stay in character and sustain attitude throughout your performance. Even though you may have the first line memorized, be prepared to adjust your delivery if asked to do so by your director.

USE MUSIC AS INSPIRATION

This advanced technique isn't for everyone, and it's definitely not a quick fix or something that will work in all situations. However, for those sessions when you have some time and are stuck trying to figure out a delivery attitude or you are facing a challenge developing a character, delivery tempo, or rhythm, this idea may serve you well.

The general idea is to use *music* as a tool for developing your interpretation or character. Music can be a powerful motivator and an inspiration at an emotional level. The ancient Greeks considered music as the study of invisible, internal, hidden objects. Your interpretation of a script is largely based on your personal, internal, hidden responses to the words in the script, so it's only natural that music can help trigger those responses to help you discover energy, attitude, emotion, and much more.

To use this technique, find a piece of music that fits the mood, tone, or energy of the script. Instrumentals will often work best, but a song with vocals may give you some ideas for phrasing or voice characterization.

As you listen to several pieces of music in your search for the most effective interpretation of your script, you may find that each piece of music you listen to will give you additional ideas. Let the music be your guide for making a variety of choices with your delivery. Test your performance choices by rehearsing the script as the music plays in the background. And, finally, rehearse your script without music to confirm that your choices are strong and effective. Of course, the music you listen to will never be used in a production, but you may discover that just by listening to music your creativity will be inspired to result in a much more effective performance.

Microphone Technique

Microphone technique is a subtle but powerful way of enhancing your character or the emotional impact of your delivery. Mic technique refers to how you use the microphone to your advantage while in the booth.

MICROPHONE BASICS

Before you can use a microphone effectively, it is helpful to first have a basic understanding of how these marvelous instruments work. The basic purpose of a microphone is to convert acoustical energy (sound waves) to electrical energy that can be manipulated and recorded. There are several designs for each of these types of microphones, *dynamic* and *condenser* being the most popular.

- *Dynamic* mics use a moving coil attached to a diaphragm (much like a loudspeaker in reverse) to convert acoustic energy to electrical energy. Dynamic mics are relatively inexpensive and rugged. Sound quality is generally better with the more expensive models. Simply plug it in to the appropriate equipment and start talking.

- *Condenser* mics use two fixed plates very close to each other, but not touching. A constant voltage is placed across the two plates, provided by a power supply (usually from a battery or external power source). As sound waves strike one plate, a change in the electrical energy is the result. Condenser mics are more expensive, far more sensitive, and more fragile than dynamic mics. The sound quality of a condenser mic is generally cleaner and "crisper" than that of a dynamic mic.

Microphones come in two primary pickup patterns: *omnidirectional* and *cardioid* (*unidirectional*). Of these, the most common type of microphone for recording is the cardioid. Omni and cardioid mics can be either dynamic or condenser. A third, less common, mic design is *bidirectional*.

- *Omnidirectional* mics will pick up sound equally from all directions and are not very common for high-quality voice recording. They are, however, usually the least expensive and most rugged.

- *Cardioid* mics (also called unidirectional mics) come in a wide variety of designs, but virtually all of them pick up sound best from directly in front of the mic. The sound pick-up reduces or fades as you move off-

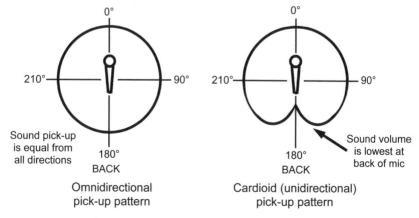

Figure 7-1: Basic microphone pick-up patterns

axis of the front center of the mic. The back of the mic is the point of maximum sound rejection.

- A *bidirectional* mic is a single mic that has the pick-up pattern of two cardioid mics placed back to back. With a bidirectional mic, maximum rejection is from the sides at 90° off-axis.

MICROPHONE PLACEMENT AND COPY STAND POSITION

In a recording studio environment you will generally be standing in front of a music stand (copy stand) with a microphone on a boom at about head level. Adjust the copy stand to eye level so you can see the entire script without having to tilt your head down. Tilting your head can affect your sound by constricting your throat and cause you to move off-mic.

Studio microphones are very sensitive and often have a "pop" screen positioned between the mic and your mouth. The pop screen serves two purposes: 1) it prevents blasts of air from hitting the microphone's diaphragm, and 2) it prevents condensation of moisture from your breath from building up on the microphone diaphragm. Over time, moisture from your breath can affect the microphone's diaphragm, dulling its sound. If properly positioned, a pop screen will not be needed for preventing breath pops, but still may be advisable for blocking condensation.

Studio microphones are usually *cardioid* (directional), and most engineers position the mic off to the side or perhaps in front of the performer, above the copy stand at about forehead level. The acoustics of the voice booth are *dead*, meaning there are no reflected echoes. The result is a very clean sound.

Microphone placement is simple for a single performer, but becomes more critical when there are several performers in the same studio, each with his or her own mic. In this case, the engineer strives to obtain maximum separation from each performer to minimize how much of each actor's voice is picked up by the other microphones.

As a starting point, position yourself so your mouth is about 6 to 8 inches from the mic. You can easily estimate this by extending your thumb and little finger; place your thumb against your chin, and the mic at the tip of your little finger. This is not a critical distance, and your engineer may adjust the mic closer or further from you. If you are working in your own home studio, you should experiment with different mic positions to discover the best placement for your voice. You may want to change the mic placement depending on the sound you want for a specific script.

WORKING THE MICROPHONE

Microphones really don't care where they are in relation to your mouth. Six inches off to the left will pick up your voice exactly the same as six inches directly in front of you or six inches above your mouth (at about eye level). You should always position yourself so you are talking across the microphone and never directly into it. Speaking directly into the mic can blast the diaphragm. Although this is rarely harmful to the mic, the resulting "popping" sounds can be a serious problem for the recording and cannot be fixed later on. In some cases, even use of a pop screen may not completely eliminate breath pops from an incorrectly positioned microphone.

As you physically move closer to a studio microphone, your voice increases in lower frequencies (bass) and the overall tone of your voice will be more intimate. This phenomenon is called *proximity effect* and is a common characteristic of all directional microphones. As you move away from a studio mic, the mic picks up more of the natural ambience of the room. This results in a more open sound, which is cooler and less intimate. Don't be afraid to experiment, but do let the engineer know what you are doing because he or she will need to adjust recording levels accordingly.

While performing, keep your head in a constant relationship to the microphone. The rest of your body can move as much as you need, provided you aren't making any noise. But your head must remain relatively stationary. If your position drifts too far off-mic, your voice will appear to fade in and out. This drives engineers crazy because the overall volume of your performance is constantly changing. Even with the best equipment, moving off-mic is extremely difficult to deal with simply because a change of just a few inches can result in a very noticeable change in the *room tone* or ambience picked up by the mic.

NEVER BLOW INTO OR TAP A MICROPHONE

Studio microphones are delicate and *very* expensive. Blowing into a microphone can cause severe damage. When testing a mic or giving a level to the engineer, always speak in the actual volume of your performance. When the engineer asks you to read for levels, consider it an opportunity to rehearse your performance.

Tapping the mic, although not usually harmful, is annoying to most engineers. It's good to keep engineers on your side; they control how you sound and have complete power in the control room. Remember basic studio etiquette—don't touch the equipment, unless, of course, it's yours!

LET THE ENGINEER POSITION THE MICROPHONE

When working in someone else's studio, always let the engineer adjust the mic to where you will be standing or sitting. Do not move or adjust the mic yourself. The same goes for the pop screen. After positioning your mic and returning to the control room, the engineer will ask for your level, and may ask you to physically change your position relative to the mic. You may be asked to *move in* on the mic (move closer), or *back off* a bit (move a bit away from the mic). These physical adjustments should be minor, and are intended to produce the right sound for your voice. If you are popping, you may be asked to change the angle of your face in relation to the mic, or to turn slightly off-mic to prevent your breath from hitting the mic.

In your personal home studio, you will, of course, have complete freedom to position your mic so you will sound your best. Experimentation will reveal the best mic placement for your home studio.

HOLDING THE MICROPHONE

You will rarely need to hold the mic during an actual session. However, it may be necessary for some auditions. If it ever happens to you, you need to know how to properly hold the mic for the best sound.

The correct handheld mic technique is to hold it vertically or at a slight angle, with the top of the mic at chin level, about an inch below the lips and slightly away from the chin, not touching the face. In this position, you will be speaking across the top of the mic rather than directly into it. Talking across the mic minimizes breath pops. You can test for proper mic placement with this exercise: Say "puh, puh," expelling a blast of air with each "puh." Slowly raise a finger from below your chin up to your lips and you will know where to position a mic to avoid being hit with your breath.

If you need to hold the mic, do not play with the cord. Just let it hang. Wriggling the cable can result in handling noise that can adversely affect your recording, even though you may not hear anything.

Using Headphones

Can you record your voice without using headphones? Of course you can! But would you want to or, more importantly, should you? There are good reasons to argue both sides of the question of whether or not to use headphones when recording your voice. Your headphones are every bit as much a tool in your studio as your microphone and, as with your microphone, there is a correct way to use your headphones, and an incorrect way. And whether you use headphones or not may depend on the type of voiceover work you are doing.

Unlike music recordings, the end product of most voiceover work is heard over speakers rather than ear buds or headphones. Monitoring under headphones removes room acoustics from the listening experience and, since music is commonly listened to under headphones, many music recording engineers are beginning to mix under headphones to create the best possible sound for the listener.

In voiceover work, headphones serve a similar, yet slightly different purpose. By wearing headphones, you will be able to clearly hear yourself as you are delivering you lines. This auditory *foldback* of your voice will accurately reflect how your microphone hears you, and will often allow you to hear subtle mistakes that might go unnoticed if you aren't wearing headphones. You will be able to hear any flaws in your room acoustics, and it will also allow you to effectively apply certain microphone techniques for achieving warmth, presence, or avoiding breath pops.

Another benefit of wearing headphones—especially if you are recording in a professional studio—is that the producer, director, and engineer will be able to communicate with you. Many studios are not set up to allow for talkback over speakers, so headphones become a critical aspect of communication. Some voice talent feel that headphones are a distraction and prefer to work without them, but the simple fact is that there is often no other option for communication between the control room and booth.

A third benefit of headphones happens when you are performing in sync with, or need to time your performance to, a playback. Headphones allow you to hear what you are working with, without the microphone picking up the playback audio.

WHAT KIND OF HEADPHONES ARE BEST?

There is no rule that says you can't take your favorite headphones to a session. It's done all the time. The important thing is that you treat your headphones as another tool for use during your performance of voiceover.

Keep in mind that your headphones need to accurately represent your voice as it will be recorded. This is the only way you will truly know that you are sounding your best. Ear buds and many consumer headphones will emphasize lower frequencies, often producing a very warm and sometimes "boomy" sound, which might sound very nice, but may not necessarily be an accurate representation of your recording. These, of course, can be used to monitor your recordings, but you should at least be aware of the consequences.

Comfort is another important aspect to keep in mind when selecting your headphones. As your voiceover work increases, you may find yourself spending many hours at a time under headphones. The last thing you want is to have sore ears or a headache caused by uncomfortable headphones.

One final consideration is listening volume. When using ear buds or some consumer headphones, the tendency is to turn up the volume. Monitoring at a too loud a level can cause your ears to fatigue, requiring you to turn up the volume. Loud monitor levels can also result in high frequency hearing loss, which can adversely affect the way you hear your recordings. If you don't already have a favorite pair of comfortable headphones, you might want to put a sample recording of your voice on an MP3 player and take it with you as you test different models of headphones. You don't need to spend a lot of money on headphones, but you should be satisfied that the headphones you choose are comfortable and will accurately represent your voice recordings.

Your Signature Voice

As you master voiceover performing techniques you will begin to discover that you fit into one of two basic categories as a voice talent: 1) you may find that you can easily create a wide variety of vocal placements and character voices, or 2) your vocal instrument is limited in range and you are learning how to create a variety of delivery styles that fall within the predominant sound of your voice. In either case, you will ultimately discover a fundamental sound or style that you are most comfortable with when performing different types of copy. This style is often referred to as your *signature voice*. It's the "voice" that gets you booked and it may be different for different genres of voiceover work. If you have the ability to create many unique character voices, you may not have a specific signature voice for your character voiceover work, but you still may have a signature voice for other types of voiceover work. This will be covered in greater detail in Chapter 10, "The Character in the Copy."

Your *signature voice* is your marketing base. When a talent buyer books you based on your demo or an audition, they believe you will be able to deliver what they need to make their project work. What the talent buyer hears may or may not be what you consider to be your signature voice—or even what you think you do best. You may be booked based on you signature voice or from a variation presented in an audition, but during a session you may be asked to take your performance in an entirely different direction. That new direction may be radically different from what got you the job, and it may even take you into unfamiliar territory.

It is unwise and unprofessional to think that you are being hired only for the *one* "style" you think you do best. As a professional voice actor, it is expected that you have the talent and ability to make adjustments with your performance and delivery style. What you think you do best—no matter what that may be—means very little to a producer. Learn how to get past your ego! You must be able to adapt your delivery style, attitude, emotional subtext, vocal placement, dynamics, and characterization to what your client needs. This does not mean that you need to be able to do wacky character voices, or take your performance to something that is beyond your abilities. But it does mean that you need to develop the skills to perform with a variety of attitudes and dynamics. And you need to do this without your ego getting in the way.

All of the techniques in this chapter are intended to help you develop range and variety in your performance. As useful as these techniques may be, their effectiveness will be compounded when you learn some basic acting skills.

---

[1] Keyes, D. (1995). *Flowers for Algernon*, Keyes, Daniel. Harcourt (reissue).
[2] Here is the proper punctuation for the line on page 91:

That that is, is. That that is not, is not. Is that it? It is.

8

Voice Acting 101

It's Not about Your Voice...
It's about What You Can DO With Your Voice!

As valuable as they are, the Seven Core Elements of an Effective Performance in Chapter 6 are just the first steps for creating an effective performance. There are many other things you can do to build on these core elements to improve the effectiveness of your delivery. The techniques you learned in Chapter 7 and the tools in this chapter will help you to develop versatility and expand the range and variety of your vocal performance. As you begin to master these techniques, you will find your delivery becoming increasingly effective. You will also discover that you will become better able to handle a wider range of delivery styles, emotions, and attitudes.

It is important to understand that the techniques you use are *not* your performance. Techniques are there to support and assist you in achieving the objectives of your performance. It's much like building a house: the hammer, saw, nails, and boards are only tools and components that are used to build the house. Acting techniques are the tools and component parts of your performance. They are the tools you use as you play your instrument.

Understanding how to use a technique in and of itself is of only limited value. You also need to know how to apply the techniques you use in the broader scope of your performance. This chapter covers concepts and performing principles that are basic to theatrical acting, and any or all will be of tremendous value to you as a voiceover performer.

Commit to Your Choices... and Adjust

All acting is based on initial choices and adjustments made to those choices as a performance develops. As you work with a piece of copy, you will be making lots of decisions and choices about who your audience is, who your character is, what the back story is, and many other aspects of your performance. It is important to commit to these choices in order to be consistent throughout the recording session.

Of course, as new choices are made to enhance your character or performance, you must commit to these also. In some cases, you may find that the choices you have committed to no longer work as well as you or the director might like. You may find it necessary to completely change or revise some of your choices. That's OK! Your choices are not engraved in stone. Learn how to explore a variety of choices and be flexible as a performer so you can make adjustments quickly, without thinking. As new choices are made, commit to them to maintain a consistent performance. As you discover and commit to the best choices that develop the character and strengthen the delivery or emotional impact of the message, you will be creating realism and believability in your performance.

The process of working through your performance to make valid choices is called *woodshedding*. This will be covered in detail in Chapter 9, "Woodshedding and Script Analysis."

BE IN THE MOMENT

This is basic acting. You must be focused on your performance. You cannot be thinking about what you are doing later that afternoon and expect to give a good performance. You also cannot be in the moment if you are struggling to get the words right or dealing with interpretation or worrying that your client might not like what you are doing. If you are even, in the slightest way, *not* focused on the copy and your performance, you will sound as though you are reading. To be in the moment, you must become comfortable with the words to the point where they become yours, and you are not thinking about what you are doing. Being *in the moment* is what Core Element #6, *Forget Who You Are and Focus*, is all about.

Being in the moment means that you understand on an instinctive level, your character, who your character is speaking to, the message in the script, your character's intentions, and innumerable other details. It also means that you speak the words in the script with a truth and honesty that comes from the heart of the character. A good way to be and stay in the moment is to practice the techniques in this chapter. Mastering this skill can take some time, so don't be discouraged if you find yourself drifting out of character or starting to think about what you are doing. Keep working at it and it will come. Some actors will spend many years developing this skill.

Your best and most real performance will be achieved when you are truly *in the moment* of the scene taking place—aware of what you are doing, but not consciously thinking about it.

BE YOUR OWN DIRECTOR—MASTERING SELF-DIRECTION

You need to learn how to look at your performance objectively, as if observing from a distance. This director in your mind will give you the cues to keep your performance on track. *Self-direction* is not only a valuable skill that you can use constantly—even when there is a director on the other side of the glass—but it is a skill that becomes absolutely essential when you record voice tracks from the comfort of your personal home studio.

When you are wearing your "director" hat, you need to be listening for all the little things in your delivery that are, and are *not* working. Look for the important words in the copy that need to receive importance or value. Look for the parts that need to be softened. Look for places to pause—a half -second of silence can make all the difference. Listen for the rhythm, the pace, and the flow of the copy. As the director, you are your own critic. Your goal is to constructively critique your performance to increase your effectiveness in communicating the message.

The process of looking at your performance objectively is part of Core Element #6, *Forget Who You Are and Focus.* It can be quite difficult if you are working by yourself. The difficulty lies in the fact that if you think about what you are doing as you perform, you will break character. Your "director's" listening process needs to be developed to the point where it happens at an unconscious level, yet you still have a conscious awareness of what you are doing as your character. The best way to learn this is to work with a voice coach or take some classes to learn what directors look (or listen) for and how they work with performers to get the delivery they want. Watch and learn as others are directed. Observe how the director focuses the performer on the particular part of the copy that needs improvement.

Record your practice sessions and have a skilled director listen to your recordings to give you suggestions on what you can do on your own. As you gain experience, your performance and self-direction become as one, and you will soon instinctively know how to deliver a piece of copy.

LISTEN AND ANSWER

An actor's job is to respond. And the best way to have a believable response is to listen. Be aware of what is going on in the copy so you have an understanding of the story and can respond appropriately. Don't just read words on the page.

- Listen to your *audience* so that your response is appropriate.
- Listen to your *character*, and to the other performers if you are doing dialogue copy. Interact with what is being said. Be real! Respond to the message emotionally and physically. Remember that acting is reacting. Listen to yourself as you deliver the lines, and observe your internal response to the words you are saying. Then react or respond accordingly. This technique can give life to an otherwise dull script.
- Listen to the director in your mind to stay on track. Learn to think critically to constantly improve your performance.
- Listen to the producer or director to take your performance where it needs to go. Your performance needs to reach the producer's vision.

- Listen to your body to find the physical tension and emotional energy needed for a believable delivery. Without physical energy, there is little more than just words.

It is only by careful listening that you will be able to respond appropriately and ultimately get out of your own way to *forget who you are.*

MAKE IT THE FIRST TIME EVERY TIME

Be spontaneous, every time! Use your acting and imagination skills to keep the copy, and your performance, fresh. Each performance (or take) should be as though the character is experiencing the moment in the script for the first time. You may be on take 24, but your character in the copy needs to be on take 1—for every take. Use your imagination to create a clear visualization of a scene, character, or situation to help make your performance real and believable take after take.

In the preface to the book *Scenes for Actors and Voices*, Daws Butler is quoted from one of his workshops[1]:

I want you to understand the words. I want you to taste the words. I want you to love the words. Because the words are important. But they're only words. You leave them on the paper and you take the thoughts and put them into your mind and then you as an actor recreate them, as if the thoughts had suddenly occurred to you.

Learn how to be consistently spontaneous! This doesn't necessarily mean that every time you deliver a line of copy it must sound exactly the same—that will depend on your choices, any adjustments you make, and the direction you receive. What this means is that you need to be able to deliver each line of your performance as though it was the first time your character ever thought of those words.

TELL THE STORY EFFECTIVELY

Don't just read the words on the page. Play the storyteller—no matter what the copy is. Search for an emotional hook in the copy—it's in there someplace—even in a complex technical script. Find a way to close the gap between the performer and the audience. Find a way to connect with that one person you are talking to, on an emotional level.

Your emotional connection may be in the softness of your voice. Or it may be in the way you say certain words. It may be in the way you carry your body as you speak your lines. Or it may be in the smile on your face. Make that connection, and you will be in demand.

The late Don LaFontaine (1940–2008) was once asked what he did as he performed. His answer was, "I create visual images with a twist of a word." It is the little shift of inflection or subtlety in the delivery of a word or phrase that makes the difference between an adequate voiceover performance and an exceptional voice acting performance. Effective storytelling is using the subtleties of performance to reach the audience emotionally and create strong, memorable visual images.

FIND THE RHYTHM IN THE COPY

Consider voiceover copy in terms of a musical composition. Music has a range of notes from high to low, being played by a variety of instruments (the voices). The tempo of the music may be generally fast or slow (the pace), and the tempo may fluctuate throughout the composition. The music also has a range of loud-to-soft (dynamics). These elements combine to create interest and attract and hold the listener's attention. Voiceover copy works the same way.

All copy has a *rhythm*, a *tempo*, and a *flow*. Rhythm in voiceover copy is much the same as rhythm in music. There are many pieces of music that run two minutes, but each has a unique rhythm. Many times, the rhythm changes within the composition. Rhythm in voiceover copy is as varied as it is in music. Some copy has a rhythm that is smooth, classy, and mellow. Other copy has a choppiness that is awkward and uncomfortable.

Some of the factors that affect rhythm in voiceover copy are pacing, pauses, breaths, the subtle emphasis of key words, and even diction and intonation. In dialogue, rhythm also includes the timing of the interaction between characters. Find the rhythm in the copy and you will win auditions.

Rhythm is something that can only be found by making the copy your own. You cannot get into a rhythm if you are just reading words off a page. Make the words your own by knowing your character, and you will be on your way to finding the rhythm. You might find it interesting to record yourself in a conversation. You may discover that you have a rhythm in the way you speak, which is quite different from the rhythm of others in the conversation.

A conversation has several things going on at once: There is a rhythm to the words, a tempo or pacing, and the interaction between the people having the conversation. Listen for pauses, people talking at the same time, the energy of the conversation, and the way in which certain words are emphasized. Observe how they move their bodies, especially when expressing an emotion or feeling. All these elements, and more, go into creating the rhythm of a conversation.

An excellent way to study vocal rhythm is to watch classic black and white movies from the 1940s. Many of these films feature some incredible character actors with interesting voices who use rhythm, tempo, phrasing, and vocal texture in powerful ways.

UNDERSTAND THE BIG PICTURE WITH SIX "WS" AND AN "H"

Look for the basic dramatic elements of a story as you study a script. These are the basic journalism five Ws—who, what, when, where, and why; and, of course, the ever popular "how." As an actor, it's very helpful to add a sixth W to define the environment in which the story is taking place—the "weather." The more details you can discover, the more accurately you will be able to portray a believable character in the story.

Here are some examples of what you can ask as you work your way through the six Ws and an H:

- Who is your character?
- Who are the other characters in the story?
- Who is your character speaking to?
- What does your character want or need at this moment in time?
- What is your character responding to?
- What is the plot of the story?
- What is the emotional relationship between the characters?
- What is the conflict?
- What complications arise?
- What events brought your character to this situation?
- When does the story take place?
- When does the peak moment happen?
- Where is the story taking place?
- Where, geographically, are other characters or objects in relation to your character?
- Why is your character in the situation he or she is in?
- Why does your character behave the way he or she does?
- Weather: Is the environment cold, hot, steamy, dry, wet, dusty, or cozy? Allow yourself to feel the temperature and other conditions of the environment so you can fully express the feelings and emotions of the story.
- How is the conflict resolved, or not resolved?
- How is the message expressed through the resolution or nonresolution of the conflict?

Ask a lot of questions! By understanding what is taking place, you will discover your role in the story. A dramatic story structure with a definite plot is most often found in dialogue scripts. However, many single voice scripts have a plot structure that evolves through the course of the story.

Unfortunately, many small-market and lower-end scripts are written solely to provide intellectual (or logical) information. Information-based copy, also known as spokesperson copy, rarely has much of a story or plot, and thus there is little or no conflict. With no conflict to be resolved, it can be very challenging to find an emotional hook. Industrial copy often falls into this category. Even with no plot, you still need to determine the audience, back story, and character, and you need to find a way to bridge the gap between performer and audience. Building that bridge can be a much greater challenge than it is with a plot-based story script. However, an emotional connection can still be made with the audience through effective characterization, and a "twist of a word."

ASK "WHAT IF... "

While working with a script, you will often find that you deliver the copy in a style that is very comfortable for you, but which may not be the most effective for your client. As you develop your performance, you'll begin to make choices that will affect the many aspects of your delivery and how the words will be perceived. At some point in time, you settle on your choices, and that's how you will perform the script.

But "what if" you slowed down on part of a line that you hadn't considered before? Or sped up on a different line? Or maybe delivered a line with changing tempo or rhythm? "What if" you lowered your volume to a point just above a whisper? "What if" you gave one word in each sentence a great deal of value and importance? "What if" you chose a different audience? "What if" your character had a different posture? "What if" your character wanted a different outcome? "What if..."

Just because you think you've got a "killer" delivery for a script doesn't mean that what you've come up with is what the director is looking for. If you get yourself stuck with one delivery, you may be in trouble when the director asks you for something different. By asking "what if..." throughout your woodshedding process, you'll come up with lots of options which will prepare you for anything the director might throw at you. Ask "what if . . ."

CHARACTER DEPARTURE

One of the best examples of "what if" is a called *character departure*. As you work with each line of a script, deliver the line in completely different ways. There should be absolutely no similarity between your various departures. If your initial choice is to speak a line slowly, run the line again very fast and with a different attitude. Then run the line again with a varying tempo with another attitude or emotional subtext. If, through your woodshedding process, you choose a vocal characteristic or character voice for the line, run it again with a completely different voice.

It only through exploring these departures from your initial choices and the other "what ifs" that you will be able to make the strongest choices for delivering your copy.

MAKE UP A LEAD-IN LINE (PRE-LIFE)—CD/8

Here's another trick to fool your brain when searching for the proper inflection of a line of copy. A *lead-in line* is simply a short statement of a possible *back story* that will give your character *pre-life* before the first line of copy. Before delivering your first line, you say something that would be a logical introductory statement, or lead-in. You can say it silently, or out loud. If you say the line out loud, leave a beat of silence before your first line of copy so that the editor can remove the unwanted lead-in line.

For example, if you are reading copy for a spokesperson commercial, you might want to have a lead-in line that sets up who you are talking to. Let's say you have determined that your audience is men and women in their thirties and forties, self-employed, and financially well-off. You have set your character as someone who is equal to the audience, so you won't be patronizing; however, you will be conveying some important information.

Here's the copy:

> Traffic! Seems like it's getting worse every day. If your
> daily commute feels like being trapped in a parking lot,
> the answer to your problem is just around the corner.
> Take the New Bus. It's not the ride you think it is.

For a *lead-in line*, you might set up the copy by putting yourself in the position of talking to your best friend, John. Rather than starting cold, set a visual image in your mind of a conversation between you and John. Deliver your lines starting with:

> *(Silently: I learned something really interesting today, and you know, John…)*
>
> Traffic! Seems like it's getting worse every day…

Your lead-in line (*pre-sentence*) sets up a conversational delivery that helps you to close the gap and communicate your message on an emotional level. This approach works for all types of copy in any situation. The *lead-in line* can be anything from a few short words to an elaborate story leading into the written copy. Generally, the shorter and more specific, the better.

MAKE UP A LEAD-OUT LINE (AFTER LIFE)

Your character lives before the first word of the script and continues to live after the last word of the script. Just as a *lead-in line* will give your character *pre-life* to help you to find the energy, attitude, and proper manner for responding, a *lead-out line* will help you to maintain your character beyond the last word spoken. And, occasionally, a *lead-out line* can help you determine the appropriate mood, attitude, or emotion for a line.

A common problem many beginning voice talent experience is that as they near the end of a script, their delivery begins to fall off, and any character they've created loses believability. There can be several reasons for this, but the most common is simply the way our brains work. Most voice talent will be reading about six to eight words ahead of the words being spoken. As the eye reaches the end of the script (or in some cases a line of copy), the brain sees its job as being done, so it relaxes and waits for the mouth to catch up. The result is a fade out in energy and delivery.

When you create a *lead-out line*, you are giving extra life to your character. The lead-out line needs to be something that is appropriate for the context of the story. It can be used to set the tone and emotional attitude for a line of copy or the end of a script.

Find an interpretation and deliver the following line of copy, first only by itself...

Please don't park the car over there.

Now, using each of the following lead-out lines, deliver the same line of copy again. Just hold the after-sentence in your head—don't verbalize it. Notice how the intention of each lead-out line can completely change your delivery of the copy just by the thought you hold in your mind:

Sweetheart!
You idiot!
I don't want it to get wet!
You'll wake the family!

Use *lead-in* and *lead-out* lines to help maintain your character and lock in the attitude and emotional subtext of your delivery.

BILLBOARDING KEY WORDS AND PHRASES

Generally, if a client or product name appears in a script, you will want to do something in your delivery that will help give it some special impact. There may also be a descriptive phrase or clause that needs some special treatment. Giving a word or phrase that extra punch is often referred to as *billboarding*. Typical methods for *billboarding* a word or phrase include: leaving a slight pause before or after you speak the words, slowing down slightly, changing your body language or facial expression, changing the inflection on the word or phrase, or even reducing the volume of your voice. All of these have the effect of giving more value and importance to the word or phrase you are *billboarding*—but only if you have the appropriate thoughts behind the words.

Emphasis is usually what directors will ask for when requesting extra punch on a word. Most people incorrectly interpret the word as meaning "to get louder," "accent," or "punctuate" in some way. By definition, the word *emphasis* means to change the intensity to add importance or value, specifically in terms of adding vocal weight to specific syllables. If you deliver a word by only making it louder, or "punching" the word, it will sound artificial and unnatural. There must be a thought in your mind in order for an *emphasized* word or phrase to have any meaning. Without the thought, it's just a louder word. When a director asks you to emphasize a word, change the way you think to interpret the request as asking for you to give greater importance and value to the word or phrase.

If you *billboard*, or place extra emphasis, on too many words or phrases, your delivery will sound artificial and forced, losing believability and credibility. Experiment with different ways to give value and importance to names, places, and phrases in a script. You will soon find one that sounds right. As a guide to help with your delivery, underline words you feel are important. As you work on your delivery, you may discover that underlining only the syllable that should receive value, rather than the entire word, can completely change the meaning or create a regional delivery. For example: defense could be spoken as <u>de</u>fense or de<u>fense</u>.

PERSONAL PRONOUNS—THE CONNECTING WORDS

First person personal pronouns—I, me, we, us, our, you, and your—are all words that listeners tune in to. These are *connecting words* that help the voice actor reach the audience on an emotional level. Use these words to your advantage. Take your time with these words and don't rush past them.

In some copy, you will want to give these words a special importance for greater impact. Most of the time, you will want to underplay the personal pronouns and give extra value to words that are the subject of a sentence. For example, the sentence—"It's what you're looking for!"—could have value placed on any of the five words, or a combination of two or more. The contractions—"it's" and "you're"—could even be separated into "it is" and "you are." Each variation gives the sentence a unique meaning. Read the line out loud several different ways to see how the meaning changes. Placing the greatest value on the word "you're" may not be appropriate if the context of the script is all about searching for exactly the right product. In that case, the word "looking" would probably be the best word to receive the greatest importance. Experiment with this phrase by changing the context to discover a delivery that sounds best for you.

It's what you're looking for.

A general rule-of-thumb is that when you emphasize or "punch" personal pronouns, the meaning shifts from the subject of the sentence to the individual being addressed. In the example above, the most important part of the line in the context of the story might be the aspect of finding that special thing everyone is "looking" for. By placing importance on the word "you're," the focus of the intent is shifted to the person and moved away from the action of looking. The result for the listener is that the meaning can be unclear, confused, or in some cases just doesn't make any sense.

There are certainly many situations in which the pronoun is exactly the proper word that needs to receive importance and value. However, this is usually only a valid choice when the individual being addressed is the subject of the intention for that line of copy.

WORK THE CONSONANTS

Bringing life to a script will often mean giving value and importance to certain words and phrases. But you can achieve similar results on a smaller scale when you *work the consonants*. Rather than emphasize an entire word, limit the emphasis to only the primary consonant in the word. This approach will help a word "pop" giving it a crisper edge in the context of a sentence. To do this, simply give the consonants a bit crisper articulation. The trick with using this technique is to find the correct amount of emphasis or articulation. If you hit the consonant too soft, the word can get lost in the mix. If you hit the consonant too hard, it can sound artificial.

This technique works well with copy that is descriptive, or which must be delivered quickly. There may not be enough time to spend with specific words and working the consonants will often achieve the same result. Also, working the consonants may be helpful for discovering a strong delivery for a line of copy. For example, in the following line of copy, the strongest delivery will be one which enables the listener to taste the food.

Crispy duck lumpia, basil scented prawns

Working the consonants in a way that lets you taste the food, will help the listener taste the food. For fast copy, working the consonants can help your delivery "cut through" the mix when music and sound effects are added. Deliver the following line first in a conversational style, then by giving the consonants just a bit more articulation or emphasis at a faster pace, and observe the difference.

The greatest deals of the decade at our grand opening sale.

Notice how *working the consonants* almost forces you into a certain delivery style. If your delivery needs to be conversational, this technique may not be appropriate as it can produce a choppiness or insincere delivery style. Use care when applying this technique. It may not work for every script, but this is definitely a technique worth keeping in your back pocket.

BUILDING TRANSITION BRIDGES

A copy *transition* is a *bridge* between concepts within a line, between subjects within a script, or between characters, and it can take many forms. It may be a transition of a character's mood or attitude. Or it may be a transition in the rhythm or pace of delivery. It might be a transition from a question asked to an answer given. It could even be a transition between concepts or products within a list.

In a script these transitions may be indicated by an ellipsis, a comma, a hyphen, a colon, a semicolon, or even no punctuation at all. As an actor, you are at liberty to include transitions wherever they feel appropriate as you perform a script.

Transitions help "hook" the audience and keep their attention. Look for transitional phrases in the script and decide how you can make the transition interesting. Avoid keeping your delivery the same through all the transitions as you read a script. Give each transition a unique twist. Change your physical attitude, movement, mental picture, or use some other device to let your audience know that something special has happened, or that you have moved on to a new idea.

Sometimes all that is needed is a slight change in your facial expression or body posture. Sometimes a shift in volume, importance, back story, or who you are talking to will create the *transition bridge*. And sometimes, a simple pause in your delivery will do the trick. Experiment with different techniques to find out what will work best for the copy you are performing. In time you will develop a style that sets you apart from other voice talent.

USING CONJUNCTIONS

How do you handle the *conjunctions* "and," "but," and "or" when they appear in a script? These three words are loaded with opportunity for creating transitions and building interest through your performance. "And" is an additive word used to connect two or more things: "We have small and medium and large sizes." "But" indicates opposites: "Oranges are sweet but lemons are sour." And "or" connotes a comparison between two things: "Do you prefer red or blue?" These little words can be stretched, emphasized, sped up, slowed down, charged with emotion, or thrown away.

How you handle these words when they come up will largely depend on your interpretation of the copy, the character you choose, and countless other choices. The challenge with conjunctions is that many voice talent will emphasize the conjunction in a misguided attempt to make whatever follows appear more important. What often happens is that the listener only hears the emphasized conjunction and what follows actually loses value.

There are many occasions when giving value, or emphasis, to the conjunction will enhance the meaning of the phrase. On the other hand, there are just as many, if not more, occasions when it will be more effective to deemphasize the conjunction. Be cautious when emphasizing conjunctions. The only way you'll truly know what works best in the context of a script is to test the phrase in several different ways.

DEALING WITH LISTS

Lists are common in all types of voiceover work, but the way a list is handled may differ depending the context of the script, the character speaking, or the genre of the voiceover work.

Some aspects of working with lists have been covered earlier. For example, in commercials, a list will most often be delivered with varying inflections that allow each item to stand alone, yet be still be tied to the list as a whole. However, a list in a promo will often be delivered with each item given a downward inflection to create a sense of intensity and dramatic impact.

The ideas of substitution and adding or subtracting a word or two, discussed on pages 97 and 98, are extremely useful when working with lists. If time permits, adding conjunctions between items of a list can add impact and eliminating conjunctions from a scripted list can create a sense of authority and drama.

CONTRAST AND COMPARISON

A common writing technique is to present a comparison between two or more items, or to contrast the positive versus negative or other aspects of a topic. In almost all cases, a contrast and comparison will be followed by a benefit as to why one or the other is better. When you discover a contrast or comparison in a script, your job is to make the difference very clear to the listener. Here's an example:

> Most digital cameras require expensive, hard-to-find batteries. The new Sigma Solar camera doesn't use batteries—it uses the power of light. So you'll never have to worry about a dead battery again.

This script contains both a contrast and comparison. The comparison is between regular digital cameras that require batteries and the new solar camera. The contrast is between expensive, hard-to-find batteries and solar (light) power. The benefit is stated in the last line.

A contrast and comparison is best delivered by understanding the meaning of all aspects of the comparison and the ultimate benefit being discussed. When you understand the benefit you can create an emotional connection to that benefit and an appropriate thought that corresponds to that feeling. Use the feeling and thought as you speak the words for both parts of the contrast and comparison to create a believable delivery. If you don't truly understand the comparison or why the contrast is important, your delivery of the copy will be flat and emotionless.

A good technique for getting to the contrast or comparison quickly is to tail off the unessential parts of the script.

Most digital cameras require... batteries. The new Sigma... uses the power of light.

Once you've got the essence of the contrast or comparison, it's a simple matter to experiment with a variety of delivery options.

THE TELL

The biggest challenge with voiceover copy is to create, in the mind of the listener, a sense of authenticity, truth, reality, and knowledge through the performance. Without this, you stand a good chance of losing credibility with the listener and all your well-intended efforts will be for naught.

If you've ever played a game of poker, you know that a player can reveal their position through a simple unconscious gesture or facial expression. Another player who can read this *tell*, and knows what it means, may be able to maneuver the game to his advantage.

Voiceover has its *tells* as well, and many of them have been discussed earlier in this book. Voiceover *tells* are those performance characteristics that affect the believability of the character and credibility of the message.

The critical *tell* in fast-paced commercials is a catch-breath, or short, audible breath between phrases. The catch-breath, along with a pause are both also common in long-form narration. These breaks in the continuity of the delivery are a clear indication to the listener of a lack of confidence, expertise, or knowledge in the presenter. It doesn't take much for a listener to know what this *tell* means. In fact it is understood on a subconscious level and can result in an instantaneous loss of credibility. The result is often a performance that shifts from a believable communication to one that sounds like the performer is reading the script or simply doesn't care.

The best way to avoid this *tell* is, first, to be aware that you are doing it, and second, to master the ability to create a character and performance style that does not allow for this *tell* to take place. Training, practice, and study with a competent voiceover coach or director are the best ways to eliminate all tells from your performance.

Physical Attitudes to Help Delivery

M.O.V.E.

Remember Core Element #5 from Chapter 6? That pesky element of *energy* is so important that it deserves some additional discussion here.

Be physical! Body movement is an expression of emotion, and your expression of emotions or feelings is the result of the thoughts you hold in your mind. When you verbalize those thoughts the meaning of the words will communicate through the tone of your voice. Move your body in whatever manner works for you to effectively get to the core emotion of the message. Your **M**ovement **O**rchestrates your **V**ocal **E**xpression! M.O.V.E.

Try the following using this simple phrase: "You want me to do what?" Begin by standing straight and stiff, feet together, arms at your sides, head up, looking straight ahead with an expressionless face. Now, say the phrase out loud—without moving your body, arms, or face—and listen to the sound of your voice. Listen to the lack of expression in your voice. Listen to how totally boring you sound.

While keeping the same physical attitude—and still without moving, say the same phrase again and try to put some emotion into your reading. You will find it extremely difficult to put any emotion or drama into those words without moving. When you begin to communicate emotions, your body instinctively wants to move.

Now, relax your body, separate your feet slightly, bring your arms away from your sides, and loosen up. Think of something in your past that you can relate to the phrase and recall the physical tension or feeling you originally felt. Say the phrase again—this time moving your arms and body appropriately for that original feeling. Listen to how your physical attitude and facial expression change the sound of your voice. Try this with different physical positions and facial expressions and you may be amazed at the range of voices you will find. A big smile will add brightness and happiness to the sound of your voice. A frown or furrowed brow will give your voice a more serious tone. Tension in your face and body will communicate stress through your voice.

It's a mistake to stand in front of the microphone with your hands hanging limp at your sides or stuffed in your pockets—unless that physical attitude is consistent with your character in the copy. Start your hands at about chest level and your elbows bent. This allows you the freedom to move your hands as you speak.

The way you stand can also affect your voice performance. Although body stance primarily communicates information visually, it can also be very important when creating a character. Body language, just as facial expression, translates through the voice. For example, to make a character of a self-conscious person more believable, you might roll your shoulders forward and bring your arms in close to the body, perhaps crossing the arms at certain points in the copy.

Physical changes help to create a believable character who is somewhat self-conscious, a bit defensive, perhaps unsure of the situation and who may even be shy and focused on how she or he is perceived by others. Your body posture assists in framing the attitude and personality of the character. The following are some typical body postures that will help you understand how body stance can affect your performance. If used unconsciously, these postures can have an adverse affect on your performance because they will have a direct impact on your speaking voice. However, when consciously applied to a character or attitude, these and other body postures can be used to enhance any voice performance:

- **Arms behind back ("at-ease" stance)**—This body posture reflects nervousness and implies that the speaker doesn't know what to do with his or her hands or is uncomfortable in the current situation. Clasping the hands in back or in front of the body tends to minimize other body movement and can block the flow of energy through your body. This in turn may result in a "stiffer" sound with a restricted range of inflection and character.

- **Straight, stiff body with hands at the side ("attention" stance)**— Standing straight and tall, with chest out, head held high and shoulders back implies authority, control, and command of a situation. This projection of power and authority can be real or feigned. This stance is sometimes used as a bluff to create an outward image of authority to cover for an inward feeling of insecurity. This body stance can be useful for a character who must project power, authority, or dominance over a situation.

- **Arms crossed in front of the body ("show me" stance)**—Crossed arms often represent an unconscious feeling of self-consciousness and insecurity, creating an attitude of defiance or being defensive. Crossed arms can also imply a certain level of dishonesty.

- **Hands crossed in front of the body ("Adam and Eve" stance)**— As with the at-ease stance, this posture implies that the speaker doesn't know what to do with his or her hands. This stance, with the hands crossed like a fig leaf, is commonly perceived as an indication that the speaker has something to hide. This stance can be useful in helping create a character who projects suspicion.

- **Hands on the hips ("mannequin" stance)**—This posture makes the speaker appear inexperienced or unqualified. Hands on the hips also blocks the flow of energy through the body and limits the performer's ability to inject emotion and drama into a performance. This stance can be used to create an attitude of arrogance.

Don't be afraid to be physical in the studio. Remember, movement orchestrates vocal expression! A simple adjustment of your *physical energy* can make a huge change in your performance. I have seen voiceover performers do some of the strangest things to get into character. The basic rule is "whatever works—do it." I once worked with a voice actor who arrived at the studio wearing a tennis outfit and carrying a tennis racket. Throughout the session, he used that tennis racket as a prop to help with his character and delivery. I've seen other voice actors go through a series of contortions and exercises to set the physical attitude for the character they are playing. A friend of mine was working a dialogue script and her male dialogue partner was having trouble getting into the right delivery. To get into the proper attitude, the two of them actually lay down on the studio floor as they delivered their lines. Your analysis of the copy can give you a starting point for your physical attitude. When you've decided on your physical attitude, commit to it and use your body to express yourself.

Many people are self-conscious when just starting in this business, and that's normal. However, when you are in the "booth," you really need to leave any judgments you may have about your performance outside. If you are concerned about what the people in the control room think about you as you are performing (rather than what they think about your performance), you will not be able to do your best work. It comes down to taking on an attitude of "I don't care" when walking into the booth or studio. It's not that you don't care about doing your best, or making the character in the copy real and believable. You must care about these things. But you cannot afford to care about what others think of you and what you are doing as you perform to the best of your abilities. And besides, as you may be starting to realize, it's not you delivering those words anyway—it's really your character who's speaking!

If getting to your best performance means moving your entire body and waving your arms wildly are appropriate for your character, that's what you need to do. You can't afford to worry that the people in the control room might think you are crazy. The engineer and producer certainly don't care! They are only interested in recording your best performance as quickly as possible, and I guarantee they've seen some pretty strange things.

Usually you can perform better if you are standing, but in some cases, being seated may help with your character. If you sit, remember that a straight back will help your breathing and delivery. If possible, use a stool rather than a chair. Sitting in a chair tends to compress the diaphragm, while a stool allows you to sit straight and breathe properly. If a chair is all that's available, sit forward on the seat rather than all the way back. This helps you keep a straight back and control your breath. Most studios are set up for the performers to stand in front of the microphone. Standing allows for more body movement and gives you a wider range of motion without being restricted.

Your physical attitude is expressed through the relaxation and tension of your face and other muscles. All human emotions and feelings can be communicated vocally by simply changing physical attitudes. Often, the copy expresses a specific emotion or attitude. Find a place in your body where you can feel the tension of that emotion or attitude—and hold it there. Holding tension in your body contributes to the realism and believability of your character. Focus on centering your voice at the location of the tension in your body and speak from that center. This helps give your voice a sense of realism and believability.

A tense face and body will communicate as anger, frustration, or hostility. A relaxed face and body result in a softer delivery. Try reading some light copy with a tense body; you will find it very difficult to make the copy believable. You can make your delivery friendlier and more personable simply by delivering your lines with a smile on your face.

Tilting your head to the side and wrinkling your forehead will help convey an attitude of puzzlement. Wide-open eyes will help create an attitude of surprise. Practice reading with different physical attitudes and you will be amazed at the changes you hear. Your physical attitude comes through in your voice.

Another physical gesture that can make a big difference in your delivery is something commonly referred to as *air quotes*. When a word or phrase needs special emphasis or needs to be set apart from the rest of the copy, simply raise both hands and use your index and middle fingers to simulate making quotation marks in the air surrounding the words as you speak. The mere gesture almost forces you to say the words differently by separating them from the rest of the sentence with a distinctive shift of attitude. Air quotes are best used in moderation and must be part of the fluid physicality of your performance. The challenge with using air quotes is to maintain the authenticity of your character and the context of the phrasing. This gesture may not be appropriate for all copy, and excessive use of air quotes can result in a delivery that sounds choppy and artificial. If you do nothing more than this single gesture, you will hear a difference, but when you use air quotes in combination with other physical movement, the effect can be profound.

Your Clothes Make a Difference

Wear comfortable clothes when recording. Tight or uncomfortable clothing can be restricting or distracting. You do not want to be concerned with shoes that are too tight when you are working in a high-priced recording studio. Stay comfortable. The voiceover business is a casual affair. With the increase in home studios for voiceover work, you can now even record in your jammies and no one will know the difference. I even know of one voice actor who will occasionally record—how shall I say this—in the all-together. He says it's a very freeing way to work.

Another note about clothing: A studio microphone is very sensitive and will pick up every little noise you make. Be careful not to wear clothing that rustles or "squeaks." Nylon jackets, leather coats, and many other fabrics can be noisy when worn in a recording studio. Other things to be aware of are: noisy jewelry, loose change, cell phones, and pagers. If you do wear noisy clothing, it may be necessary for you to restrict your movement while in the studio, which can seriously affect your performance. Maybe my friend who records in the all-together has something!

If you are recording in your home studio, you'll need to not only be aware of clothing noise, but also the many other potential noise sources around your home, both inside and out. Dealing with environmental noise will be covered later in the section on home studios.

The Magic of Your Mind:
If You Believe It, They Will!

One of the objectives of voice acting is to lead the listener to action. The most effective way to do that is to create believability through a *suspension of disbelief.* You suspend disbelief whenever you allow yourself to be drawn into a story while watching a movie or play or read a book. You are fully aware that what is taking place in the story really isn't real. However, as you experience it, you suspend your disbelief and momentarily accept the appearance of the reality of what is happening in the story.

Suspension of disbelief in voiceover is essential for creating a sense of believability in the message. The audience must believe you, and for that to happen, *you* must, at least momentarily, believe in what you are saying.

Use your imagination to create a believable visual image in your mind for the message you are delivering. The more visual you can make it, the more believable it will be for you and for your audience. On a subconscious level, your mind does not know the difference between illusion and reality. Just as your physical attitude affects the sound of your voice, if you create a strong enough visual illusion in your mind, your words will be believable.

Creating a visual illusion is a technique used by most great actors and virtually all magicians. For a magician to make the audience believe that a person is really floating in the air, he must momentarily believe it himself. The performer's belief in what is taking place contributes to establishing the suspension of disbelief in the audience. If the magician is focused on the mechanics of his illusion, he will not give a convincing performance.

If you are focused on the technical aspects of your performance you cannot possibly be believable. The technical aspects and techniques of your voiceover work must become completely automatic to the point where you are not even aware of them. The words on that script in front of you must come from within you—from the character you create. Only then will you be able to successfully suspend disbelief. This is what's meant by the phrases "making the words your own" and "getting off the page."

Visual imagery is a powerful technique for creating believability when delivering any type of copy. Read your script a few times to get an understanding of what you are saying. Then, set your visual image and let your character come in and be the storyteller, the expert, the spokesperson, the salesperson, the eccentric neighbor, the inquisitive customer, the kooky boss, and so on. By allowing your character to take over, you automatically shift your focus from the technical aspects of reading the copy to the creative aspects of performing and telling the story.

A visual image helps give life to your character, reason for its existence, an environment for it to live in, and motivation for its words. Visualization helps make your character believable to you. If the character is believable to you, its words become true, and the message becomes believable to the audience. To put it another way: If you believe it, your audience will.

Trends

A considerable amount of voiceover work is in the form of advertising as radio and television commercials. The advertising industry is generally in a constant state of flux simply because its job is to reach today's customers in a way that will motivate them to buy the current "hot item." In order to do that, advertisers must connect on an emotional level with their audience. And, in order to do that, the delivery of a commercial must be in alignment with the attitude and behavior of the target audience. Each generation seems to have a unique lifestyle, physical attitude, slang, and style of dress. These constantly shifting *trends* are reflected in the advertising on radio and TV. In other words, what is "in style" today may be "out of style" tomorrow.

As a voice actor, it is important that you keep up with the current trends and develop flexibility and versatility in your performing style. You may develop a performing style that is perfect for a certain attitude or market niche, but if you don't adapt to changing trends you may discover that your style is no longer in demand. During the mid 1990s, the Carl's Jr. restaurant chain ran an advertising campaign that featured a very droll, flat, almost monotone voiceover with a very definite lackadaisical, yet sardonic attitude. The key phrase of the campaign was "If it doesn't get all over the place, it doesn't belong in your face." The delivery style became a trend. Suddenly there were commercials everywhere that had a similar delivery style. The trend lasted for a few years, during which a handful of voice talent who could effectively perform in that style did quite well, financially. But when the trend had run its course, that flat, monotone delivery style vanished from the advertising scene almost overnight. Those voice actors who were at the top of their game during those few short years found it necessary to adapt and follow the current trends if they were to continue to be successful in their voiceover careers.

Probably the best way to keep pace with current trends is to simply study radio and television advertising that is on the air today. Listen to what the major national advertising producers are doing in terms of delivery attitude, pace, and rhythm. Observe the energy of the music and how the visuals are edited in television commercials and notice how the voiceover works with or against that energy. Look for commonalities among the commercials you study, and you will begin to notice the current trends.

One thing you will notice is that most locally produced advertising does not follow national advertising, or at best, is several months behind.

You don't really need to do anything about these trends, other than to be aware of what they are and how they might affect your performance. That awareness will prove to be another valuable tool for you to use when you audition or are booked for a session. Use it to your advantage.

---

[1] Butler, D. (2003). *Scenes for Actors and Voices.* BearManor Media.

9

Woodshedding and Script Analysis

The Director in the Front Row of Your Mind

As you study voice acting you will develop instincts as to how to develop your character, deliver your lines, and create drama in your performance. These instincts are good and necessary for a professional performer. However, if left at the level of instincts, they will limit your ability to find the nuance and subtlety of the performance—those seemingly insignificant things that make the drama powerful, the dialogue interesting, or a comedic script hysterical rather than just humorous.

All voiceover copy is written for the purpose of communicating something—selling a product or service, providing information, education, or expressing an emotion or feeling. No matter how well-written, it is not the words in and of themselves that convey the message, it is the *way* in which *the words are spoken* that ultimately moves the audience. It is the details of the performance behind the words—the nuance—that allow a performer to bring a script and a character to life. And behind every performer, there is a director. Somewhere in your mind is a director. You may not have realized it, but that director is there. Allow your director to sit front row, center in your mind—in a big, overstuffed chair—so he or she can objectively watch your performance to keep you on track and performing at your best.

Voiceover copy is theatrical truth—not real-life truth—and your internal director is the part of you that gives you silent cues to keep you, or rather, your character, real. As you work with copy, you will find a little voice in your head that tells you, "Yeah, that was good" or "That line needs to be done differently." The director in the front row of your mind is the result of critical thinking. He or she is the part of you that keeps you on track, helps you stay in the moment, and gives you focus and guidance with your performance. Think of this director as a separate person (or part of you) who is watching your performance from a distance, yet close enough to give you cues.

Over time, your internal director and your performance will become as one—a seamless blending of director and performer resulting in a truly professional dramatic artist, without any conscious effort. This mastery of self-direction is the level to strive for. This is what theater is all about. This is what you, as a voice actor, can achieve with any type of copy you are asked to read.

But there is a catch! As with most things in life, you must learn to walk before you can run. When performing voiceover, a mastery of performing skills is only the beginning. You must also learn how to dig deep into a script to uncover the truth that is hidden behind the words. This is a process commonly known as *script analysis*, or *woodshedding*.

"Woodshed" Your Copy

Although this chapter will include some review of concepts covered earlier, everything here is intended to help you discover a process of your own for quickly and effectively uncovering the details in any script so you can practice in a way that will bring the character to life. Once mastered, your personal process for *woodshedding, or script analysis,* may happen as quickly as reading through a script.

The term *woodshedding* comes from the early days of American theater. As theatrical troupes traveled to new frontiers in the early West, the only place they could rehearse and work out their performances was in a woodshed. The term stuck and its still in common use today.

From the moment you first read any script, you will instinctively come up with a way to speak the words. Sometimes your gut instincts and choices will be dead-on accurate. At other times, you may struggle with a script as you try to figure out what it's all about and your character's role in telling the story. The character you create may ultimately be defined as simply an "announcer" or spokesperson doing a hard-sell sales pitch or, perhaps, a "friendly neighbor" telling the story about a great new product he has discovered. In other cases, the character you need to define may have a complex personality with a range of emotions. For almost every script, you'll need to do some sort of basic analysis to uncover the information you need for an effective performance. The process you use may be very simple, or it may be a complex analysis of every detail in the script. As your skills develop, you will most likely change they way you *woodshed* a script.

Let's review some of the key elements of copy that can help determine your character, attitude, emotion, and other aspects of your performance. For a more complete explanation, please refer to Chapters 6, 7, and 8.

- **The structure of the copy** (the way it is written) — Is the copy written in a dialect style? Is the wording "flowery" or expressive in some way? Is the copy a straight pitch? What is the pace of the copy? What is the mood of the copy? What is the attitude of the character?

- **Know the audience**—Knowing the target audience is a good way to discover your character. Experienced copywriters know that most people fit into one of several clearly defined categories. The words and style they choose for their copy will be carefully chosen to target the specific category of buying public they want to reach. Specific words and phrases will be used to elicit an emotional response from the target audience. Your character may be defined in part by the words spoken to convey a thought, or his or her attitude may be clearly expressed within the context of the copy.

- **What is the back story** (the moment before)?—What happened before the first word of copy? The back story is the specific event that brought your character to this moment in time and to which he or she is responding. This may or may not be obvious in the script. All voiceover copy has a back story. If a back story is not defined within the context of the script, make one up.

- **Who are the characters?**—Who is your character and how do other characters, known or unknown, interrelate with your character and each other (as in a dialogue script)? This interaction can give solid clues about your character.

- **What is the scene?**—Where does the story in the script take place? What is the environment? Temperature? Understanding the big picture of the script will reveal a tremendous amount of information that will help you discover the most effective performance.

- **What does your character want?**—Your character has a specific purpose for speaking the words in the script. What is the underlying want and need of your character, and what is ultimately achieved by the end of the script?

- **How does your character behave or move?**—The writing style or context of a script will often reveal how your character moves and behaves as he or she responds to various to other characters or situations occurring in the story.

- **What is the conflict?**—What happens in the copy to draw the listener into the story? Where is the drama in the story? How is conflict resolved or left unresolved? Is the conflict humorous or serious? How is the product or message presented through the resolution or nonresolution of conflict? It may take some digging, but when you discover the conflict, your performance will be much more interesting and compelling.

There are many other clues in the copy that will lead you to discover the character. As the performer, you may have one idea for portraying the character, and the producer may have another. If there is any question about your character, discuss it with the producer.

Creating a Performance Road Map: Analyzing and Marking a Script

One of the first things you should do as you begin working with a script is to quickly analyze it; *woodshed* it, searching for clues to help you create a believable character and effective delivery.

As you begin working with voiceover copy, you may find that it will take you a few minutes to make the choices about your character and other aspects of the copy. However, as you gain experience, you will be able to do a thorough woodshedding in the time it takes you to read the copy a few times.

The Script Analysis Worksheet on pages 132 and 133 can be used when working with any piece of copy. The worksheet is essentially another tool you can use when breaking down a script to define the *Seven Core Elements* of a performance. If you find a sequential, linear process beneficial, you may find the worksheet helpful.

Once you've done this process a few times, it will become automatic and you won't need the worksheet any longer. By answering the questions on the worksheet, you can quickly learn everything you need to know about a script and your character. If an answer is not clear from the copy, then make it up. You won't be graded on your answers, I promise. The answers you come up with will give you critical information you can use in developing effective characters and delivery. They are simply a way for you to make practical choices for the script you are performing. For you to maintain a consistent performance, it is important that you stick with the choices you make in your script analysis. If something isn't working for you, of course, you can change your mind. But any new choices or changes should only be made to make your performance and your character more real and believable.

TO MARK OR NOT TO MARK

Through experimentation, you will find a form of script analysis that works for you. You may find that it is very helpful to mark your script with notes, lines, and boxes designed to chart your path through a performance. Or you may find it unnecessary to mark your script, and instead only make minor notations as needed. Whatever works for you is what you should use.

If you find you are paying too much attention to your notations as you read a script, you are probably over-analyzing the text. This can result in a delivery that is unfocused and sounds like you are reading. As you develop your personal process for script analysis and notation, and your performing skills improve, you will most likely find you need to mark your script less and less.

Regardless of your individual process, or how much you mark your script, the basic process of *woodshedding* will remain the same. As you analyze a script, you will want to look for key words and phrases that reveal attitude and emotion, and give clues about your character and how your character responds to information, situations, and other characters. Notice the context of the copy and how the message is presented. Look for places where you can add variety by using the dynamics of pacing, energy, attitude, tone of voice, and emotion. Look for natural breaks, shifts of attitude or emotion, and transitions in the copy. Look for *catchphrases* that reveal something about your character's attitude, emotion, or feelings.

By the time you read a script through once or twice, you should be able to make some solid choices on how you intend to perform it. You should know who the one person is you are speaking to (the *audience*); who you are as the speaker (your *character*); and what you are responding to, or why you are speaking the words in the script at this moment in time (your *back story*).

Marking your script with specific notations can help you create a map of how you will deliver it. These markings are your personal cues to guide you through an effective performance of the copy.

Practice marking magazine or newspaper articles or short stories and you will quickly find a system that works for you. In a short time, you will refine your system to a few key markings which you can use regularly to guide you through almost any script.

Here are a few suggested markings and possible uses. Adapt, modify and add to them as you like:

- Underline (_____)—emphasize a word, phrase, or descriptive adjectives
- Circle (O)—key elements of conflict in the script
- Box (□)—the peak moment in the copy—put a box around the words or phrase at that point in the copy
- Highlight (▓▓▓▓) or different color underline—resolution or nonresolution of conflict
- Arrow pointing UP (➚)—take inflection on a word up
- Arrow pointing DOWN (➘)—take inflection on a word down
- Wavy line (~~~)—modulate your voice or inflection
- Slash or double slash (//)—indicate a pause

One of the most common markings is to simply underline a word that needs to be emphasized. This works fine in most cases, but there may be times when you want to make sure you say a word correctly. Try underlining only the syllable of the word that needs emphasis. For example: <u>de</u>fense or de<u>fense</u>. Another important thing about script marking is that, although you certainly should understand its proper use, it's a good idea to

Script Analysis Worksheet

Answering the following questions, based on the copy, will help you discover the audience you are speaking to, your character, and any special attitude you need to incorporate into your performance.

Who is the advertiser or client? _____

_____

What is the product or service? _____

_____

What is the delivery style?
- ☐ Fast and punchy   ☐ Conversational/friendly   ☐ Relaxed/mellow
- ☐ Single voice   ☐ Dialogue/multiple   ☐ Character/animation
- ☐ Authoritative   ☐ Business-to-business   ☐ Narration

Who is the advertiser/client trying to reach (target AUDIENCE)? Determine the age range, income, gender, buying habits, and any other specific details that become apparent from the way the script is written. Who is the "other person" you are talking to? Visualize this individual as you perform the copy.

_____

_____

Find important key words or catchphrases where the use of dynamics of loudness or emotion will give value and importance. Look for the advertiser's name, product, descriptive adjectives, and an address or phone number. These elements may need special attention during your performance. Underline or highlight the words or phrases you want to make important.

What is the message the advertiser/client wants to communicate to the target audience? What is the story you are telling through your performance? What is the USP (unique selling proposition)?

_____

_____

How does the story (plot) develop? For dialogue copy, find the setup, the conflict, and how the conflict is resolved or not resolved. Discover how the plot flows. Are there any attitude changes with your character or others? Plot development is critical to effective dialogue copy. Determine your role in the plot and how your character develops.

_____

_____

Use arrows ↗ ↘ to indicate copy points for changes in inflection or attitude.

What is your role (CHARACTER in the story) in terms of how the story is being told? Do a basic character analysis to define your character's age, lifestyle, clothing, speaking style, attitude toward the product or situation in the script, etc. What are your character's motivations? What are your character's WANTS and NEEDS (DESIRES) at *this moment in time*? What happened immediately before the copy to which your character is responding (BACK STORY)? Be as detailed as you can in order to discover your character.

How does your character relate to any other characters in the script, or to the audience in general? Is your character an active player in telling the story (as in a dialogue commercial), or is your character that of a narrator imparting information to a captive audience (as in a single-voice "spokesperson" commercial)? What can you do to create a bond between your character, other characters in the script, and the audience?

What can you do to make your character believable? Any special vocal treatments or physical attitudes?

Does your character have any unique or interesting attitudes, body postures, or speaking characteristics (speaks slowly, fast, with an accent, squeaky voice, etc.)? If so, identify these.

Study the copy for pauses that might be used to create tension or drama, and for places to breathe. This is especially important for industrial copy, which frequently contains long, run-on sentences with technical terminology. Mark breaths and pauses with a slash mark (/).

Find the rhythm of the copy. All copy has a rhythm, a beat, and timing. Discover the proper timing for the copy you are reading. Dialogue copy has a separate rhythm for each character as well as an interactive rhythm.

Look for transitions in the script (similar to attitude changes). These may be transitions from asking a question to providing an answer (common in commercial copy), or a transition between the attitudes of your character.

Look for key words you can give importance to, and that will connect you with the audience. Personal pronouns, such as "you," "our," "my," and "I," may be written into the script or simply implied. If connecting words are implied, find a way to make that implied connection through your performance (without actually saying the words).

reduce your markings as your performing skills develop. A heavily marked script may not only be difficult to read, but may also require a great deal of thought as you follow your roadmap. The more you must think about what you are doing, the less you are truly in character.

The degree to which you mark your script may vary from project to project, but it will certainly help to have a system in place when you need it.

Woodshed Your Script to Be More Believable

Just as you have a personality, so does the character written into every script. The character for a single-voice script is often simply that of an announcer or spokesperson delivering a sales pitch of some sort, or communicating basic information. But, even this announcer has a personality that is appropriate to the copy. Scripts written for dialogue or comedy have multiple characters that are often more easily defined. For all types of copy, finding the personality of the character allows you to give the character life and helps make your performance believable. Remember, making your performance believable is what voice acting is all about.

The best way to effectively communicate a scripted message is to create a believable character telling a believable story. To be believable, your performance must include variety, tension, conflict, and sincerity. It must also be easy to listen to and in a style that the audience can relate to. To be believable, you must know your character and develop a performing style that is conversational and real.

CHARACTER ANALYSIS

The role you play in a voiceover performance may be defined simply by the manner in which the words are written, or the context may be vague leaving it up to you to create something. Scripts written for specific or stereotyped characters occasionally have some directions written on the script, something like: "read with an English accent," "cowboy attitude," or "edgy and nervous." Many times, producers or writers will be able to give you additional insight into their vision of the character. It will then be up to you to create an appropriate attitude and voice for that character.

In theater, this process of defining the attitude and personality of a character is called a *character analysis*. As a voice actor, you need to know as much about the role you are playing as possible. The more details you include in your character analysis, and the more you understand your character, the better you will be able to take an attitude and personality to "become" that character for your performance. Or, to put it another way, the more you understand the character in your copy, the easier it will be for you to find those emotions, attitudes, and personality traits within you that you can use to create your character and bring life to the words in the script.

As you have seen, there are many clues in copy that will help you discover the character and his or her personality. The target audience, the mood or attitude of the copy, the writing style, and any descriptive notes all give you valuable information. As with other parts of the woodshedding process, the process of character analysis is something that will become automatic in time. Once you know what to look for, you will soon be able to define your character after reading through the copy once or twice.

Voice acting does not usually require the same sort of in-depth, detailed character analysis that might be necessary for a theatrical performer. However, to be believable, you do need to have a good idea of the character you are portraying. Here are some things to look for and consider as you read through your copy to discover and define your character:

- Who is this character talking to? (target audience)
- What is the environment for the copy? (mood)
- What is the character's age? (young, old, middle-aged)
- How does the character stand? (straight and tall, hunched over, arms crossed, hands on hips, etc.)
- Where is the character from? (geographic region, country)
- Does the character speak with an accent or in a dialect? (If so, what would be the country of origin? A poorly done dialect or accent can have negative results unless done as a parody or characterization.)
- How would the character dress? (business suit, or casual)
- What do you know (or can guess) about the character's economic status? (financially well-off, struggling, etc.)
- What is the overall mood or attitude of the copy? (fast-paced, slow and relaxed, romantic feel, emotional, aggressive, etc.)
- What is the pace of the copy? (Slow-paced copy often calls for a relaxed type of character while fast-paced copy demands a character with more energy.)
- From the context of the script, what do you know about the way your character moves? (energy)
- What is the product or service for which the copy is written? (The subject of the copy often dictates a specific type of character.)
- What is the character's purpose, or role, in the script? (protagonist, antagonist, delivering the message, part of a story script, comedic role, that of straight-man)
- What life events or actions brought the character to this moment in time? (theatrical back story)
- What is your character responding to? (back story)
- What does the character want from telling the story? (desires)

Finding answers to questions like these will help you develop a visual image of your character that will help you to instinctively know what is needed to deliver the copy effectively and believably. You will know, for example, if the character needs to speak quickly or slowly, with an accent, or with an attitude.

Creating a visual image of your character and the environment she finds herself in will help to develop the necessary tension for drama. The tension here is not between characters, but rather a physical tension located somewhere in your body. It is this tension that will allow you to bring energy to the words and give life to the character in the copy.

Discovering the character in the copy may appear to be a lengthy process, but, in fact, it happens quickly once you know what to look for.

FIND THE BACK STORY

All copy has a back story, also known as "the moment before." There are two definitions for back story: the first is theatrical back story, which refers to the life experience of the character that brought him or her to the moment of the story. The second definition of back story is what we use in voiceover: that is, the specific event or action to which our character is responding.

No matter how you define it, the back story is the result of the wants and needs of the character that provides the motivation for the words, actions, and reactions to what happens in the environment of the story.

In theater, the back story is frequently unveiled during the course of the performance. With voiceover copy, there is rarely enough time to reveal the back story or provide much character development. A radio commercial must tell a complete story with a beginning, middle, and an end—and with fully developed characters from the outset—all in a very short period of time.

In a dialogue script, you will often be able to figure out the back story with ease simply because the interactions between characters are written into the script. It is these interactions and responses that reveal clues to the back story and the relationship between characters.

It can be more of a challenge with a single-voice script in which there may be few, if any, clues that reveal what brought your character to the point of speaking the words in the copy, or even why, or to what your character is responding. If a back story is not clear from the copy, make one up! After all, you are an actor and you do have permission to pretend.

The idea is to create a believable motivation for your character that brings him or her to the particular moment in time that is taking place in the script. The back story will reveal your character's wants and needs at this moment in time, and that information will help guide you in your delivery. The fastest way to do this is to figure out what your character is responding to with those first few words of the script.

Define the back story and what the character wants in just a few words. Keep it concise, believable, and real.

UNVEIL THE CONFLICT

Conflict is an essential part of dialogue copy, and can also be present in a single-voice script that tells a story. Conflict rarely occurs in information-based copy in which the message is more of a sales pitch or instructional in nature than a story. Conflict creates drama, and drama holds interest.

A dialogue script without conflict will be boring and uninteresting. On the other hand, a dialogue script with a well-defined conflict can be funny, emotional, heartwarming, and informative—all at the same time. Look for the primary conflict in the script. Usually, this will be some difference of opinion, a crisis, an impasse, or some other obstacle. Define this primary conflict in a few concise words.

Once you have defined the primary conflict, look for any complications that support or exaggerate it. These are often secondary or minor conflicts that serve to add meaning and importance to the primary conflict.

Follow the development of the conflict to reveal its peak moment, which is the climax—the key moment in a commercial. It will usually be found immediately prior to the resolution or nonresolution of the conflict.

During the course of developing the conflict, the advertising benefit (*unique selling proposition*) should be revealed. The *peak moment* often is the point in the copy where the advertiser's name is mentioned or the purpose of the commercial is revealed.

DISCOVER THE RESOLUTION OR NONRESOLUTION OF THE CONFLICT

In commercial copy, it is through the resolution or nonresolution of the conflict that the message is expressed. Sometimes ending a commercial with an unresolved conflict can actually create a memorable impression in the mind of the listener. An unresolved conflict leaves the end of the story up to the listener's imagination, and that can be a very effective motivation for action. For example, a radio commercial we produced for the high-end toy store, Toy Smart, presented a conflict between a mother and her "child." As the story developed, the mother tried to coax her "child" to eat his green beans with less than satisfactory results. This conflict resolved when the "child" turned out to be the husband who said "I'll be happy to eat all the green beans you want, as long as you put them with a T-bone steak!" However, at the very end of the commercial, the husband had one more line, which left the conflict in a state of nonresolution: "What do I get if I eat all my brussels sprouts?" This left the resolution of the conflict to the imagination of the listener and created a memorable impact moment in the commercial.

Look for details in the copy that give clues as to how the message is actually communicated. Are there a series of gags, jokes, or a play on words that lead to expression of the message? Do characters in the copy shift roles (reversals)? Is there a list of information that ends with an unusual twist? Does the story take place in an unusual location? Is there something in the story that appears to be out of context with what is taking place? Is there a personality problem or physical limitation with one or more of the characters? How are these resolved—or not?

MAKE THE COPY YOUR OWN

As you analyze a script, remember that there are no right or wrong answers to the questions you ask, and there are no good or bad choices. Use your imagination and bring something of yourself into the copy. The idea is to create a believable character and situation for the copy you are reading. Bringing your personal experience into the character you create will aid in making him or her real to the listener.

Use what you learn from the copy and the tools at your disposal to make the copy your own. If you have a naturally dry and sarcastic style of speaking, you may be able to apply that trait to your character to make it unique. If you have a bubbly speaking style, that trait might give a unique twist to a character. Don't be afraid to experiment and play with different approaches to performing a character.

On the surface, "making the copy your own" may appear to be a contradiction. After all, according to Core Element #6, *Forget Who You Are and Focus*, one of our objectives is to get out of our way to allow the character to become real. But bringing part of your own personality or attitude to your character can actually make it easier to create an interesting and compelling performance.

CREATE TENSION

When making copy your own, it is important to be specific when defining a scene or character and to commit to the choices you make. Using specific terms creates a tension in your body that you can use in your voice. Without tension you will be unable to create drama, which is essential for capturing and holding the attention of the listener.

To create tension in your body, begin by observing your feelings and emotions as you read the copy. Allow your senses to be open to experience whatever sensations might appear and make a mental note of where that sensation occurred in your body. As you begin to add life to your character, recall the memory of the sensation you just experienced (*sense memory*). Focus on placing your voice or performance at that place in your body. This technique may be somewhat difficult to master at first, but keep working at it—the result is truly amazing once you have the knack of doing it.

WORK BACKWARDS

To quickly get an idea of the copywriter's intent, the target audience, the client's message, and some solid clues about your character and the story in the copy, try looking at the last line of the script first. The end of a script is where the resolution or nonresolution of conflict occurs and is usually the point where a character's attitude or true motivation is revealed. It is also where the most important part of the client's message usually resides. By working from the bottom of the script to the top, you will be able to learn important information that you can use to quickly create a basic character and attitude. Then use other clues in the copy to more fully develop your character.

FIND EMOTIONAL HOOKS

These are the words or phrases that carry an emotional impact. Call on your past experience to recall a memory of a similar emotion (*sense memory*). Notice that the memory of the emotion creates a certain physical tension someplace in your body (see "create tension" on page 138.) Observe the tension's position in your body and what it feels like. Hold this tension or sensation as you deliver the copy, reexperiencing the emotion or feeling. Now speak from that place in your body, fully expressing the tension. This technique helps to make your performance more believable and your character more real.

LOOK FOR QUESTION MARKS IN THE COPY

Question marks are opportunities for dramatic punctuation. I'm not referring to the punctuation mark—?. I'm referring to words or phrases in the copy that give you the opportunity to ask a question. If the copy specifically asks a question, you should make that clear with your performance. Question marks that do not ask questions are usually found in sentences that describe or explain something. Someplace in the sentence there will be an opportunity to answer the unasked question.

Find those spots and figure out your own answers to the questions. This woodshedding technique can be incredibly useful to bring your character to life because the answers you come up with are part of the character's knowledge or history, which helps make the character real. Here's a 30-second radio script with places noted where question marks present opportunities for discovering information noted in parentheses:

Have you ever started a relationship (*What kind of relationship?*) **–
and then discovered the truth?** (*What truth? And HOW DOES IT
FEEL to discover that kind of truth?*)
I was thinking about working with an agent to sell my home,
(*What kind of home?*) **but then I found out about their high**

commissions! *(How high?)* **Not my idea of a great relationship.**
(What is a great relationship?)
Then I discovered MyOpenHouse.com! *(How does it feel to make a great discovery?)* **I can get my home listed with an agent,** *(What is that like?)* **and <u>save</u> up to 40% on their commission.** *(How does it feel to save that much?)* **It's like the best of both worlds –**
professional help, *(What does "professional" mean to you?)* **and a really low commission.** *(How does that feel?)*
MyOpenHouse.com. Now that's a relationship I can live with!
(How long will this relationship last?)

You can take this process as far as you like, even to the point of asking questions about every word in the script. As you choose the answers to the unasked questions, you will be creating the foundation of your character's attitude and personality, and creating a context for your performance. Commit to the answers you come up with and use them as tools for giving your character life. However, be prepared to modify your answers as your character develops and as you receive direction from the producer.

LET GO OF JUDGMENTS AND INHIBITIONS

An important part of the woodshedding process is to experiment with your choices out loud, exactly the way you intend to perform the lines. This means you can't hold back just because you are afraid of what someone nearby might think. Always keep in mind that you are an actor, and as an actor, your job is to perform. And in order to create a great performance, you must rehearse the way you will be performing.

Be careful not to make the mistake of rehearsing and woodshedding silently or at a whisper. Unless you test your woodshedding and script analysis out loud, you can't possibly know exactly what your performance will sound like. Your delivery might sound great in your head, but the minute you start performing on mic, it will almost always come out of your mouth sounding completely different from what you had in mind.

One of the keys to success in voice acting is to let go of any judgments, inhibitions, and concerns you might have about what you are doing. Leave your ego outside. Allow yourself to become the character in your script. If your delivery needs to be loud, go someplace where you can be alone to work on your performance.

The director in the front row of your mind is not there to judge you, but should be considered a coach and an advocate whose sole purpose is to make your performance better. There is an important difference between being critically analytical about your performance and judgmental.

Judgmental thinking would be:

- "The way I delivered that last copy was just horrible! I'll never be able to do these lines right."
- "I just can't get into this character!"
- "I can't do this kind of copy!"
- "I shouldn't feel embarrassed when I do copy like this."

Analytical, or critical, thinking would be:

- "I didn't like the way I delivered the copy—it just didn't seem real."
- "I know I can be more effective than that last read."
- "What can I do to make my character more believable?"

Judgmental thinking usually approaches the subject from a negative point of view, stops you in your tracks, and prevents you from discovering the solutions you need. Critical (analytical) thinking is constructive and helps move you toward solutions that will make your performance more believable. Of the two, judgmental thinking comes naturally to most people, while critical thinking is a learned skill.

When you leave your ego, judgments, and inhibitions in the car, you'll be open to critically analyzing your script to achieve the best possible performance. Chapter 10, "The Character in the Copy," will give you some tools and techniques for doing this.

TAKE THE "VOICE" OUT OF "VOICEOVER"

While woodshedding and rehearsing, don't just read your copy. Have a conversation with the listener. Talk *to* your audience, not *at* them, always striving to motivate, persuade, or move the listener to action. Remember that even if you are the only person in the booth, the *other* person is always there. Visualize the perfect person to hear the message, and talk *to* them. Talking *at* your audience will sound like you are either reading the script, selling the message, or acting. All of these perceptions are ineffective and ultimately result in the listener disconnecting from the story.

Only by taking the voice out of voiceover—in other words, creating a completely believable and compelling conversation—will you be able to draw the listener into your story. Although we refer to the craft as voice acting, or voiceover, the reality is that you are a storyteller. Remember:

- Use drama (*emotional hooks*) to attract and hold attention.
- Talk in phrases, not word by word.
- Don't read—tell, don't sell.
- Don't act—be authentic and real at all times.
- Let the content and subtext of the copy determine your dynamics.

- Have a conversation with the listener.
- Talk out loud to yourself to find hidden treasures in your delivery.
- Experiment with different attitudes, inflections, and emotions.
- Take out the punctuation marks in the script to make the copy flow more naturally and conversationally.
- Have a mental attitude that allows you to create a feeling of reality and believability. If you believe your character is real, your listener will.

USE MUSIC TO INSPIRE YOUR PERFORMANCE

Music can be a powerful woodshedding tool for helping you discover the emotional content, attitude, pacing, timing, and overall delivery style for a piece of copy. Experiment by rehearsing a script while playing an instrumental. Focus on matching your delivery to the mood, tempo, rhythm, and tone of the music. You'll quickly discover that if you let the music guide you to your character, everything about your performance will change depending on the music you are working to. An upbeat music track will result in more smile, a quicker pace, and a brighter performance. A slower, dramatic music track will result in a more intense, dramatic, and emotional performance. By testing your performance against a variety of musical styles, you'll be better prepared to make valid choices for your performance when you record your auditions and paid session work.

Of course, the music you rehearse with will never actually be used, so you can feel free to use your favorite instrumental CDs or downloaded files. Movie soundtracks are excellent for this technique because of the wide range of emotions and dramatic content. If you want to work with the same type of music that commercial producers use, you can visit any of the numerous online music libraries. Although these music libraries sell their music downloads, there is no charge for auditioning, or listening to the music. It's relatively easy to select a genre and start listening to music as you work with your script.

Because your job as a voice actor is to provide dry voice tracks to your clients, I don't recommend purchasing any library music. Of course, if you have the talent for providing complete production services, having some library music on-hand can certainly be a benefit.

There are literally dozens of online music libraries and the easiest way to find them is to simply enter an Internet search for the keywords: *production music library*.

INTERRUPT – ENGAGE – EDUCATE – OFFER

Regardless of the type of voiceover script you might be working with, your job as a voice actor is to effectively communicate the message, often attempting to reach the listener on an emotional level. The challenge is in figuring out how to do that.

We can borrow a basic concept of marketing and apply it to our woodshedding process to result in a powerful tool for creating an effective delivery. Whenever we want to communicate something to someone else, we need to do four things: 1) we need to get their attention, 2) we need to keep their attention, 3) we need to give them the information, and 4) we need to give them an opportunity to respond or act on what we've said.

In marketing, this is the process of *interrupt, engage, educate, and offer*. This process should not be confused with advertising. Although an advertisement might include these four components—and some of the best ads do—advertising is more about creating an interrupt and making a message memorable through repetition. Marketing is more about creating a unique aspect to the message that will make it memorable without repetition.

So, how do we apply this concept to voiceover? Glad you asked!

As you peruse your copy, woodshedding for the various elements discussed so far, take a close look at the first sentence or two. How can you speak those words in a way that will instantly take the listener's mind off of what they are thinking and swing their attention towards you? Your interpretation of those first few words creates the *interrupt*—and there may be dozens of ways to do it! You might achieve it with a whisper, an emotional subtext, through tempo, or by speaking with an attitude in your tone of voice. Every script will be different, and there may be only a few ways that will work well with any given script. Many of the techniques explained earlier are specifically intended to help you to create a powerful Interrupt.

Now that you've got the listener's attention the real work begins. In order to keep them listening you've got to *engage* them in the message. A well-written script will help, but the real secret to successful engagement is in the nuance and subtlety of your interpretation. You can't just be reading the words. And if you sound at all like you are acting, or in any way phony, all credibility and believability will be lost. This step of engaging the listener requires a deep understanding of your character's role in telling the story. This component is critical to an effective voiceover performance because it gets the listener involved and invested in the story.

Once the listener is invested in listening to the message, important information can be delivered to *educate* them. This part of a script is usually pretty obvious. It's the description and price details, or the explanation of how something works. It's often nothing more than raw, uninteresting information. But you can't let it sound like that. You've put a lot of work into getting your listener invested in what you have to say. Don't throw it all away now! By the time you get to the educational part of your story, your delivery needs to have evolved in such a way that the flow from interrupt through engagement and into education is imperceptible.

The final step in this process is the *offer*. This could be a tag delivered by a different voice, or it could simply be an address or phone number. The idea of the offer in marketing is to provide a safe and low-risk way for the audience to take the next step in the sales process. In advertising, this is referred to as a *call to action*. As with the other three components, the way you deliver the offer will be directly dependent upon the context of the script and your choices in how the story will be told.

These steps of *interrupt, engage, educate*, and *offer* must be positioned in that order for the communication to be effective. A properly written script will use this structure and may actually repeat it throughout the script with multiple interrupts, engagements, educational sections, and offers—but they will always follow that sequence.

Many inexperienced copywriters don't understand basic marketing and advertising concepts and will leave one of the components out completely, or worse—begin the script with the offer. This sort of poorly-written copy is all too common. As a voice actor, it will be your job to bring the words to life—regardless of how they are written. When you master the various ways to incorporate *interrupt, engage, educate,* and *offer* into your delivery style, you will be far ahead of most other voice actors who will still be struggling with their basic interpretation.

It is only by woodshedding a script that you will be able to discover the most effective punctuation, phrasing, attitude, character, emotion, subtlety, nuance, and the meaning of words in the context of a story. You can't change the words in a script, but as a voice actor, you have a tremendous amount of flexibility in determining how those words might be spoken. And that's what the process of woodshedding and script analysis is all about.

Tips for Woodshedding

- Develop your woodshedding skills so they become automatic.
- Look under the surface to discover the subtlety and nuance of copy.
- Don't settle on your first choices.
- Always experiment and test different options for delivery of a line.
- Explore emotion, attitude, pacing, rhythm, tempo, and so on. to reveal alternative choices.
- Look for key words and catchphrases.
- Mark your script with a pencil. It is inevitable that at least some of your choices will change.
- Be careful not to over-analyze your script. Over-analysis can result in a flat delivery.
- Find a way to deliver the first line in a way that gets the listener's attention and evolve your telling of the story using the *interrupt,*

10

The Character
in the Copy

How Will You Play the Role?

Are you a voice actor? Or are you a voice talent? There's a big difference! These seemingly similar references to our craft are, in reality, radically different approaches to performing and working with a script.

When you are performing a voiceover script as a voice actor, you are playing a role, no different than if you were playing a part in a stage play or movie. That's why this craft is called *voice acting*! Unless you are telling your own personal story, the words and situations are not yours—they are those of a character who may be substantially different from you. To play the role of any character believably requires training and developing the ability to detach personal beliefs and attitudes from those of the character being portrayed. This is the essence of all acting.

THE TWO TYPES OF ACTORS

Whether you work behind a microphone, on stage, on television, or in film, there are two distinctly different approaches to performing and creating characters. One is where the actor develops a strong and highly identifiable performing style that is at the foundation of every role. The style may be one of a specific voice characteristic, physical appearance, performance rhythm, body movement, or underlying attitude. I refer to actors in this category as *celebrity actors*. When these actors perform, we have no doubt in our mind that we are watching that person perform. We become involved with their performance, in part, because their acting style is completely appropriate for the roles they choose to play. In other words, no matter what the role, their characters are believable, largely because there is some aspect of the character role that is very similar to the actor. Some film actors I would place in this category are Jack Nicholson, Christian

Slater, Adam Sandler, Tom Cruise, Keanu Reeves, Cameron Diaz, and Jennifer Lopez. Many highly successful voiceover talent frame their performance through an interpretation of each script based on skill and instincts developed over many years. Although many of these performers have the ability to create a wide range of vocal styles, emotions, and attitudes, their performance comes more from who they are, than by creating a character for each role they play.

Some acting courses teach that the actor should bring as much of his or her self to the performance as possible, and design their performance on how *they* would handle the situations, based on personal experience and interpretation. If you are merely "being you" as you perform a script, even on an extended level, then your performance may risk sounding like **_you_** doing the words, and there may, or may not, be anything unique or special about your performance. Now, you may be an excellent reader with a talent for interpreting or spinning a phrase, or you may posses an incredible vocal resonance and command when you speak, but if you are personally attached to the words of the script, you are not truly an actor.

The other approach to performing is one in which a wide range of acting skills and abilities is developed which allows the actor to literally create many different emotions, attitudes, and personalities that are outside of who they really are. Actors who have mastered this approach literally become the character they are playing. As we watch or listen to them, we see the character they have created, not the person they are. I refer to these actors as *character actors*. I consider Jim Carrey, Jodie Foster, Drew Barrymore, Tom Hanks, Dustin Hoffman, Meryl Streep, and Robin Williams all excellent examples of actors who truly become the characters they are playing. In the world of voiceover, many of the best known and highest paid voice talent have developed the ability to create a variety of uniquely different voices and personalities for the characters they play.

Both approaches to voiceover work are completely valid, and both offer potential for success. However, it's important to understand the differences because your individual abilities may direct you to follow one path or the other. Not everyone working in voiceover is a voice actor. For example, if you have very strong personal or religious beliefs, you may discover that it is very difficult to separate yourself from those beliefs in order to create a believable character that has opposing beliefs or attitudes.   No matter how hard you might try, you may not be able to create a sense of truth as you speak the words. If this is true for you, then the path of mastering performing skills as an actor may not be for you. You must follow a different path, with different training that will give you the skills to base your interpretation and performance of a script on who you are. You will need to develop a deep understanding of your innermost self, and you will need to learn how to tell a compelling story from the perspective of *you* as the story teller. You will need to learn how to be a masterful reader of stories, rather than a creator of characters who tell their stories.

The ability to read a script with a powerful interpretation is no less a skill than that of a voice actor creating and playing a believable character. In voiceover, performers with these heightened reading and interpretive skills commonly refer to themselves as voice artists or voice talent. Those who develop the skills for creating compelling characters can accurately refer to themselves as voice actors. Both performing styles are common, but the trend in most areas of voiceover has been moving toward voice acting.

THE DILEMMA

A common dilemma with performers just learning the craft of acting is the thought that they are "lying" or being "untruthful" when they perform the role of a character who expresses thoughts, ideas, beliefs, or opinions that may be radically different from their own, or they feel guilty when they are getting paid to read a script for a product they don't believe in. By definition, the term "actor" simply means *playing the role of a character*. There is nothing in the definition that implies that the performer is lying, cheating, or being dishonest in any manner. In fact, the underlying precept of all forms of acting is that it is the actor's job to create a believable reality of the moment for the character he or she is playing. The dilemma occurs because the neophyte actor is confusing their personal beliefs with those of the character they are playing. Without a disconnect of personal beliefs, it is extremely difficult to create a believable and compelling character. This disconnect is essential and necessary in all forms of acting, including voice acting. And, in some situations, it can be difficult to achieve.

So, does this mean that, as a voice actor, you should take any job offered to you, regardless of the message or its ultimate purpose? Of course not! Your personal beliefs, ethics, and philosophy should certainly be major factors in choosing your performance material. All scripts are not right for all voice actors. Even if you are a highly skilled reader, there will be many scripts that cross your desk that are not appropriate, either for your style of delivery, or in their content. Ultimately, it is up to each individual performer to learn how to choose the jobs they will accept. As with anything else in life, some choices will be better than others.

Your Best Tool Is Your Own Personality

Whether you approach your voiceover work as a voice actor or as a voice talent, the best tool you have to define a character or discover an interpretation is your own personality. When you know yourself, you can tap into parts of your personality to give life to the character in the copy.

Personality analysis is a subject that has been studied for thousands of years. Hippocrates developed a system of defining personality traits, which placed individuals into four separate personality types with dominant (sanguine and choleric) and recessive (melancholic and phlegmatic) traits. The Hippocrates system of personality analysis was very restrictive in its definitions of personality types but it did provide a basic structure within which people could be placed.

More recently the psychologists of our world have developed highly refined methods of determining specific personality types. Some of their studies have shown that personality is largely a result of the chemical makeup of the brain. Cultural upbringing and conditioning further contribute to personality development.

There are many excellent books available that will help you discover some fascinating aspects of your personality. Many of these books are written as aids to improving relationships or developing self-awareness. Three excellent personality books are: *Please Understand Me: Character and Temperament Types* by David Keirsey and Marilyn Bates (1984); *Are You My Type, Am I Yours* by Renee Baron and Elizabeth Wagele (1995); and *Dealing with People You Can't Stand* by Dr. Rick Brinkman and Dr. Rick Kirschner (2002). Another approach to understanding personality types is through the *Enneagram*. There are many books on this subject, among them, *Personality Types: Using the Enneagram for Self-Discovery* by Ross Hudson and Don Richard Riso (1996). These books look at personality types from different points of view and offer some fascinating reading.

An advertiser's understanding of who buys their company's products is crucial when it comes to a marketing campaign. Your understanding of yourself is equally necessary when it comes to creating a character that will effectively communicate the message in the advertiser's copy. The best way for you to learn more about yourself is to ask questions and find the most appropriate answers. Based on your answers, you will be able to determine some of your dominant and recessive personality traits.

Most studies of personality type start with several basic categories, then divide those into subcategories. Every person has characteristics in several categories, but certain areas are dominant, and others are recessive. The following simple questions will give you an idea of some basic personality differences.

- Do you respond to problems emotionally, or do you think about them before responding?
- Do you have a strong need to express yourself creatively, or do you prefer quiet activities?
- Do you avoid unpleasant emotions (including fear), or are you inclined to take risks?
- Do you rely on your instincts for information, or do you rely on what you see and hear?
- Do you seek approval from authority figures, or do you rebel?

- Do you play the role of a nurturer, or do you treat others in a detached manner?
- Do you express anger readily? Are you accommodating and out of touch with your anger, or do you see anger as a character flaw?
- Do you prefer literal writing or a more figurative writing style?
- Are you more realistic or speculative?
- Do emotions impress you more, or do principles?
- Are you attracted to creative, imaginative people, or to more sensible, structured people?
- Do you tend to arrive at events early, or are you generally late?
- Do you do things in the usual way, or in your own way?
- Do you feel better having made a purchase or having the option to purchase?
- Do you operate more from facts or from principles?
- Do you find it easy to speak to strangers, or is this difficult?
- Are you fair-minded or sympathetic?
- Do you prefer planned activities or unplanned activities?

Your answers to these and other questions will only scratch the surface of your personality. When you gain an in-depth understanding of who you are, you will be ahead of the game when it comes to creating a marketable style. When you understand yourself, you will be able to tap into some of the core elements of your own personality as you create a unique character. Discovering the essence of who you are is the first step in developing acting skills that will allow you to create believable and compelling characters.

THE TWO BOXES

In our *VoiceActing Academy Workshops*, I teach the concept of *The Two Boxes*. You and the character you are playing each live in a box. You are very comfortable within the walls of your box. Your box contains all of your life experience, belief system, habits, behaviors, attitudes, emotions, feelings, knowledge, wisdom, and more. The box your character lives in contains all the same stuff you have in your box, only it's those of the character—not yours. The character's box may be larger or smaller than your box, and the character you will portray is very comfortable within the walls of his or her box.

You need to understand the real you and how you exist in your box before you can fully understand how your character exists in his or her box.

As an actor, your job is to climb out of your box and into the box of your character. You bring everything from your box with you except the box itself. You separate yourself from the confines of the walls of your box as you enter the box of your character. Everything you bring with you is available as tools that can be used to help bring the character to life.

If your character's box is larger than yours, you need to be aware of this in order to allow yourself to behave believably as the character. Learning how to do this may be uncomfortable at first, but that's only because your comfort levels are relative to existing within your box and you've not yet grown comfortable in a bigger box. The path to becoming comfortable in the character's box is through the use of the many acting and performing techniques you've learned from this and other acting books, acting classes, and improvisation classes.

Once you've climbed into your character's box, you need to let go of the real you and experience how the character you are portraying lives and behaves. There will always be a part of you there to make your character real. In a very real sense, when you create a character, you are tapping into that part of you where the character lives.

This is the commonality between you and the character. It's the stuff you brought with you from your box that also exists as the same, or very similar, stuff in your character's box. Those things in your character's box that are different from anything in your box must be created through your performance. But in order to achieve this, you must know what they are, and have some way to create them. This is why basic training in acting and performing technique is an essential part of the study of voice acting.

A mastery of this process results in a truly believable character that you can create on demand without thinking about what you are doing. This is what Core Element #6, *Forget Who You Are and Focus* is all about.

Socio-cultural Awareness and RISC 3-D

You're about to learn a process to help you create a compelling and believable character for virtually any script, based on the same techniques major advertisers use when developing powerfully effective advertising.

The corporate business world uses highly refined methods of personality and social analysis to define the demographics (statistical data) of the marketplace for selling products and services. These studies define the buying attitudes and purchasing habits of consumers and aid advertisers in reaching their desired market.

There are several companies whose entire business is based on analyzing the buying trends of different types of people. By understanding a buyer's motivations, advertisers can write in specific words, phrases, or style. For TV commercials and print advertising, editing techniques and use of color, font style, and other visual elements are used—all of which are "hot" buttons designed to trigger a buying impulse in the viewer, or reader. In radio commercials, similar hot buttons are triggered through a careful choice of words and phrases, use of appropriate music and various production techniques. In every case, the desired result is to reach the audience on an emotional level and to motivate the audience to take action.

Today, advertisers are faced with a marketplace of "occasional" consumers who are no longer characterized by predictable buying habits and who no longer exhibit strong brand loyalty. The key objective of marketing socio-cultural research is to identify the links between personal motivations and buying behavior in order to understand the consumer and why she is attracted by certain propositions and not by others. Simply studying consumer behavior is not adequate, nor is analyzing buying habits in terms of age or class. To understand modern society, it is necessary to look much deeper at the socio-cultural diversity of society and find the trends and characteristics that can make the difference between commercial success or failure.[1]

The Research Institute on Social Change (RISC), started to monitor social change in Europe in the early 1980s and developed the RISC socio-cultural segmentation system in 1983. The RISC system was extended to the United States in 1989 as RISC Ameriscan. Marshall Marketing, a full-service market research and communications consulting firm based in Pittsburgh, PA, has been working with RISC in the U.S. since 1996. The RISC 3-D program was launched in 2000, which Marshall Marketing utilizes to help local, regional, and national advertisers understand and adjust to the purchasing behaviors of present and future consumers. The RISC 3-D program is quite extensive, but there are some specific elements that, when put to use by a voice actor, can have a powerful effect.

Through a series of studies, on both national and local levels, a probability sample of people is surveyed with a carefully developed questionnaire. The questions don't ask for opinions, but rather register facts and preferences about the individual. The results of the survey, as processed through RISC's proprietary algorithm, capture the person's socio-cultural characteristics.

To more easily view the results, a chart is created that takes on the appearance of a 3-dimensional compass (Figure 10-1).[2] The vertical axis is linked to attitudes of change. At the north are people who see change as a positive force in their lives and are open to change (Expansion). To the south are people who prefer stability, structure, and consistency (Stability). The left-to-right horizontal axis of the compass (east-to-west) relates to the balance between the individual and society. To the east are those who are more independent and seek immediate pleasure (Enjoyment); to the west are people with strong ethics who are more community oriented (Responsibility). The front-to-back horizontal axis relates to an individual's attitudes toward Flexibility (front) or Structure (back).

Respondents are scored on each of approximately 40 socio-cultural characteristics. Their scores result in a specific placement within the three dimensions of the compass, and can be represented as an arrangement of 10 "cells" in multidimensional diagrams (Figures 10-1 and 10-2).[3] Individuals positioned close to each other tend to have shared values and similar preferences, while those at opposite extremes have little in common.[4]

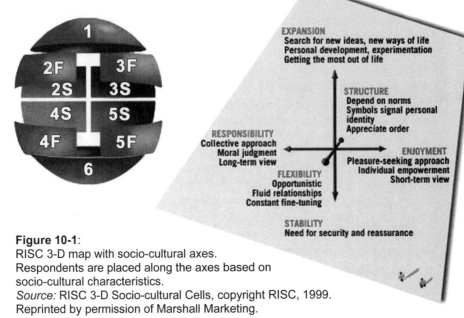

Figure 10-1:
RISC 3-D map with socio-cultural axes.
Respondents are placed along the axes based on
socio-cultural characteristics.
Source: RISC 3-D Socio-cultural Cells, copyright RISC, 1999.
Reprinted by permission of Marshall Marketing.

A basic understanding of how advertisers target their message will be beneficial to you as a voice actor. Knowing what the cultural and social norms are for any specific demographic group will give you some much-needed information to aid in the development of a believable character. For example, let's say that, based on the copy you are given, you can determine that your audience is a person who is outgoing, youthful, interested in experiencing new things, and likes to live on the edge. You make this determination based on your analysis and interpretation of the words and phrases in the copy. With this information you can now make reliable choices and adapt your character and performance energy to something your audience can relate to, thus creating a sense of believability.

For the audience described here, you would most likely need to perform with considerable energy and excitement in your voice. A slow, relaxed delivery probably would not be an effective way to reach the audience, unless the script was specifically written for that attitude.

The Ten Socio-cultural Cells & Your Character

Each of the 10 cells on the RISC 3-D map (Figure 10-2) represents a grouping of socio-cultural attitudes, beliefs, preferences, motivations, and buying habits. Advertisers use these socio-cultural "cells" to aid in targeting their advertising and marketing plans. All aspects of a campaign, including words, visuals, colors, music, and sound effects, are carefully chosen to match the targeted cell's characteristics. The closer the match, the more likely it is that the message will reach the target audience.

These 10 RISC 3-D socio-cultural cells can also be useful in developing a believable and compelling character, simply by working the process in reverse. Understanding the motivations, attitudes, and belief system of your audience will enable you to tap into those parts of your own personality and bring them into the character you are creating. And, if those characteristics are not a predominant part of your personality, this understanding will give you a functional guide you can use to make valid choices that will bring life to your character.

When you create a believable character, an emotional connection can be made with the audience, giving the message a stronger impact. Similarly, by understanding your audience and adapting the traits of the RISC socio-cultural cells, you can develop a unique performing style for every script that will elicit an instinctive response.

The following pages separate the 10 RISC 3-D cells into their socio-cultural profiles, key attributes as defined by RISC, and other useful information to help you understand your audience and create a believable character. As an exercise to develop your acting skills, use the charts that follow as a guide to create a variety of characters with different attitudes. Find a paragraph in a book or newspaper and read the same copy from the attitude of a character in each of the 10 cells. Allow your mind and body to take on the characteristics, body posture, belief system, and attitudes described for each cell and observe how each character you create can be unique.

You can learn more about Marshall Marketing, consumer research, and strategies for advertisers at **www.MarshallMarketingUSA.com**.

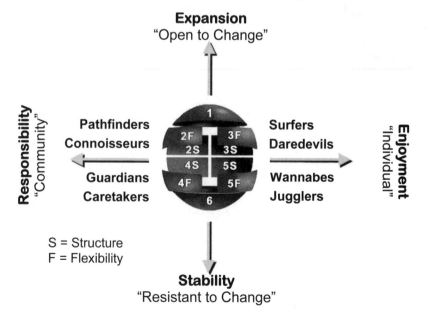

Figure 10-2: RISC 3-D map with cell names.

Cell 1 "Enthusiasts"

Descriptive traits: Active. Curious, Energetic, Hungry, Independent, Insecure, Instant Gratification, Obsessed, Stressed

Profile: Enthusiasts try everything. They expend their considerable but unfocused energies in an intense pursuit of something different, something else, something that will fill their needs. Their choice of activities is eclectic and nondiscriminating. Human contact and constant activity make them feel alive. They are deeply concerned with maintaining and increasing their levels of energy and vitality, in support of their level of activity. Enthusiasts do not like to be controlled or limited in any way; they are individualists, but need connections to others. They are not loners.

© RISC, 1999

Cell 2F "Pathfinders"

Descriptive traits: Activist, Capable, Charismatic, Experts, Intuitive, Practical, "Professors", Responsible, Tolerant

Profile: Pathfinders accept the challenges of creating new approaches in unfamiliar circumstances, seeing the challenges on a grand scale but finding pragmatic solutions. They are mature, capable, experienced, knowledgeable and self-confident. They believe they are best able to figure out the right path to the future and encourage others to follow their strong convictions. They are intuitive, sensitive to the feelings of others and respect differences (even if they don't agree). They prefer to influence rather than to give orders.

© RISC, 1999

Source: RISC 3-D Socio-cultural Cells, copyright RISC, 1999.
Reprinted by permission of Marshall Marketing

Cell 2S "Connoisseurs"

Descriptive traits: Adaptable, Chameleon, Energy, Fast-action, Image, Networking, Opportunist, Overachievers, Situational ethics

Profile: Connoisseurs know what is right and want to help others do the right thing. They operate within a known framework of norms and structures, which they seek to reinforce and strengthen. This framework is their way of dealing with a world in which change is the only constant; their way to manage change is to adapt to the existing rules. They recognize and rely upon links between past, present, and future as a means of understanding and managing that change.

© RISC, 1999

Cell 3F "Surfers"

Descriptive traits: Adaptable, Chameleon, Energy, Fast-action, Image, Networking, Opportunist, Over achievers, Situational ethics

Profile: Surfers "surf the net," "surf the waves," and operate on the surface. They find opportunities in the lack of system they perceive in the world and take control of their surroundings by whatever means are available. They are not loners and can find personal gain in creating good for the group, but never lose sight of their own objectives. They are charismatic, easy to get along with. They are constantly testing the limits, rethinking and reorganizing situations as necessary. They live life in the fast lane.

© RISC, 1999

Source: RISC 3-D Socio-cultural Cells, copyright RISC, 1999.
Reprinted by permission of Marshall Marketing

Cell 3S "Daredevils"

Descriptive traits: Advantage seekers, Appearance, Empowerment, Extreme, Flashy, Image, Pleasure, Recognition, Status, Thrill-seekers

Profile: Daredevils find opportunity within structure, which empowers their individual pursuits and provides the basis of their self-image. Clothing, labels, symbols of rank and status are "uniforms" which help them to define themselves; these badges also announce their position as members of "the club." They live life on the edge, needing to prove themselves through risk-taking and unorthodox behavior. For Daredevils, competition is a core value, and the recognition of their success and status by others is a deeply-felt need.

© RISC, 1999

Cell 4F "Caretakers"

Descriptive traits: Careful, Cooperative, Empathetic, Hands-on, Harmony, Cohesion, Religious, Responsible, Sensible, Teachers

Profile: Caretakers want to preserve what is good in the status quo for the benefit of others, but are tolerant and to not impose their views. They rely on relationships rather than orders. They believe in their own skills and visions, but enlist the support and cooperation of others. They creatively encourage shared success. They know and observe the customs and norms that provide social cohesion and do not seek to stand out or to succeed at the expense of others.

© RISC, 1999

Source: RISC 3-D Socio-cultural Cells, copyright RISC, 1999.
Reprinted by permission of Marshall Marketing

Cell 4S "Guardians"

Descriptive traits: Conformist, Defensive, Duty, Followers, Loyal, Obedient, Obligation, Trusting

Profile: Guardians see change as a possible source of trouble, and do what they can to avoid it. They worry about things going wrong and like to make plans to prevent mishaps—for themselves and for others. Close personal relationships are important to them. They know the rules, follow orders, and believe in leaders, although they don't want to be the leaders. They are happiest when everything goes according to schedule, with no surprises. Consistency, tradition, and routine are their watchwords.

© RISC, 1999

Cell 5F "Jugglers"

Descriptive traits: Carefree, Fast track, Happy-go-lucky, Materialist, Practical, Risk takers, Self-confident, Self-indulgent

Profile: Jugglers are the original multitaskers; they like to try things, to experiment with new sensations. They do not, however, expect any permanent change to result from this experimentation. They desire freedom (but not independence), and avoid all limitations, restraints, and schedules. They enjoy relationships, but focus on having fun rather than on community. They have two goals: money (which brings both possessions and respect), and personal expression—and, they often use technology to achieve them, as well as for entertainment. Their only worry is that something beyond their control may go wrong and limit their free activity.

© RISC, 1999

Source: RISC 3-D Socio-cultural Cells, copyright RISC, 1999.
Reprinted by permission of Marshall Marketing

Cell 5S "Wannabes"

Descriptive traits: Consumption, Enjoy Life, Hardworking, Hedonism, Materialistic, Results-Oriented, Security, Survival, Vanity

Profile: Wannabes want to be... either Jugglers who seem to have everything they want, or Daredevils who seem to have all the fun. Wannabes themselves are good at finding what they want and work hard to get these "prizes." They don't really understand the world—but as long as things don't change, their energy and activity let them achieve the entertainment and distractions they seek. They are loners, to some extent, and have no strong sense of connection to others—except as sources of recognition. They dislike plans and schedules, and make them only out of a sense of survival.

© RISC, 1999

Cell 6 "Contenteds"

Descriptive traits: Careful, Cooperative, Homebodies, Local, Loyal, Predictable, Religious, Small pleasures, Togetherness

Profile: The Contenteds are quiet, comfortable people, for whom familiarity and consistency are preferable to spontaneity and novelty. They live their lives in small groups—at work, at play, and in the family. Some anxieties exist, usually concerning potential dangers and difficulties. But Contenteds have a strong sense of personal pride and competence, and feel entitled to self-expression and fulfillment.

© RISC, 1999

Source: RISC 3-D Socio-cultural Cells, copyright RISC, 1999.
Reprinted by permission of Marshall Marketing

Theater of the Mind

Voice acting is *theater of the mind*! You do not have the advantage of props, flashy lighting, or scenery. All you have are the words on a piece of paper, and your individual creativity. From the words alone, you must create an illusion of reality in the mind of your audience. In order for you to create a believable illusion, you need to know what is going on in the mind of the character you are playing. You also need to know your character's role in the story, and his or her relationships to other characters, objects, and the product or service. To learn what is going on in the character's mind, you need to analyze the script. (See *Woodshed Your Copy* on page 128).

Analyzing, or woodshedding, a voiceover script is very much like reducing a play to its essential parts. The more information you can discover in the copy, the easier it will be to create a believable performance. Single-voice spokesperson copy is frequently information-based and may not require much analysis. However, dialogue copy and story-based scripts are short theatrical pieces and must be thoroughly understood to be effectively performed.

Although analyzing a script is helpful in understanding its component parts, it is important to realize that *overanalyzing* a script can kill spontaneity and cause the voice actor to place too much focus on technique and thinking about what he or she is doing. Remember, to be effective, technique must become automatic and occur without any conscious effort. Study a script just long enough to discover what you need to know, then put the script down and let your instincts and training do the rest.

To create effective theater of the mind, your performance must reflect real life, exhibit some sort of tension, contain something the listener can relate to, and have a sense of honesty and a ring of truth. These are all elements of good theater and should be incorporated into any voice acting performance, regardless of the type of copy or the length of the project.

When creating a character for your performance, keep in mind the following basic elements of good theater:

- Interesting characters with wants and needs "at this moment in time"
- A story or sequence of events that builds and leads to a climax
- Conflict in one or more forms
- Resolution or nonresolution of the conflict, usually in an interesting or unexpected manner
- Closure in which any loose ends are satisfactorily resolved

Uncover these elements in a voiceover script and you will be able to understand your character better.

An Exercise for Creating Real Characters—CD/9

Visualization is a powerful technique that can help bring your characters to life, and this exercise will do just that! The first time I used this exercise in my workshop, the result was amazing. We witnessed a total transformation and the student, who was having difficulty finding the proper voice and attitude, was able to create a completely believable character that she did not know existed within her. As I've mentioned before, for a character in a script to be "real" to a listener, everything about the character must flow through you just as if you were the character.

Once understood, the following visualization process can take as little as only a few seconds to a minute or so. However, as you learn this technique you may want to spend some additional time relaxing your body and mind prior to doing this exercise. Of course, in an actual session you won't have much time for a lengthy visualization, but by then the process should be second nature.

Define your character in as much detail as you possibly can, including physical appearance, clothing, hair, posture, mannerisms, and other features. Begin by thoroughly woodshedding your script and making choices for your audience, back story, and character. Visualize this character in your imagination. This character description and image will become important later on, so don't skimp on the details.

You may find it helpful to do this as a guided visualization by listening to track 9 on the CD or recording the script yourself. Take your time with this, and don't rush it. The clearer and more vivid the visualization, the better the results, and the more believable your character will be.

At first glance, this visualization may seem a bit unusual. However, when you give it a try, you may be surprised at what you are able to come up with, not only with physical changes, but also with the sound of your voice that results from creating a believable character.

Visualization Script for "Creating a Character"

With your character in mind, close your eyes and take a slow deep breath through your nose. Fill your lungs completely. Exhale slowly through your mouth to relax. Repeat with another long deep breath… and slowly exhale. Don't forget to keep breathing.

Imagine yourself standing in front of the microphone, or in the voiceover booth. See yourself in your imagination—it's as though you are observing yourself from across the room. Create the image of yourself as clearly as you possibly can, in whatever manner works for you. When you have a sense of seeing yourself standing in the room, take another long deep breath… and slowly exhale.

Now, imagine the character you will be playing coming into the scene in your imagination. See the character walking in. Notice how the character is walking. Observe the posture and physical movement. Notice what the character is wearing: What do the clothes look like? What kind of shoes is your character wearing? Is your character wearing glasses or jewelry?

As you observe this scene, see yourself look at your character's face. Notice any facial details, color of the eyes, appearance of the skin. Does your character appear to present any sort of attitude or have interesting facial expressions? When you have a clear image of the character in your imagination, take a long deep breath… and slowly exhale.

Now, as you are observing the two of you in the room, imagine seeing the real you step out of your body and come to the place from where you are observing. As the real you steps out of your body, imagine the character stepping into your body. Everything about the character is now reflected in your body: The character's posture, the way the character stands and moves, the character's physical appearance, facial expressions, and mental attitude. Everything about the character is now expressed through your body, mind, and voice.

Allow yourself to fully experience this transformation. Notice any tension in your body. Be aware of how you feel as this character—physically, mentally, and emotionally. When you have a sense of the transformation, take a slow deep breath; exhale, maintain the physical, emotional, and mental state; open your eyes, and begin speaking the words in your script.

Making Your Characters Believable

FIND THE MUSIC

There is *music* in your daily conversations, and there is a great deal of music in any voice acting performance. Some of the basic elements of music are pitch, tempo, rhythm, volume, quality, and intonation—all of which are present in every sentence you speak.

It is the music of your performance that will convey the subtlety and nuance of the meaning behind the words. Find the music in the way your character speaks and you will create a believable character. Chapter 14, "Character Copy," explains in detail how to find and sustain interesting and musical character voices.

STUDY OTHER PERFORMERS

Study film and television actors. Observe how they deliver their lines and interact with other characters. Listen to the dynamics of their voices. Notice that most actors use a lot of vocal variety and inflection. They also move and express emotion physically as well as verbally. Mimic what you see other actors do and how they speak so you can get the experience of what they are doing. Study their techniques and apply them to your style. You will soon find the point where your stretch becomes uncomfortable. To grow as a performer, you need to find a way to work past that boundary.

STRETCH YOUR BOUNDARIES AND BE WILLING TO RISK

Be willing to experiment and risk moving beyond your comfort zone (Core Element #7, *Gamble*). Practice the techniques to develop the skills that will make moving outside of your comfort zone easier. Don't worry about how you will appear or sound to anybody! As a voice actor, your job is to perform the copy and your character in the best manner possible. Leave your inhibitions and self-conscious attitudes outside the studio door.

Stretch beyond what feels comfortable. It is better to stretch too far than not far enough. It is easier for a director to pull you back after setting a character that is too far out there, than it is to stretch you further. Remember, there is no right or wrong way to perform. Each performer is unique and different techniques work better for different performers. Do what works best for you to make your performance real and believable.

As you stretch your abilities, you will probably feel uncomfortable at first. Remember to be nonjudgmental and to not worry about how well you are doing. Each of us has an individual concept of some point at which we feel we would be going too far, or over the edge. Practice taking yourself just a little bit over that line until you begin to feel uncomfortable. Then take yourself a little bit further. The more you move beyond the point of discomfort, the faster you will develop the ability to create any character.

You must be willing to risk total failure. Intend to perform to the best of your abilities. Become the character and do whatever it takes to make the character real. Remember that you are uniquely you, and that you are interesting just as you are. Also remember that the people you are working with have insecurities of their own and may actually know less about the business than you do. Know that you know what you are doing. If you never risk, you can never learn. Use each audition or session as a learning experience. Keep an attitude of always being in training.

MAKE EVERY TIME THE FIRST TIME

Make each and every performance seem as if it is the first time. It is very easy to get sloppy by take 27. Take 28 should sound as fresh and real as take one—only better. Unless your director tells you otherwise, you should maintain the same energy and attitude for each take. Use the director's guidance as a tool to help you focus in on your best performance. Add a little spin to a word, or shift your emphasis here or there with each take, but keep your energy and attitude consistent. This becomes very important to the editor who needs to put the final project together long after you have gone. Variations in your performance energy can stand out very clearly if you are not consistent, and make the editor's job a nightmare.

CHARACTER HAS PRECEDENCE OVER COPY

As you learn how to create believable characters, you will discover that your characters may want to say and do things in a certain way. After all, if the characters you create say things exactly the way you do, what would be the purpose of creating the character?

To be real, each character you create must have its own personality, mannerisms, thought processes, and speaking style. Every subtlety and nuance of your character contributes to its believability. Forcing your character to say or do something that is not appropriate will instantly take your performance out of character.

Scripts are often written with a specific attitude, phrasing, or delivery style in mind, yet what the writer had in mind may not be what ends up being recorded. Provided your delivery is *in character*, it is perfectly acceptable, for example, to contract or uncontract words that aren't written that way. This is closely related to the idea of changing or removing punctuation marks to create a more natural and conversational delivery. Your objective as a voice actor is to bring your character to life, and you should strive to do whatever is necessary to create that reality. You can't change the words in the script, but you can change the way you say the words. The way your character speaks will always have precedence over the written script. This will be true until the director tells you otherwise.

ACT PROFESSIONAL

Play the part! Enter a studio with the attitude of a professional there to do a job and with the confidence that the character you create will be exactly right. Be friendly, cooperative, and ready to work. Making money does not make you a professional. Acting professionally makes you money. When you act like a pro, the people hiring you will believe that you are a pro and they will respect you. Remember that this business is all about creating believability in the mind of the audience. When you enter a studio, your first audience will be the people who hired you. Make them believe you are good at what you do and prove it with your performance.

Become the child you once were! Pretend! Play! Have fun!

Tips for Performing All Types of Copy

When you become the character in the copy, you will be believable to the audience, and a suspension of disbelief will be created. When the audience suspends their disbelief in what they hear, they become more open to the message. This all starts when you discover the character in the copy.

- Don't overanalyze your copy. Overanalyzing can cause you to lose spontaneity and cause your delivery to become flat and uninteresting.
- Rely on your instincts, and trust the director in your mind to guide your delivery to keep it on track, conversational, and real. Don't read.
- Tell the story. All scripts tell a story. Storytelling is always about relationships. To be believable, make the relationships appear real.
- Make your character believable and unique by adding something of yourself. Let your imagination run wild. If you believe in the reality of your performance, the audience will believe.
- Don't become so focused on your character that you lose sight of the story, the drama, and the relationships between characters and conflict.
- Internalize the wants and needs of your character, both physically and emotionally. Find the place in your body where a tension develops. Hold it there to *set your character*, and deliver your lines from that place. Use your imagination to create a vivid reality of the scene, situation, relationships, and conversation in your mind.
- Play it over-the-top on your first time through. Use more attitude, dynamics, or energy in your delivery than you think may be necessary. It's much easier to pull back than to push you further out.
- Underplay, rather than overplay. Louder may not be better. Pull back, speak more softly, and be more natural. Remember, "less is more."
- Keep your posture in a stance consistent with the character and the choices you have made in regards to how your character stands, moves, and behaves. Maintain this attitude throughout your performance.
- Find the music in the copy. All copy has a tempo, rhythm, dynamics, and other musical qualities. Speak as quickly and as quietly as you would if you were talking to someone in a real conversation.
- Stay in the moment. Pick up cues. Interact with other performers. Don't let your lines become separated from those of the other performers. Listen to yourself, the director, and other performers and respond appropriately (*listen and answer*).

---

[1] "RISC AmeriScan: Western Socio-Cultural Scan", RISC 1999 (1-2).
[2] RISC 3-D map, scan, and other materials copyright RISC, 1999.
  Used by permission of Marshall Marketing and RISC International.
  www.MarshallMarketingUSA.com
[3] "RISC AmeriScan: Western Socio-Cultural Scan", RISC 1999 (3).
[4] *Marshall Marketing RISC 3-D Guidebook*, RISC, 1999 (5).

11

Single~Voice Copy

Single-voice copy is written for a solo performer who will deliver the entire message, with the possible exception of a separate tag line which may be voiced by a different performer. Because there is only one character speaking, any interaction is implied, either between the character speaking and an unheard second character or between the performer and the listener.

Reading or "announcing" copy will rarely result in an effective communication. Both styles direct the performance inward and imply that the performer is speaking to himself or herself. To properly interrupt, engage, and educate a listener, a voice actor must speak to a one-person audience with the expectation of receiving a response. Only then can the *offer* be made with the expectation that the listener will take action. See Chapter 9, page 144 for more about *interrupt, engage, educate, offer.*

When discussing the target audience for any given script, the producer may speak in terms of the demographic audience she is wanting to reach, for example: women, 25–35. However, your performance must be focused on speaking to only one person. Knowing the broader demographic can help, but the fact that you and the producer are referring to two different things can be somewhat confusing. Remember, if you ask the producer specifically who you are speaking to (the one person), she will probably not know what you are talking about. Figuring out your one-person audience is your job, not the producer's.

The target audience of a single-voice script can usually be determined pretty easily; however, sometimes it can be a challenge to define the character speaking. Well-written copy that clearly tells a story makes the character easy to define. Poorly written copy that contains only information in the form of facts and figures can make this difficult.

Consider single-voice copy as a story you are telling. Find your storyteller and commit to the attitude and style choices you make. Deliver the copy from a set point of view by finding the *subtext* (how you think and feel) behind the words you speak and express it through your voice. Study your script closely to determine if it was written to match a current trend.

One key to effective single-voice delivery is to use the basic dramatic principle of having a conversation with another person. Make the other person the ideal person who needs to hear what you have to say. Another key is to find the appropriate attitude or style. Both of these can be effective interrupts and engagements when done properly. Make your conversation natural, believable, and candid, speaking to only one person at a time. *Shotgunning*, trying to speak to several people at once, tends to make your delivery sound more like a speech than a conversation, although that may be appropriate for some types of copy.

In single-voice scripts, as in others, there can be many different written references to the performer, such as VO, ANNCR, or TALENT—and all may be used interchangeably. You may also see references to music, SFX (sound effects), and even directorial cues, which are not to be read by the performer. The format may be single- or double-spaced and may or may not include a separate column for video and other instructions. Read everything on the page in order to fully understand the message and your character's role. Then quickly run through your Seven Core Elements to set your choices.

Tips for Performing Single-Voice Copy

- You are a storyteller, and stories are always about relationships. Find the relationships in the story you are telling.
- Analyze the copy for character, mood, attitude, conflict, rhythm, and so on. (See Chapter 10, "The Character in the Copy.")
- Look for the message, image, feeling, or unique quality that the advertiser wants to communicate to the listener. What separates this product or service from its competitors?
- Find the subtext, thoughts and feelings behind the words.
- Determine who the one perfect audience is and why she should be listening to what you have to say.
- Speak conversationally, having the expectation of a response. Talk *to* the other person, *not at* him or her.
- Determine the creative strategy that will enable you to build dramatic tension and allow for expression of the message. Use sense-memory techniques to locate tension in your body and speak from that place.
- Find a way to deliver the first line of copy in a way that will interrupt the listener's thoughts and bring them in to listening to your story.
- Be careful not to telegraph the message or send a message of "here comes another commercial."

Single-Voice Scripts

As you work with the following scripts, you might find it interesting to read through the script before reading the copy notes. Come up with your interpretation for attitude, pacing, character, and performance, and then read through the notes to see how close you came to what the producers of these projects intended. After working with a script, play the corresponding audio track on the CD that comes with this book to hear the actual commercial.

PUBLIC AWARENESS CAMPAIGN—CD/12

Title/Media: "Expectations" Regional Radio :60
Agency: CA Attorney General Office
USP/Slogan: "Think through it… Don't do it."
Target Audience: Parents of teenagers
Style: Teenager, dry, remorseful
Character: Male,
VO Talent: Jon Allen
Copy Notes: This radio commercial is intended to create an awareness of the potential ramifications for permissive parents who allow their teenage children to drink alcohol at a party, under the premise that it is "OK for teens to drink if they are supervised." The challenge with a script and story as emotionally loaded as this, is keeping the character and delivery "real."

There are certain things you expect from your parents. You expect them to love you. You expect them to protect you, and you expect them to know what's best for you. But even though teenagers expect these things from our parents, we still "push" 'em whenever we can. But there were things I didn't expect. Like them giving in when I wanted to have a party at the house and serve alcohol. I never expected them to do that! I mean, I'm only 17. I expected them to say "no way". I wish they had. I wish I hadn't pushed so hard. Because, if they'd said no, my best friend, Shari, might still be alive. If they'd said no, the police might never have traced the alcohol back to our party.

And if they'd said no—my mom and dad might still be together. I always expected them to be.

TAG: Never provide alcohol to a minor. Think through it… Don't do it.
A message from the California Attorney General's Office and the California office of traffic safety. To learn more, go to safestate.org/TRACE

Copyright © The Commercial Clinic. All rights reserved.

MY OPEN HOUSE—CD/13

Title/Media: "Numbers" Regional TV :30
Agency: The Commercial Clinic; **www.commercialclinic.com**
USP/Slogan: "Where buyers and sellers come together."
Target Audience: Home owners looking to buy or sell a home.
Character: Spokesperson – female; Tag – male
VO Talent: Penny Abshire **Tag:** James Alburger
Copy Notes: This is a typical TV script with separate columns for video and audio. It's a very good idea to read, and thoroughly understand, what will be happening visually as your words are spoken. Many voice talent overlook the visual column and deliver a voiceover performance that is inappropriate. This wastes time in the studio and inevitably requires that the director explain the visuals—something he should not have to do.

This spot is a short "story" using the *interrupt, engage, educate, offer* structure. The story is written in a way that anyone who has bought a home can relate to. The delivery is compelling in that it quickly establishes a meaningful relationship with potential home-buyers or sellers. Notice the use of tempo and rhythm to convey the message's emotional context. Also notice the conversational delivery of both the main voice track and the tag.

VIDEO	AUDIO
	NO MUSIC:
Rapid shots of numbers and related symbols. Numbers fades to BG as words fly forward, ending with a big question mark.	Numbers hurt my brain… you know? When I was buying my house I was overwhelmed with all the numbers. Fees, commissions, listing, mortgage – I just didn't understand 'em all.
Text graphic full screen "Give it to me Straight" dissolves to sequence of words, ending in 40% down…	So I asked 'em to give it to me straight, in words I could understand. And they did! And I like the words… 'Free Listing'… 'No money down'… 'No closing costs.' And one number I did understand... save up to 40% on real estate commissions.
Diz: moving video of home with "sold" sign in foreground.	My brain feels better now.
Diz: Full screen graphic with URL and logo.	Tag (:05): Bringing buyers and sellers together… painlessly… MyOpenHouse.com.
	Copyright © The Commercial Clinic. All rights reserved.

EAST CAMBRIDGE SAVINGS BANK—CD/14

Title/Media: "It's Not Rocket Science" Local Radio :52/:08
Agency: Direct
USP/Slogan: "Connecting with customers."
Target Audience: Adult men and women
Style: Friendly attitude projecting a sense of understanding
Character: Spokesperson – female
VO Talent: DB Cooper
Copy Notes: This radio commercial was part of a larger campaign consisting of print, radio, and television. The intention of this spot is to make the point that ECSB provides more personal service than other banks. Experiment with different ways of delivering the first sentence in order to grab the listener's attention. Then test different ways to tell the story so that it builds to the conclusion at the end of the second paragraph and resolves to the solution in the third paragraph. Because the message is about "… real, knowledgeable, people…" the delivery must be relaxed and conversational.

When you've made your choices for your delivery, listen to how DB Cooper delivered the actual voice track. Notice how she uses inflection, phrasing, and a hint of "attitude" to clearly imply how her character feels about the way other banks treat their customers "Do it BY yourself", and how she feels about the positive experience to be had by doing business with ECSB.

At many banks, things are becoming more and more "Do It Yourself." They talk about banking on the web and on your Smartphone, and how you'll never have to actually SPEAK with them.

But what's getting lost is that "Do It Yourself" often means "Do It BY Yourself." Service has become an afterthought.

At East Cambridge Savings Bank, they know there will be times when what you want from your bank isn't on the Internet or automated phone line. That's why they have people there. Real, knowledgeable people with the experience and inclination to help you.

Call 1-866-354-ECSB. Someone will answer. It's not rocket science. It's people. And there, people come first.

Copyright © East Cambridge Savings Bank. All rights reserved. Used by permission.

SMALL BUSINESS SUCCESS—CD/15

Title/Media: "Small Business" Internet Radio :60
Agency: Sun Marketing
USP/Slogan: "Growing your business"
Target Audience: Adult entrepreneurial business owners
Style: Very conversational
Character: Independent business owner seeking success
VO Talent: James R. Alburger
Copy Notes: The advent of easily accessible Internet radio has created a new frontier for advertisers. Target audiences are often highly niched, according to the type of show, thus making it relatively easy for an advertiser to effectively reach her customers quickly. Notice that the first sentence of this script immediately attracts the attention of the intended audience. Also notice how the delivery tells the story through a combination of shifts in phrasing and repositioned punctuation.

You got a small business?

Well, then you know how tough it can be… marketing, finding new customers, and especially just staying focused on the day-to-day details of running your business.

Even though my business was doing OK, it wasn't where I knew it could be. I was getting a bit discouraged. Then I heard about this little book called "Growing Your Business" by Marc Le Blanc.

Wow! I still can't figure out how such a small book could make such a big difference in my business! It only took about an hour to read, and the things I learned… well, all I can say is I'll be using Marc's ideas for a long time to come. Why? Because they work!

I learned how to really focus on what I need to do to attract more customers and how to be more successful by creating a plan for generating more business. Guess that's why Marc named his website small business success.

If you want to be more successful with your business— and who doesn't—you should check out Marc Le Blanc's website at smallbusinesssuccess.com.

No spaces—just… small business success.

Copyright © The Commercial Clinic. All rights reserved

12

Dialogue Copy:
Doubles and Multiples

Types of Dialogue Copy

THE CONVERSATION

As with single-voice copy, dialogue copy involves a conversation between two or more characters. The primary difference with dialogue copy is that your one-person audience is present in the script.

Unlike most single-voice copy, dialogue usually involves a story with a specific plotline and interaction between two or more characters. It is important for you to understand the whole story, not just your part in it. If you limit your understanding to just your role, you may miss subtle details that are vital to effectively interacting with the other characters, or for creating the dramatic tension that is so necessary for giving the characters life and making them real to the listener. When two or more characters are having a conversation, I refer to it as *interactive dialogue*.

Another form of dialogue is one in which the characters are not talking to each other, but are instead speaking directly to the audience. A conversation is still taking place, but in this case, it is more one-sided with each actor sharing a portion of the overall delivery. I refer to this type of dialogue as *shared information*. This is very similar to a single-voice delivery, except that in this type of performance each character is interacting primarily with the audience, but must also respond appropriately to the other character sharing information.

In the area of multi-voice ADR and looping, the dialogue lines may be adlibbed and a scene may involve many voice actors.

Regardless of the structure, all dialogue requires excellent listening and performing skills. Interactions between characters must be believable and timing must be correct for a dialogue performance to be accepted by the listener.

COMEDY

Comedy is a very popular form of dialogue copy. It is not the words on the page that make a script funny; it is the intent behind the words. In part, comedy is based on the unexpected—leading the audience in one direction and then suddenly changing direction and ending up someplace else. Comedy is often based on overstating the obvious or placing a totally serious character in a ludicrous situation. Comedy can also be achieved by creating a sense of discomfort in the mind of the audience.

Think of a comedy script as a slice of life—with a twist. Playing lines for laughs doesn't work. Laughs come only when the audience is surprised.

Rhythm and timing are essential with dialogue. A natural interaction between characters, overlapping lines, or stepping on lines, gives a more real feeling and helps set the rhythm and pace of the story. Pauses (where appropriate), and natural vocal embellishments can add naturalness.

Ask the producer or director before taking too many liberties with any copy; this is especially true with comedy dialogue. If the producer understands comedy, you may be given the freedom to experiment with your character and how you deliver your lines. Ultimately, your character should have precedence over the copy and certain ad-libs or other adaptations may be necessary to create the illusion of reality. Say your lines in a natural, conversational way, appropriate to the situation, and the comedy will happen.

To be effective, comedy dialogue must have a sense of reality, even if the situation is ludicrous and the characters are exaggerated. The following tips and suggestions will help you perform comedy copy effectively.

Tips for Performing Comedic Dialogue Copy

- Be real. Keep your character spontaneous and natural. Use a back story or lead-in line to help get into the moment.
- Find the dialogue rhythm. The rhythm for a comedic script will be different from that of a serious script.
- Humanize your character by adding natural sounds, such as "uhh," "yeah," "uh-huh," "mmm," etc. These sounds help give the feeling of a real, natural conversation. Ask before making copy changes.
- Find the *subtext*—what's going on behind the words—this is especially important with dialogue. If your character is that of a normal person in a ludicrous situation, you need to have a subtext of normalcy. If your thoughts anticipate the punch line, it will be communicated through your performance.
- Stay in the moment, listen and answer. Respond authentically to the other characters or situations that occur, expecting a response.

Dialogue and Multiple-Voice Scripts

DONATE LIFE—CD/16

Title/Media: "What are you waiting for?" Radio :60
Agency: AM Strategies
Writer/Production: Elliot Rose; **www.am-strategies.com**
USP: "Spend a minute, save a life"
Target Audience: Adults
Style: Very conversational, young and old share their stories
Characters: Young boy—happy. Older woman—somewhat sad. Tag.
VO Talent: Young boy—Ethan Rose, Woman—Stephanie Riggio,
Tag—Jack Dennis
Copy Notes (as described by the writer): This commercial was produced as a demo for Donate Life California, an organ and tissue donor registry. I wanted to create a sense of drama, of urgency, of hope, of happiness, and sadness in this spot. I wanted to shake people up and make them realize that the difference between life and death all comes down to a very simple task: marking the box that says "I want to be an organ donor" on their driver's license renewal, or registering online. For the talent I cast my six-year-old son, Ethan, and a wonderful actress from Los Angeles named Stephanie Riggio. Unfortunately for various reasons, the spot never ran. But it's one the most meaningful projects I've ever had the honor of working on.

Boy:	My name is Trevor. My dad is alive because he got a new heart.
Woman:	My name is Ellen, and my husband, Jerry, died waiting for a liver transplant.
Boy:	It took a long time to find a new heart for him.
Woman:	My Jerry was on a waiting list for over a year.
Boy:	But then, we got a call from the doctor.
Woman:	One day I got a call from the Hospital.
Boy:	It was good news.
Woman:	News wasn't good.
Boy:	They found a new heart.
Woman:	They couldn't find a liver donor in time.
Boy:	Now I have my daddy every day.
Woman:	And now, all I have is my memories.
Boy:	My daddy is alive because somebody registered as a donor.
Woman:	My Jerry is gone because not enough people register as donors.
Boy:	And I really love my daddy.
Woman:	And I really miss my Jerry.
Tag:	The only difference between these two stories is you. So, please register as an organ and tissue donor. It

only takes about a minute and you can do it online or at the DMV. Close to 7,000 Californians will die waiting for organ donations. What are you waiting for? Spend a minute… save a life. Register online at Donate Life California dot org.

Copyright © AM Strategies. All rights reserved.

TOWNE CENTER PLAZA—CD/17

Title/Media: "Lost in the Mall" Radio :60
Agency: Enjoy Development
Production: The Commercial Clinic; **www.commercialclinic.com**
USP: "In the middle of things"
Target Audience: Men and women who shop
Style: Real person, conversational
Characters: Wife helping her husband navigate his way through a mall.
VO Talent: Penny Abshire, James R. Alburger, Ross Huguet
Copy Notes: This is one of several spots created for a radio advertising campaign to promote a small shopping mall. The client's requirement that every store in the mall must be mentioned in the commercial created a real challenge, not only in the writing, but also in creating a delivery that would tell the story without sounding like it was selling anything. The purpose of this commercial was to bring awareness to the mall as a whole, and especially to stores that had traffic problems, which included the food court. A typical, relatively uninteresting, spot for this type of client might simply provide a list of stores and talk about the great food at the restaurants. However, we chose to take listeners on a virtual walk through the mall, which resulted in a strong visual image and better client identification.

Husband:	I know I'm supposed to meet you…
Wife:	At Magnolia, Jack. For a romantic dinner…
Husband:	Yeah! And that would be where?
Wife:	Right next to Fisherman's Market & Grill!
Husband:	And that would be where?
Wife:	Just cross from Kokopelli Café. Are you lost?
Husband:	No. Where are you?
Wife:	I'm at Magnolia, Jack.
Husband:	Hey, isn't that where I'm supposed to meet you?
Wife:	(laughing) Yes – where are you?
Husband:	At Pier 1 Imports.
Wife:	Okay, now just walk toward Michael's, see it?
Husband:	Yes, there's Evan's Eye Wear (I need to get my eyes checked).

Wife:	Then Crowfoot Travel...
Husband:	And Uncle Don's Hobbies – and Gordon Fancy?
Wife:	No hon, that's Uncle DAN'S Hobbies and GARDEN Fancy.
Husband:	I really do need to have my eyes checked!
Wife:	Yes, you do...
Husband:	Yummmm, something smells... Oooh! Cookies by Design!
Wife:	AFTER DINNER, honey – keep walking!
Husband:	Okay – I'm at Fisherman's Market & Grill. There's Adolph's Bakery and Kokopelli Café and...
Wife:	See the woman standing in front of Magnolia, waving?
Husband:	I see a cute blonde on a cell phone.
Wife:	That would be me, Jack.

Copyright © The Commercial Clinic. All rights reserved.

HOLIDAY MAGIC—CD/18

Title/Media: "Holiday Magic"
Agency: Gelderhead Productions; **www.holidaymagiccd.com**
Production: Gelderhead Productions; **www.gelderhead.com**
USP: "A gift for children of all ages"
Target Audience: Children
Style: Multivoice dialogue
Characters: Jeffrey, the surfer San Diego reindeer; Kate, the adventurous reindeer from Milwaukee; Carlito, Jeffrey's cousin from the Wild Animal Park; Santa Claus.
VO Talent: Jeff Gelder (Jeffrey); Heather Martinez (Kate); Greg Dehm (Carlito); Phillip Tanzilo (Santa)
Copy Notes: The Holiday Magic CD is an independent project to brighten the holidays for children under care at Children's Hospitals. The full-length audio CD includes songs and stories submitted by voice actors, singers, and other professionals woven together by a story about the adventures of Jeffrey the surfer San Diego reindeer. Each year's CD features a different story requiring a cast of 10 to 15 voice actors. The following script is a good example of the current trend in multivoice production in which four voice actors in different cities (most of whom had never met) recorded their lines independently. The dialogue was then edited, and music and sound effects added last. Each actor had to determine the attitude and personality of their character, and how to create a believable conversation—all while recording alone in their home studio.

Part 4, Kate, Carlito, Jeffrey, Santa

Kate *(panting)*:   Oh my gosh Jeffrey… that was… SOOOOOO… much fun!

Carlito *(panting):*   Aí… Dios mio… sí Jeffrey… ¡que bárbaro!

Jeffrey *(panting):*   I… can't… believe… we did it!

Santa *(overhearing):*   HoHoHo you little reindeer… You did WHAT exactly?

Carlito: Oh Santa… I hope you don't mind…

Kate: Yeah Santa really… it was just soooo aMAZing!

Jeffrey: Oh Santa it was… it was just AWEsome!

Santa: HoHoHo. OK, I'm LISTENing!

Kate: Well it was like this…

Carlito: Sí, Santa, Jeffrey had this… "idea fantastico"!

Jeffrey: Yep and we got all the reindeer to help

Santa: HoHoHo. Help… with what exactly?

Jeffrey: Well… we loaded up all the sleighs with every big speaker we could find…

Kate: … and then everyone's MP3 players – even the elves all helped!

Carlito: … and all the Holiday Magic music we could find…

Santa: And then?

Carlito: Sí Santa, and then, we flew all the sleighs all around the world playing the most happy music we could find for everyone…

Kate: … and the most amazing part…

Jeffery: … is that everywhere we went…

Carlito: … all the people started singing!

Kate: … it's true! They all started singing!

Jeffrey: … quietly at first

Kate: … but then it got louder

Carlito: … and louder

Jeffrey: … it was like the whole world was singing again.

Santa: So no wonder our workshops are running at full capacity now! Oh it is going to be a Happy Holiday for everyone after all. And you know what?

Jeffrey & Kate: What, Santa?

Carlito: Sí, Santa, what?

Santa: All that joyful singing takes away all the NoNoNo's and gives us all back our merry old HoHoHo's!

Copyright © Gelderhead Productions. All rights reserved

13

Industrials: Long-Form and Narrative Copy

Sales presentation, marketing videos, in-house training tapes, point-of-purchase videos, film documentaries, telephone messages, and many other projects all fall into the category of corporate and narrative. Frequently, these scripts are written to be read and not spoken.

Writers of industrials—corporate and narrative—copy are often not experienced writers, or usually write copy for print. There are exceptions to this, but overall you can expect copy in this category to be pretty dry. Corporate and narrative copy is often full of statistics, complex names or phrases and terminology specific to a business or industry. These can be a challenge for even an experienced voiceover performer.

As you perform a corporate or narrative script, you are still performing a character telling a story, just as for any other type of copy. You should know who your character is, who you are talking to, and what you are talking about. You also need to find a way to create an image of knowledge and authority for your character. What is it about your character that gives him the authority to be speaking the words? Is your character the owner of the company, a satisfied customer, the company's top salesperson, or a driver for one of the delivery trucks? To create an image of credibility, figure out an appropriate role for your character and commit to your choice.

A corporate or narrative script for a video project might have several performers on camera. These are often professional actors, but sometimes include employees of the business. There also may be several voiceover performers for different sections of the script. Many scripts in this category are written for a single voiceover performer, but occasionally two or more performers will alternate lines or voice different sections. There may also be some interactive dialogue sections of the script. The complexity of a corporate script will vary greatly depending on the intended purpose, the content, the length, and the budget for the project.

It is sometimes more challenging to deliver a script of this nature in a conversational manner, but it is possible. Facts, numbers, unusual terms, and complex names all contribute to a presentation more like a lecture than a conversation. However, the information is important, and the audience must be able to relate to the presentation as well as clearly understand what they hear. If the presentation of the information (your performance) is interesting and entertaining, the effectiveness of the communication will be much better.

Tips for Performing Corporate and Narrative Copy

The following tips and suggestions will help you with corporate and narrative copy in general.

- Talk *to* the audience on their level, not *at* them, even though the script might be full of facts, statistics, and unusual names or phrases.
- Take your time delivering the copy. Unlike radio or TV copy, which must be done within a specific time, there is rarely any time limitation for corporate and narrative copy.
- Be clear on the facts and pronunciation of complex words. These are important to the client and need to be correct and accurate.
- Slow your delivery or pace in sections where there is important information; speak more quickly in other parts of the script.
- If you are alternating lines with another performer, and the script is not written for dialogue, be careful to not overlap or step on the other performer's lines. Keep your delivery more open for this type of script, unless the producer specifically requests that you tighten your delivery.

Corporate and Narrative Scripts

As you work with the following scripts, you might find it interesting to read through the script before reading the Copy Notes provided for each one. Come up with your own interpretation for attitude, pacing, character, and performance. Then read through the notes to see how close you came to what the producers of these projects intended. The following scripts are reproduced as accurately as possible, including typos, grammatical errors, and awkward phrasing.

THE ART OF RISK—CD/19

Title: "The Art of Risk"—audio training program
Agency: Direct—Bobbin Beam
USP: "Get comfortable with the uncomfortable."
Target Audience: Actors and performers
Character: Educator—Motivator
VO Talent: Bobbin Beam; **www.bobbinbeam.com**
Copy Notes: This script was originally written as part of a blog post. Its intention was to help and support voice actors who were experiencing frustration with their audition submissions. The author, who also voiced this recording, delivered her script from the point of view of a very experienced, international voice talent who knows the business well. Her tone of voice, although encouraging and supportive, simultaneously reveals an underlying truth of the voiceover business: that in order to have any level of success, it is essential that we allow ourselves to take risks and explore options and opportunities outside of our comfort zone.

Notice how, in her delivery, Bobbin captures the listener's attention in the way she phrases the opening sentences. As she proceeds through the story, it is clear through the tone of her voice that she speaks from experience and has an underlying desire to help others who are experiencing what she describes.

The Art of Risk

How many auditions does the average actor perform to nail a single job? How many times do you "put yourself out there," and see nothing come of it?

Good question. Obviously the answer varies, depending on so many factors, which would be difficult to quantify into a solid statistic. It is the question as well as the answer that makes me wonder.

In general, it would be safe to assume that you are in the majority if you take the risk of performing on any level. You run the risk of not booking the job more often than not.

Risk of failure. Risk of rejection. Risk of "de-selection." Think about the Oscars: So many actors auditioned for the films and just so many got the job. Only so many films or actors were nominated, and just a few select won the "golden ticket."

Being in this business is like the supreme rollercoaster ride of your life. If you want to ride, better strap yourself in. You may have exhilarating highs one week or one day, and have all the air let out of your balloon the next.

In voiceover acting, we are at a grander disadvantage. At least on a film or video shoot, you're interacting with other human beings. Not so in voiceover, unless you enjoy the rare occasion where you're booked into a studio for a double or an ensemble gig. Even so, many times you end up perhaps with just the director, and or the engineer to record the session.

So most of the time, we work in a very isolated environment, and take our daily risks. We operate in a vacuum, and in so doing, we risk it all. We spill out our best, we think, and can still fall flat on our face.

Working through this "art" of the process is challenging at times. It can be quite painful, to risk and lose, as it can be incredibly heady in getting the recognition or landing a gig.

When we suffer losses, we must train ourselves to place them into perspective. You do this any way you can. But it helps to have practical training and experience to weather them.

Where we can get into trouble is when we allow our emotions and ego to take off on a self-absorbed "pity party." Many of us do this because we are actors, we are competitive, and have innate and trained sensitivity, combined with a healthy ego. For those who can't get this aspect of it simply give up.

That's when it's time for a break from the business. Really, take a break! Keep doing things you love and surround yourself with people who love you, and get back in touch with what truly matters.

Know it's not your fault that you've been rejected, ignored, disrespected, overlooked, under-appreciated, low-rated, or the latest industry buzzword, "de-selected."

Don't let this stop you.

It's OK to take the risk, while giving permission to others not to hire you for whatever reason that is not in your control. All you can control is your own performance and spill it out there.

And, next time, take the risk and get comfortable with the uncomfortable.

Copyright, © Bobbin Beam. Used by permission.

THE POWER OF YOUR THOUGHTS—CD/20

Title: "Passion, Profit, and Power"—audio training program
Agency: Mind Power, Inc.
USP: "Reprogram your subconscious mind to create the relationships, wealth, and well-being that you deserve"
Target Audience: Self-motivated people who desire to improve their life
Style: Friendly, professional, expert
Characters: Author/expert
VO Talent: Marshall Sylver; **www.sylver.com**
Copy Notes: Marshall Sylver is a motivational speaker, professional hypnotist, and an expert in the area of subconscious reprogramming. This

excerpt from his program "Passion, Profit and Power" is an excellent example of taking the words "off the page." Many professional speakers will produce an audio version of their seminar, or a training program. For studio recordings such as this, the material is scripted in advance and read during the recording session. The challenge for the presenter is to sound as though the words being spoken are coming off the top of the head. They must sound completely natural, very conversational, and be absolutely comfortable with the complexities of their topic—yet it is totally scripted.

With this sort of material, it is very easy to fall into a "read-y" narration style that can sound more like a lecture than a conversation. As an experienced motivational speaker, Marshall delivers his message in an intimate style that presents him as a knowledgeable expert. Notice how effectively he uses many voice-acting techniques like sense memory, pulling lines, pacing, and physicalization just to mention a few. Also notice how you, as a listener, are affected by the way in which Marshall tells the story. He draws you into what he is saying, and keeps you listening by presenting a skillful blend of intellectual and emotional content.

As you listen to the CD, observe the variety in Marshall's delivery; the way he adjusts his pacing to emphasize a point, the way he changes the pitch of his voice, and the way he creates a vivid image in your imagination. Also listen for changes in his delivery that are affected by physical alterations of his body, arms, and face.

As you work with this script, keep in mind that as a voice actor you are both a communicator and a storyteller. Notice that there are several lists in this story, as well as lots of vivid imagery. Use the techniques you've learned in previous chapters to determine your audience, back story, and character so you can create a powerful interpretation. Use sense memory, visualization, and physical movement to create a totally believable experience in the imagination of your audience.

This story is excellent for stretching your performing abilities. Become familiar with the basic story, and then tell it to different groups of friends in different ways—for one group simply tell the story without much emphasis or feeling; for another group, pull out all the stops. Notice how much more effective and real the reaction will be for the second group.

Passion, Profit and Power—script:
The Power of Your Thoughts

Every single thought that you can have possesses the ability to psychosomatically affect you. What this means is that every thought has a response in the physical body. In a moment I'm going to cause your body to respond to something simply by the thought of it.

Recently, I was driving through a citrus orchard near my home, and as I looked around me I saw thousands and thousands of totally ripe lemons. If you've ever driven through a citrus orchard, the first thing you'll notice is the scent… it smells like "sweet tarts." As I was driving through the citrus orchard, and looking at these juicy, plump, ripe lemons, I couldn't resist. I pulled my car off to the side of the road and I got out. And I walked up to the nearest tree that was laden full and heavy with the juiciest, plumpest, biggest, sour lemons I'd ever seen in my life.

I couldn't resist. I reached up to the tree… and I plucked the biggest, juiciest lemon I could find… and when I did, I smelled that scent, popping as the stem burst away from the fruit. That sour, tangy, juicy citrus scent from the lemon. I couldn't resist any more. In that moment, I reached into my pocket. I pulled out my pocket knife, and I pressed the shiny metal blade of the knife against the smooth, juicy flesh of this yellow lemon. And I began to slice through this juicy, tangy, tart, sour lemon… and as I did, the pulp and the juices from the lemon ran down the knife blade and across my hand. I put the knife aside, and I took my thumbs and I pressed them between the halves of this juicy, sour, tangy, tart lemon – and I split them apart. It was like sunshine in both hands! I couldn't wait – I tipped my head straight back – and I began squeezing half the lemon into my mouth. The sour juices, dripping, running down my chin, going back into my mouth. I couldn't wait any more… I took the other half of the lemon and I bit into it fully… tasting the sour, juicy, tangy lemon. And when I did, my jowls tightened, just like yours are tightening now.

Every thought has a psychosomatic response in the body. By virtue of the fact that you think it – you'll telegraph to your body to respond to that thought as if it were true. What you believe is true for you. Nothing else.

Copyright, © Marshall Sylver. Used by permission.

14

Character Copy

A good actor can do a thousand voices
because he finds a place in his body for his voice
and centers his performance from that place.
Charles Nelson Reilly

Vocalizing Characters

Many character and animation voices are an exaggeration of specific vocal characteristics or attitudes, which enable the performer to create an appropriate vocal sound for the character. A forced voice is rarely the most effective, is difficult to sustain, and can actually cause physical damage to your vocal cords. The most effective character voices are those that slightly exaggerate the attitudes and emotions of the character you are portraying, or that take a small quirk or idiosyncrasy and blow it out of proportion.

Over the past few years, the trend in feature film animation has been to apply very real, human characteristics and voices to animated characters. Although there is still a place for the occasional wacky voice, the vast majority of animation work has moved to a more conversational style.

ANIMATION AND CHARACTER VOICES

Character and story analysis are most important with animation voice work. Many factors will affect the voice of the character, so the more you know about your character, the easier it will be to find its true voice.

Consistency is extremely important in character voice work. When you find the character's voice, lock it into your memory and keep the proper attitude and quality of sound throughout your performance, adapting your character's voice when the mood of the script changes. The important thing here is to avoid allowing the sound of your character to *drift*. To make your character believable and real to the audience, the quality of the voice must not change from the beginning of the script to the end.

Most animation voice actors have a repertoire of several voices. A typical session may require voicing three characters before lunch, three more after lunch, and then going back for pick-ups on some of the first set of characters. This sort of schedule means character voice actors must be extremely versatile and must be able to accurately repeat and sustain voice characterizations. These demands make animation work a challenging niche to break into, and one of the most creative in voiceover.

In addition to voices and sounds for animation, character voice work can also include dialects, foreign and regional accents, and even celebrity impersonations. Special accents and dialects require an ability to mimic a sound or attitude that is familiar to a portion of the listening audience. Usually, this mimicking is a stylized interpretation and doesn't necessarily have to be 100 percent accurate, unless the character is represented as being authentic to a region or culture. Many times, the best accent is one that reflects what a community "thinks" the accent should sound like, which is often not the real thing. However, when authenticity is required to give the character believability, vocal accuracy is important. Most of the time, however, a slight exaggeration of certain regional vocal traits tends to give the character attitude and personality. Being personally familiar with the culture, region, or dialect is also helpful.

Celebrity voice impersonations are often the most challenging because the celebrities are usually well-known. The voice actor's job is to create a voice that offers recognition of the celebrity, yet hints at being just a bit different. Celebrity impersonations are usually done in the context of a humorous commercial in which some aspect of the celebrity's personality or vocal styling is exaggerated or used as a device for communicating the message. In most cases, if a producer wants an extremely accurate celebrity voice, he or she will hire the celebrity. It may cost more to hire the actual person behind the voice, but the increased credibility is often worth it and hiring the real person will circumvent any possible legal issues.

There are two critical aspects to character voice work: regardless of the character, the voice must be believable; and it must, in some way, be different from your natural voice. Achieving excellence in this aspect of the craft requires specialized knowledge and a mastery of performing skills.

THE SIX ELEMENTS OF CHARACTER VOICE

Pat Fraley is one of the most amazing voice actors I've met. He's not only a consummate performer—in just about every area of voice work—but he's also an excellent, and very generous, teacher and coach. Pat understands character voice work better than most professionals. He's one of the top 10 voice actors for animation, having performed more than 4,000 different cartoon voices. He's narrated dozens of audio books, and voiced thousands of commercials.

Just as I've broken down a performance to its *Seven Core Elements*, Pat breaks down a character voice to its six critical elements.

Pat Fraley's Six Critical Elements
of a Character Voice
© Patrick Fraley, 2006
www.patfraley.com

With any artful endeavor, you will find two words that are important to define: Form and Content. What is *form* of a character voice? The way this applies to character voice is that the *form* is the *sound* of the voice. The *content* is the thinking and feeling, the psychology of the character. The Six Elements of a Character Voice are all about the *form*. What makes up the form of a character voice? Like everything else, there's a finite amount of elements.

These are the six elements: *Pitch, Pitch Characteristic, Tempo, Rhythm, Placement,* and *Mouth Work.*

Pitch _____
(Higher or lower than your own?)

Pitch Characteristic _____
(Gravelly? Breathy? Husky? Constricted?)

Tempo _____
(Faster or slower than your own?)

Rhythm _____
(Syncopated? Plodding? Loping?)

Placement _____
(Nasal? Back of Throat? Normal?)

Mouth Work _____
(Accent? Lisp? Tight lips?)

The first element is ***Pitch***. Pitch relates to the musical notes of the character's voice. Is the character's voice higher than your own? Is it lower? Or perhaps the character goes both higher and lower, showing a wider range. Thus far we've dealt with Pitch and it has been assumed that the characteristic of the pitch was clear.

The second element deals with the specific characteristic of the pitch, ***Pitch Characteristic***. Pitch Characteristic is the dynamic of the pitch, or coloring. If Pitch were a noun, Pitch Characteristic would be it's adjective, as it describes the nature of the pitch. It is clear? Is it gravelly? Is it hoarse? Is it breathy? Is it constricted? Is it cracking? Velvety?

The third element is ***Tempo***. Tempo refers to the character's general rate of delivery. There are three possibilities: (1) Does the character speak faster than you? (2) Slower than you? (3) Or does the character vary or have a wider range of tempo than you?

The fourth element is **Rhythm**. Vivid characters go about getting what they want, and they go about getting it in the same way, over and over.

There is a pattern to their behavior. This pattern shows up in the way they speak. Character Rhythm is defined as a repetitive pattern of emphasis in the way the character speaks which emerges from the thinking and feeling of the character. It's kind of a vocal thumb print.

The fifth element is **Placement**. Placement refers to where the voice seems to be coming from or where it's placed. When I think about placing my voice in my nose the sound takes on a whole different dynamic. I feel a lot of air coming through my nose. It sounds a whole lot different than when I place my voice in the back of my mouth. How about if I think of creating the voice in my throat? Distinct sounds? The trick is to learn how to *stay* in the placement all the time for any given character.

Because placement happens in and around the mouth, it has a kind of relationship with our sixth and final element, **Mouth Work**. Mouth Work refers to anything done in and around the mouth to effect the character. The kind of effect that comes to mind is accents. But also, having tight lips effect the way we sound, stretching your mouth to one side and talking, or the way a character pronounces their "S"s.

So that's all the ingredients it takes to bake up the form of a character voice—no less, no more: Pitch, Pitch Characteristic, Tempo, Rhythm, Placement, and Mouth Work.

Finding Your Voice

Begin creating a character's voice by doing a thorough character analysis to discover as much as you can about him or her (or it). Based on the copy, make the decisions and commit to who you are talking to and if your character has any special accent or attitude. Finally, decide where in your body the character's voice will be placed.

Visualize the voice coming from a specific location in your body and work with the copy until the voice feels right, using choices about the character's physical size and shape to help you localize the voice. Use the "sweep" (Exercise #10, page 49) to find a suitable pitch for your character voice. Once you have found a pitch, use placement, pitch characteristics, tempo, rhythm, and mouth work to create a unique sound. Here are some possible placements:

top of head (tiny)	under tongue (sloppy)	nose (nasal)
behind eyes (denasal)	diaphragm (strong)	chest (boomy)
top of cheeks (bright)	loose cheeks (mushy)	throat (raspy)
front of mouth (crisp)	back of throat (breathy)	stomach (low)

Practice different voices with different attitudes. Use computer clip art, comic strips and other drawings to get ideas for character voices. You may find that a particular physical characteristic or facial expression is needed in order for you to get the proper sound and attitude. Remember: *Physicalize the moment and the voice will follow.* Record a variety of voices and characters, then listen to what you've recorded. If your character voices all sound the same, you'll need to work on your character voice skills.

The Character Voice Worksheet

In animation, you must be able to recall a voice on demand. The Character Voice Worksheet on the next page is just one of many good ways to document the characteristics of each voice you create. The worksheet is divided into four parts: (1) Accessing and recalling the character voice, (2) Placement, (3) Physical characteristics, and (4) Other notes.

Accessing and recalling the character voice:

- Give your character a name for quick recall
- Define age, gender, attitude, physical attributes, energy level
- Create a *key phrase* that will allow you to return to the character. You will know the phrase is correct when it sounds natural.

Placement:

- Determine where in your body you are positioning the voice.
- Choose appropriate pitch, pitch characteristics, tempo, rhythm, and mouth work to contribute to the reality of the voice.
- Understand the character's emotions and feelings.

Physical:

- Determine how your character stands and moves in space and time. Experiment with facial expressions, physical gestures, and speaking quirks, including how your character laughs.

Other Notes:

- Include any additional information that will help you recall your voice.

Each voice you document is what Pat Fraley terms a *starter*. This is a core voice from which you can create many more simply by adjusting some of the characteristics. Start with a defined voice, then experiment with changing the pitch and altering pitch characteristics. Change the tempo and rhythm as you work with the voice, or see what happens as you adjust the voice placement or slightly modify the mouth work—and voila, you've got a new character.

Penny Abshire demonstrates how she creates and documents her characters on track 10 of the CD.

The Character Voice Worksheet

© James R. Alburger, all rights reserved

Sketch or pic of character

Character name: _____

Age: _____ Height: _____

Gender: _____ Body type: _____

Character source (where did you get the idea?):

_____

Describe primary energy: _____

Key phrase: _____

Appearance (hair, clothing, etc.): _____

Placement (location of voice in your body):

Vertical pitch: _____

____ Abdomen   ____ Chest   ____ Throat   ____ Eyes

____ Adenoid   ____ Nasal   ____ Face   ____ Top of head

Horizontal placement: _____

____ Front of face/body   ____ Centered   ____ Back of head/body

Pitch Characteristics: _____
(raspy, gravelly, smooth, clear, smoky, edgy, nasal, de-nasal, nervous, breathy, tight, etc.)

Vocal Dynamics—Phrasing/Pacing (musicality of your character's voice):

Tempo: ____ Fast   ____ Slow   ____ Moderate   ____ Varying

Rhythm: ____ Smooth flow   ____ Staccato   ____ Melodic

Attitude—Tone of Voice: _____

Emotion: _____

Volume (loud/soft/varied): _____

Physicalization (how does your character move in time and space?):

Stance: _____   Walk: _____

Quirks: _____   Laugh: _____

Body: _____   Hands/arms: _____

Mouth work: _____   Dialect/accent: _____

Associated color, sound, or taste: _____

If your character was real, who or what might it be like?

_____

Other notes: _____

_____

Tips for Character and Animation Copy

Character voice work can be challenging, but lots of fun. Use the following tips and suggestions to help find your character's voice.

- Understand your character and the situation. In animation, you must often use your imagination to make up what you are reacting to.
- Discover who the audience is and understand how the audience will relate to the character.
- Maintain a consistent voice throughout the copy and be careful not to injure your voice by stretching too far. It is better to pull back a little and create a voice that can be maintained rather than push too hard for a voice you can only sustain for one or two pages.
- Be willing to exaggerate attitudes or personality traits for the sake of finding the voice.
- If a drawing, photo, or picture of the character is available, use it as a tool to discover the personality of the character.
- Find the place in your body from which the voice will come.
- Experiment with pitch, pitch characteristics, tempo, rhythm, placement, and mouth work to discover the most appropriate vocal delivery for your character.
- Experiment to take on the physical characteristics of the character.

Character and Animation Scripts—CD/21

Find a suitable voice and delivery for each of the following characters. Use *Pitch, Pitch Characteristics, Tempo, Rhythm, Placement, and Mouth Work* to create a believable voice. Do a quick "A-B-C" (*Audience, Back Story, Character, Desires*, and *Energy*) for each line of copy to discover clues about your character. Explore any relationships that may exist in the copy or in the drawing. Experiment with different voice possibilities and add layers of *emotion, attitude,* and *physical movement.*

Character Name: Scotty Snowboots

Description: He's pure, innocent and full of wonder!

Copy: *Momma… I think God has a bad case of dandruff!*

Copy: *How much more does it have to snow before I can make snow angels?*

Character Name: "Sergeant" Snyder

Description: He's a gym teacher. Wasn't a jock in high school and never will be. Makes up for his physical limitations with an overly aggressive attitude and a small amount of power.

Copy: *So it's a joke in this class that my glasses are so thick, I can see into the future! You think that's funny, do you? Well, you little maggots! I can definitely see into YOUR FUTURE! Now drop and give me 20!*

Copy: *What's with all this whining… 20 laps around the track is nothing… ya want 30?*

Character Name: Buzzer D. Kill (Buzz for short)

Description: Crotchety, ornery, angry and fed up with the injustices dealt him in life.
Southern drawl.

Copy: *I'm gettin' me somethin' to eat this time fer sure! I'm downright sick-n tired of eatin' carrots! I want me some meat!*

Copy: *So I says to him, I says, "Just how am I supposed to fly without wings – huh? Just how is that suppose to happen? You give me these long legs and a huge nose – but no wings! Great design… bud!"*

Character Name: Captain Raltoon

Description: Captain of the King's Battalion. He is loyal and steadfast. He puts honor above all else – even his personal safety.

Copy: *Meaning no disrespect, Sire, I shall not leave the palace. My place is at your side. My men will guard our borders. I will allow no harm to come to you.*

Copy: *I beg of you, Princess, do not take these threats lightly. You must take great care and be on guard. There are those among us who would prefer you do not see tomorrow.*

Character Name: Buck E. Beaver

Description: He's a cocky, arrogant and extremely confident space marauder with a heart of gold.

Copy: *Well now, honey, you'd better listen up. Either you lend a hand with the chores while you're on board my ship, or I'll push your cute little toosh out the airlock and never look back. Are we clear, sweetheart?*

Copy: *So… You say you've got a… few minutes? That could work for me…*

Character Name: E.T.A.

Description: He's an alien lost on Earth and has given up any hope at all of ever being rescued.

Copy: *If I'm stuck on this lousy planet – I might as well take advantage of all its vices… until they kill me! Pass me that red meat, butter and undercooked chicken!*

COPY: *I wonder if Viagra works on antenna?*

COPY: *I can't feel my right arm – oh well, who needs it?*

Character Name: Trinity T'Ovoli

Description: Half woman, half feline from the planet K'A<u>k</u>is. Confident, but somewhat fragile – being forced into adulthood and leadership following the assassination of her parents.

Copy: *N'Dar, my wise advisor, our kingdom is in grave danger. Rebel forces are losing ground on the battle field. Our world will end if I cannot find a way to unite these opposing factions. It is up to me.*

Copy: *Ahhh… at last a glimmer of hope. If only the Alliance will agree, this could all end tomorrow.*

Character sketches © Tim Abshire. Used by permission. All rights reserved

Production: Beginning Animation Class Exercise—CD/22
Title: Game Pilots
Writer/Producer/Director: MJ Lallo

This script and the recording on track 22 of the CD are the end result of a beginning animation voiceover class. Students were assigned a role and given only a very short time to find the voice and attitude for their character. The recording includes the first take of all lines, followed by MJ's direction, and finally the produced version of the audio.

CAPTAIN	Eagle 1 we've spotted more drones. They're flying in fast at 11 o'clock .
EAGLE 1	Copy that Captain! I got 'em on my radar. OK guys prepare for another assault.
CHEETHA GIRL	Ohh MY GOD... 6 planes are coming in at 1 o'clock.
EAGLE 1	Oh hell, they sure are... those S.O.B.s!
CAPTAIN	They're coming at us head on. Cheetah lay down some screening fire.
CHEETAH GIRL	Freaking A! They're starting another pass! I'm flying right into 'em... cover me.
CAPTAIN	I've got bullet holes my left wing. Keep those jerks off our backs!
EAGLE 1	Cheetah... They're divin' hard. Keep firing on 'em. Blast 'em back.
CHEETHA GIRL	AHHHHH yeeeeeeeeesssss... two of 'em down. Way to go. Captain... they're dropping like flies now! Eagle 1... come in Eagle 1... Do you copy...
EAGLE I	Copy that... Ahhhhh... I'm losing oil pressure in engine 2. 0h no that last plane is coming at me... 2 o'clock. Hit his tail! Damn Hit him again!
CAPTAIN	It's like playing a game of ENDWARS! Hitting him hard... bingo he's down!
EAGLE 1	Man that was close. You guys are awesome!

CHEETAH GIRL & CAPTAIN & EAGLE 1 (victory improv)

We're clear... ohhhhh yeah... way to go... nailed it... owe ya brewski. (improv)

Copyright © MJ Lallo. All rights reserved

15

Imaging, Promo, Trailer
Voiceover for the broadcasting and film industries

Radio Imaging, Branding, Signature Voice

The term *imaging* refers to the niche area of voiceover work that specifically promotes a radio station's sound or marketing image. The best imaging voice actors come from radio station production departments where they learn first-hand what station imaging is all about. An imaging production is usually fast-paced with lots of rapid-fire, short liners, station call letters, station slogan or USP, punchy music cues, catchy sound effects, and processed voice tracks that define a station's "sound" within its format.

An imaging voice talent will be voicing a station's IDs, promos, sweepers, and liners, and in some cases may even be handling the production. If you don't know what those are, imaging may not be for you.

If imaging is something you are interested in, you'll need to know how radio stations promote themselves, and how your voice will be used as an identifiable part of that promotion process. You'll also need to understand the differences between the various radio formats, and you'll most likely need to produce a separate demo for each format. Imaging formats include: Contemporary Hit Radio (CHR), formerly Top Forty; Adult Contemporary (AC); Hot AC; Urban Contemporary; Alternative; Modern; Jazz; Oldies; Classic Rock; Country; and News/Talk. Each format has its unique style.

GABRIELLE NISTICO (Charlotte NC)—CD/27
www.VOCareer.com & www.VoiceHunter.com

Vice President of VOCareer.com, Gabby is a coach, author, voice actor and producer. Her career began in radio at the age of 14 and moved to VO very quickly. She found her niche is radio imaging. Today she is the voice of dozens of radio & TV stations across the country and a leading expert on radio imaging.

Is Radio Imaging for You?

Many VO talent, upon hearing the phrase Radio Imaging for the first time immediately think; "This isn't for me. I have no experience in radio." It likely is "for you" and your lack of broadcasting knowledge is not a problem. If fact, in might be a plus.

Yes, to become a great radio imaging talent there is a necessary amount of radio knowledge you must absorb. However imaging is not part of the DJ job description. Radio stations want skilled voiceover actors and rarely, if ever, consider radio personalities for their imaging. So "Radio voices" need not apply. Few if any stations still use big, old-timey, phony, unnatural voices for their imaging. That sound is made fun of – profusely!

Imaging is all the pre-recorded, produced, station identifiers that you hear *between* songs on the radio. Identifying imaging from other station elements is easy. Listen for a distinct voice that stands out from the other station elements. That voice will usually be heavily effected and surrounded by lots of sound effects and noises.

Imaging also promotes, brands and enforces the overall message of the station. Stations invest a lot of time and money into their own marketing efforts and try to personify a listener's lifestyle. Imaging completes the total radio experience by creating a more personal relationship for the listener.

For most stations, imaging is an effective way to quickly deliver station related information, powerfully. Everything from new artists to events and concerts are publicized via a station's imaging. Imaging is aired so frequently that it acts almost like an audio ticker for the station.

This is the future of radio! Traditional, terrestrial radio still has a lot of life left and new forms of broadcasting like satellite and Internet are in their infancy. Voiceover talent are an essential part of the broadcasting experience and as DJs fall more and more out of fashion, forms of VO like imaging will take their place.

Imaging offers a freedom not found in other types of VO. Not only can you *change* the copy a station sends you; you can ad-lib, embellish and re-write all or part of it whenever you want!

Station's LOVE creative input from their VO talent and they welcome the copy suggestions we make.

I have a serious potty mouth, a twisted sense of humor and a pretty unconventional outlook on life. Most of the rock, country and CHR stations I work for hire me for those very reasons. It's not just my voice they are interested in, it's my personality too. That never happens—everyone else wants you to "stick to the script." Screw that—imaging breaks all the rules. It's voiceover anarchy!

Promo & Trailer

This category of voice work is for television and film promotion, which is different from radio imaging, although there are similarities. While imaging covers station IDs and programming transitions to establish a radio station's audio brand, *promotion* is generally considered to be advertising for a television station's programming products. *Promo* voiceover work usually refers to television station promos, although it can sometimes apply to radio. *Trailer* voice work, on the other hand, refers to the promotion of motion pictures.

Originally *promo* and *trailer* voiceover work were lumped into the same category because of the similarities of style and purpose for the product. However, as the voiceover world has become more and more niche oriented, the two genres of voiceover work have each become specialties in their own right. Both are essentially storytelling and the differences in style are subtle, but distinct.

Let's discuss promo voiceover first: As with radio, television stations present a specific image in their market. While radio must rely only on audio, a television station has the advantage of adding pictures and graphics to their broadcast image. If you watch much TV, you'll notice that each station has its own look and sound. The pictures, graphics, and voiceover will be consistent for just about everything the station airs to promote its programming and image. This is true for all television and cable broadcasters, from the local station to the national networks.

A network promo voice talent will generally work to an already-edited video with a music track and sound bites that set the tone, pace, and attitude of the promo. The delivery will be one that is involved with the story and characters. For local television stations, promo voiceover will often be done as *dry voice tracks* without the benefit of seeing the picture or hearing the music.

Working as a television promo voice talent can often mean the talent is "the voice," or one of the voices, of the television station. A promo voice talent will voice station IDs, program introductions, program promos (commercials for the station's programming), program VOCs (voice over credits — the "coming up next" voiceover you'll often hear at the end of a program), public service announcements, news opens, news promos, and possibly even station marketing and sales videos. As the voice of the station, you may be asked to voice anything and everything that serves to promote the station. Depending on the station and market, a television station might have two, three, or even more voice talent to handle specific promotional needs. One might voice only news promos, while another might voice the VOCs and program promos. Still another might only voice the radio commercials for the station.

Landing a job as a station promo voice can mean a lot of work on a regular basis. It used to be that you would have to travel to the station's studios to record your voice tracks. However with today's technology, you will most likely be recording at your home studio and send the tracks to the station via ISDN or an FTP website. Promo work is usually recorded on a daily basis, and most promo voice talent are "on call" in the event of a major breaking news story that needs a promo produced quickly.

Promo voice work for a television network is a completely different animal. At one time the networks hired one person as "the voice" of the network. However, today the major networks will use many different voice talent for different aspects of their promotion. There might be a specific voice talent hired for comedy promos, dramatic promos, political promos, and so on. Most voice talent who are hired to work on promos at the network level have several years of voiceover experience and understand the performing styles for voicing promos.

Although very similar to promo voiceover, trailer work does require a somewhat different form of storytelling. The copy is usually read *wild*, without the benefit of picture or a soundtrack, but the lines must be delivered within a specific time.

While a promo may be delivered with a range of delivery styles from conversational to intensely dramatic, trailer voice work will generally treat almost every word and phrase as the most important, dramatic, or impactful word or phrase ever spoken. The style is often one in which the voice actor is less involved with the characters and moves the story forward with a very dramatic, detached or almost "announcer-y" delivery. A conversational style used for a commercial will usually not be effective for promo or trailer voiceover.

Tips for Performing Imaging, Promo, and Trailer

- You are a storyteller, and stories are always about relationships. Find the relationships in the story you are telling.
- Look for the message, image, feeling, or unique quality of the program or station. This can often be determined by listening to the music and sound bites for a promo, or getting a sense of the energy of a radio station for imaging.
- Determine the creative strategy that will enable you to build dramatic tension and allow for expression of the message. Use sense-memory techniques to locate tension in your body and speak from that place.
- Find a way to deliver the first line of copy in a way that will interrupt the listener's thoughts and bring them in to listening to your story.

16

Other Types
of Voiceover Work

By now you can see that there are many facets to the craft of voiceover. To be successful in this business you need to discover what it is that you do best, and focus on developing those skills and your marketing within that niche.

This chapter covers some of the highly specialized performing areas of voice acting. Some of the top professionals in each area have contributed their thoughts on what makes their work unique and their suggestions for breaking in and succeeding.

Audio Books

Audio books are a highly specialized area of voiceover work. This kind of work is not for everyone! To be a successful audio book narrator you must have a passion for reading. Audio books are challenging projects, and the pay is often not as good as other types of voiceover work.

An audio book reader will read a book, on average, three or four times. The first time through is to get a sense of the story. The second time through is for the purpose if identifying and marking the scenes, characters, emotions, and story dynamics; and starting to develop the voice treatments for each character. The third time through is usually the recording session, or perhaps a final read to rehearse the delivery, in which case the fourth time through would be the session.

A typical audio book recording session will last about four to five hours per day. Most *readers* recommend taking frequent breaks and speaking continuously for no more than 50 minutes at a time. For the most part, the session goes nonstop, and the same process will repeat for as many days as it takes to complete the project—and that could be one or two days or up to several weeks. And even when you think your recording is complete, you may be called back to the studio months later for *pick-ups* to replace some lines or retrack a page or two. When this happens, you need to be able to match the original delivery both in delivery style and vocal quality.

Remember that the microphone hears everything. Even a slight change in position can result in a change in the quality or consistency of your performance. Wear clothing that doesn't make any noise, move silently, and learn to take shallow breaths. Page turns should be done as quietly as possible. If a sentence at the bottom of one page continues to the next, complete the sentence at the bottom of the page, pause, turn the page, then pick up with the next sentence. Don't try to do a page turn in mid-sentence. Changes in your delivery style, vocal placement, or bad edits will all result in your voice becoming a distraction to the listener.

Marketing yourself as an audio book narrator requires some special skills and knowledge. There are some companies that specialize in audio books; they arrange for the recording rights and book the performers. These are companies you can contact regarding work as an audio book narrator. Some will ask that you record at their facility, while others may be open to having you simply provide the completed and edited recordings.

Some audio book voice actors will take on the roles of both talent and producer by finding a book that they think will be marketable, arranging for recording rights from the author or publisher, preparing a production budget, and pitching the book to an audio book publisher with themselves as the performer. Developing an audio book project in this manner can create an income stream from several different areas.

One of the best ways to learn about audio books is to begin by studying acting and other performers who narrate audio books. Then find a voiceover coach who knows the audio book business, and who offers a class in audio book narration. You'll find additional information about the audio book industry at the Audiobook Publishers Association website, **www.audiopub.org**. Audio book work can be tiring and exhausting, but if you enjoy reading and storytelling, it can be very enjoyable and gratifying.

HILLARY HUBER (Hollywood, CA)
www.hillaryhuber.com

Hillary Huber is a successful voiceover talent based in Los Angeles. She has voiced thousands of commercials including national accounts for Toyota, Birds Eye, Boeing, Ford, and McDonalds. She is also a critically acclaimed audio book reader for titles including *A Light in the Piazza*, by Elizabeth Spencer, *A Field of Darkness*, by Cornelia Read, *A Map of Glass* by Jane Urquhart, and a six-book series titled *Southern Women*, by Elizabeth Spencer. Hillary is also an accomplished voice-acting coach and will often be found working with Pat Fraley as a co-instructor at his workshops. Here's what she has to say about the field of audio books.

A Bit about Audio Books

Reading audio books is the most rewarding voiceover work I have done. It offers me the opportunity to act, not just announce. I get to be multiple characters, wind a story, do it all on my own time AND I get to line my shelves with my finished work!

This fast-growing market is much easier to break into than other voiceover arenas such as commercial, promo, animation and looping. There are no agents. You forge a relationship with an audio book publisher and they funnel you work. It's that easy.

How does one go about creating that relationship? There are a couple of ways. You might try the gun-for-hire route. In this case you make a demo (generally consisting of three <u>short</u> excerpts from books demonstrating your particular skills; the demo is no more than three minutes total.) Your demo is then sent to small- and medium-sized audio book publishers with the hope that they need your talents and hire you. The larger audio book publishers are a little harder to get to.

I think a better way is to become your own producer. Most audio book publishers farm out their work. Find a book that has not been previously recorded. Record yourself reading a few excerpts from the book. Then approach the appropriate audio book publisher with an offer. "This is how I can make you money," is more attractive than "How can you make me money?"

You can research audio book publishers at **www.audiopub.org**. This is a terrific site loaded with current information about the state of the audio book market.

Reading audio books requires many of the same skills that this book teaches. You must be a good reader. You must be able to interpret themes and wind a story. Acting is critical, unless you want to read nonfiction—medical text, scientific works, self-help, and so on. The ability to change the sound of your voice to differentiate between characters is a big plus. Not all readers do this but I find a little vocal characterization enhances the read and makes the listening easier.

I record for only about four hours at a time. If you work as a gun-for-hire, and record at the audio book publisher's studio, you can expect to read for six to eight hours.

This market is not as lucrative as the commercial world. But it's so much more creative and ultimately, for me, more fulfilling. I produce all of my own work, so I am able to work around my schedule and never have to miss an audition or another job. I record in my home studio (many books are done this way) so I don't have studio and engineer overhead costs. I highly suggest you look into this wonderful, fast-growing world of audio books. It's a blast.

Video Games

Voicing for video games falls within the Character and Animation category, but there are some unique differences for this type of work. For the most part, today's video games have characters that must be presented as real people. Therefore, as a voice actor, you truly need to be able to create a believable and conversational delivery for the characters you create. You'll also need to have a stable of at least several characters who sound completely different from one another—and from you.

When you get a script for a commercial, animation, or narration, the script is complete and you can read what comes before and after your lines. It's in a very linear format, and you will sometimes have a sketch of your character. Video games are nonlinear! Your script will have only your lines, and you may or may not know what your character looks like. Each line will need to be recorded in a variety of ways to express different emotions, attitudes, energy, and so on. Also, your delivery for each line will need to be able to be attached before or after one or more other lines. This is because, when a game is played, the exact sequence of your lines will depend on how the player moves through the game. Regardless of the path a player takes, the continuity of the voiceover must be consistent and appropriate for the events taking place in the game.

Video games are usually action-oriented and often include scenes that depict violence or death. When you play a video game character you may find yourself being directed to "die" in a dozen different ways. How would your character sound if he died from being hit by a car? By being stabbed? Shot? Hit on the head? Blown up? Choking? Drowning? Each vocalization of your character's "death" must sound different from the others.

ZACH HANKS (Santa Monica, CA)
www.soundawg.com

Zach Hanks is a professional voice talent and creative director of SOUNDAWG, where he directs and casts voices for interactive games. His directing credits include the *Warhammer 40,000: Dawn of War II* series, the *Company of Heroes* franchise, and *S.T.A.L.K.E.R.: Shadow of Chernobyl.* As a voice talent, his credits include *Mass Effect 2, Dragon Age: Origins, Final Fantasy XIII, Aion,* and *Brütal Legend,* but he is best known as "Captain MacMillan," the Scottish sniper captain from "All Ghillied Up" in *Call of Duty 4: Modern Warfare.*

The Truth about Video Game Voiceover

Video game dialogue has unique demands and requires a special skill set that not every voice actor has. Contrary to popular assumption, voice acting for games and voice acting for animation are very different. In general, western animation is character acting for *comedy*, is peopled by broad *caricature* character-types, and is geared toward a young audience. The performance style is usually more along the lines of sitcom, sketch comedy, melodrama, farce, clown, and vaudeville. Interactive, on the other hand, is character acting for *drama*, is peopled by realistic characters (cultural stereotypes and archetypes notwithstanding), and is often geared toward an adult or young adult demographic. The performance style is often realism, in the vein of film genres like action, war, horror, suspense thriller, high fantasy, space opera, etc.

Video games require actors. Voice talents who do not identify primarily as "actors," such as audio book narrators, announcers, and radio personalities, are not likely to have the necessary foundational skills and training the medium requires. There are exceptions, but they are rare. Besides good acting, there are two essential qualities that all the most prolific game actors share: *endurance,* and *efficiency.*

Video game dialogue recording sessions are usually rigorous at the very least. Often, they can be exhausting, painful, and injurious. Voice actors sometimes call these games "screamers" or "throat rippers." Modern and historical war games, like the *Call of Duty, Company of Heroes,* and *Medal of Honor* franchises, can require hours of intense screaming. These sessions push the performer to the farthest limits of what even a vocally trained actor can endure. To be a commodity for video games, the actor must know the limits of their instrument and be able to perform near that limit for the duration of the session. This is a skill that comes with training and experience. Well-trained stage actors with vocal training for the stage and for singing have blown out their voices completely within just 10 lines, because they lacked game voice acting experience and pushed too hard. Even the best game voice actors routinely lose jobs because of vocal fatigue and injury. Risks include laryngitis, chorditis, burst blood vessels, fainting, soreness, and vocal nodules and polyps, to name a few. This is not a career for singers, unless you want to be the next Tom Waits.

Video games can require tens of thousands of lines of spoken dialogue. In game dialogue recording, time is money, and there's never much to waste on slow actors. Audio budgets are a very small fraction of a game's total development budget, and voice recording is a small sliver of the audio budget, so every minute counts. Actors who can deliver a keeper take in the first two reads become regulars in every casting person's stable of talent. A decent game actor can get through over 400 lines of gameplay dialogue in a four-hour session (at two takes per line).

The best can get through over 600 lines. That kind of efficiency requires an actor with an "instantaneous process," who can deliver reads at emotional extremes with only a few seconds of preparation. Many of the highest paid, Oscar-winning film actors in Hollywood can't hit that kind of a mark on demand.

The financial rewards of working on a single title are modest compared to a national SAG television commercial. As of this writing, AFTRA scale for a session is a little over $800, SAG scale is just under $800, and there are no residuals. Double scale fees are rare. Nonunion session fees in Los Angeles generally are around $150 to $200 per hour with a two-hour guaranteed minimum, but some triple-A titles offer fees comparable to union scale rates for nonunion sessions. Almost no game actors can support themselves exclusively on video game work. Rarely, a winning-lottery-ticket booking happens where an actor books a character with 10,000 lines of dialogue *and* does the motion-capture for the animators, and he makes a year's pay on a single game. Despite this, the busiest Los Angeles-based game actors can earn a five-figure bump to their annual income from video game session fees, and greatly increase their notoriety and fan base as character voice actors.

Because a career in game acting as a sole vocation is impossible at this time, game actors primarily pursue this work for the sheer joy of it. Actors get to play roles they'd never play on screen in larger-than-life epics. On-screen leading men and beautiful ingénues get to be character actors, and actors with a "face for radio" get to be romantic leads. The work demands a vast vocal and emotional range that your average primetime network guest star role does not. Games allow actors to inhabit skins that aren't necessarily even human, fight for their lives, and save (or destroy) the galaxy. And when the writing is good, even the roughest sessions are still pure joy. I am a working video game voice actor, and I am grateful every day to have the greatest job in the world.

Telephony

This area of voiceover is often combined with Industrials because most of this work is for business. *Telephony* voiceover work includes the broad range of telephone messaging. If you've recorded an outgoing message on your home answering machine or cell phone voice mail, you've done the most basic of telephony work. Many businesses will require voiceover for *message-on-hold* (MOH), *voice prompts*, and *concatenation*, among others.

A message-on-hold recording is one or more messages, usually with background music, that are heard when a caller is placed on hold. A good MOH script will contain several short messages that focus on a service, product, or benefit offered by the business. It's a vehicle for self-promotion.

Voice prompts are the automated outgoing messages that provide instructions to a caller. One type of voice prompt simply asks the caller to take a specific action, much like the message on your answering machine. Systems using these prompts will require the caller to press a key on their phone in order to proceed, or simply "wait for the tone."

The other type of voice prompt takes the idea of an outgoing message to the level of creating a virtual person having a conversation with the caller. This is known as *Interactive Voice Response* (IVR). The caller is greeted by a recorded "person" who engages the caller in a conversation that will ultimately get her where she wants to go. The caller responds by speaking the request, and the computer moves to the next prompt based on what the caller said. It's completely automated, but fully interactive. When properly produced, the voice prompts sound completely natural and may even be mistaken for a real person. Ah, technology.

Telephony voiceover has specialized software that allows for the creation of multiple prompts in short order. And there is a technology and language for telephony voice work that must be learned.

Voice prompts are increasingly the first contact a customer has with a business, and there are literally dozens of companies who are eager to provide this service.

CONNIE TERWILLIGER (San Diego, CA)—CD/23
www.voiceover-talent.com

Another aspect of voiceover work for telephony requires the voice talent to create a sequence of recorded words or phrases that can be linked together seamlessly in any order without sounding like bad edits. This process of assembling individual recordings is known as *concatenation*. Connie Terwilliger is one of many professional voice talent whose specialty is telephone messaging.

Concatenation

The trick to doing this type of work is to be consistent with each segment and remember what you sounded like when you recorded the part that comes before what you are now recording. Each line of copy must be delivered in context, and with continuity for what is before and what follows. The following script from TeleMinder is typical of this type of work.

Hello, a member of your family has an appointment…
at the Eye Clinic.
at the Allergy and Asthma Clinic.
on Wednesday (include Monday through Friday)
April 23rd (include all calendar dates)
in the morning (include afternoon)
at
8:30 (include all times at 1/4 hour intervals)
AM (include PM)
If you need to cancel or change this appointment, please call the
office at…

Track 23 on the CD, voiced by Connie, demonstrates how this works. The first part of the track is the individual lines. The second part shows how the lines are assembled by the computer system. This is a very simple example. Some complex systems might have hundreds of different short lines that could be assembled into any of dozens of possible sequences that must sound natural when heard over the phone.

As you listen to the CD, you'll notice that the audio quality is considerably lower than what you might expect. Most telephony systems require audio files that are compressed to an 8-bit format. These files are very small, and much easier to use in the telephony computer systems. The lower quality often goes unnoticed when heard over the telephone primarily because most telephones are not designed to reproduce audio much over 8 kHz. One of the most popular formats is the mu-law (μ-law) format which starts with a 16-bit .wav file recorded at 44.1 kHz. That file is then down-sampled to an 8-bit, 8 kHz file during conversion to the mu-law format. OK… that's probably more than you wanted to know.

ADR/Looping

Automated Dialogue Replacement (ADR) and *Looping* are niche areas of voiceover that are have remained hidden secrets for quite some time.

The term looping comes from the early days of film sound when a segment of the film was spliced into a continuous loop. A set of white lines was drawn on the film using a grease-pencil, with the lines converging where dialogue needed to be replaced. The actor, watching the lines converge, would begin speaking on cue, matching his original performance.

In today's complex world of film sound, the process is basically the same, except it is now done with a video playback. Audio beeps have been added to assist with the countdown. Looping has also taken on new meaning in that it not only includes the process of dialogue replacement, but it also includes recording the natural human sound effects of crowd scenes, nonscripted voiceover, and efforts (the sounds of human exertion in fights, etc.)

ADR is the process of replacing dialogue for a feature film. The original dialogue may need replacing for any of several reasons: the original location sound is unusable; the director did not like the way a line was delivered even though the rest of the scene was OK; or, profanity needs to be replaced for television or air-travel use. In most cases, the original actor will replace their own dialogue, perfectly matching their original performance. However, occasionally, the actor may not be available, so his or her lines may need to be replaced by a soundalike voice talent who will perfectly match the actor's lip-sync. Most of the major studio feature films will have 50 percent or more of the dialogue replaced—and some will ADR the entire film, discarding all of the original dialogue.

A typical looping session may include a group of up to 20 "loop group" actors. The ADR session director will assign one or more actors to a specific task, or to replace a line of dialogue, as needed. The actors may perform individually, or as a group, depending on the needs of the film.

Breaking in to ADR/Looping can be a challenge as it is a small niche of voiceover work. Improvisation and acting skills are essential, as is an ability to perfectly match lip-sync. "Loopers" must be excellent researchers in order to create the realism for specific time periods and locations.

The best way to break in to this part of the business is to find a loop group and ask to start sitting in on sessions. As a loop group director gets to know you and learn about your abilities, they may invite you to participate.

JAMES PHILLIPS (Barcelona, Spain)
www.jamesphillips.info

James Phillips has always been fascinated with sound and voice. From childhood nights scanning short wave radio frequencies, absorbing the colors of language, to formal training in radio and television at the CBS affiliate, Brown Institute of Broadcasting. He put his passion to work as an announcer first on FM radio in Miami and then in Barcelona as a voiceover talent, becoming a signature voice in English on premier corporate films, audiovisuals, TV and radio commercials, trailers and documentaries. Combining vocation and experience, he directs voice castings and film dubbing, ADRs for feature films and the original voice soundtracks for animation films.

ADR—Automatic Dialogue Replacement

There is nothing "automatic" about dialogue replacement. Whether replacing dialog lines in post-production for technical reasons (such as mic noise or unwanted background sound), adding background voices and ambience, or dubbing a released movie or series into a new language, ADR is anything but "automatic." All these processes involve voice acting and, most often, working in a professional studio as an ensemble of actors under a director. A fourth area of voice acting in film and TV involves recording the voices for animation, often prior to the animation itself.

My work in voice acting and directing involves these four areas of ADR or "looping," with a particular focus on full cast dubbing into English of foreign language films and the original English language soundtrack for animation films.

Many countries have a long-standing tradition of full cast film dubbing into their local languages. However, while dubbing a foreign language film into English has not been as widespread, the demand is growing particularly for multilanguage audio tracks on DVDs and to increase distribution and sale of foreign series and movies in markets where English is a major language.

In Spain, for example, where I live and work, more and more live-action and animation films are being produced with English as the original language to facilitate worldwide distribution. The films are then dubbed into Spanish, Catalan, Basque and Galician (the main languages of the country). Examples are *Agora* (2009, Alejandro Amenábar), *Donkey Xote* (2007, José Pozo), *Fragile* (2005, Jaume Balagueró) and *My Life Without Me* (2003, Isabel Coixet). Often the story in coproductions naturally involves casts that combine Spanish and English speaking actors, such as the TV movies *Reflections or The Lost*.

On supervising the ADRs for *Fragile*, numerous scenes called for recording the background voices and ambience with children. A main issue in working with children is scheduling, around their school hours, and complying with labor legislation. It is fascinating to watch how children often have natural (and excellent) eye-screen contact, which is a must for credible synchronization. However, children develop so fast that locating good child actors is a never-ending quest. When you least expect it, a child who artfully sells you a pastry at a sidewalk stand or does a spontaneous reading in a church service, brings the right nuance to your next ADR session.

Donkey Xote is an adventure-comedy animation feature film about Sancho's donkey that tells the "true story" of Don Quixote. All the voices were recorded without image. A rough soundtrack was then created and the animation drawn to the voices. About two years later, once the animation was completed, ADR sessions were held to accommodate scene changes or to improve the interpretation. Care must be given to ensure that the color of the new dialogue matches the old.

Full cast film dubbing begins with the translation of the original dialogues into the new target language. Once translated, the script then goes to the "adjuster", who makes modifications in order to match the lip movements as closely as possible to the new target language dialogue.

The director chosen to direct the dubbing preselects the voice actors for all the characters in the film. Once the replacement language script is adjusted and the cast of voice actors is approved, the actors are scheduled. The sessions are generally intense, with a feature film typically dubbed in five to seven days. In the past, actors were convened in groups as needed for the different scenes, and recorded together. However, with the increased use of surround sound and other mixing considerations, actors are more often than not recorded individually.

Screen-eye contact is paramount in dubbing. Voice actors must be guided more by visual cues for starts, pauses and stops, rather than by listening to the original voices. As a result, the use of headphones is generally discouraged except for complex situations. After explaining the character to an actor, rehearsal of the first loop begins. The original sound is played back over the speakers and the actor rehearses on top. If the actor is new to dubbing, sometimes I explain the need for eye contact; other times I wait for the reaction when we rehearse without sound and the sudden realization: "Whoa, I have no idea where I come in, where the pauses are, where I have to end."

You need to be able to retain the line in your mind so you can deliver it while keeping as much eye contact as possible with the screen. The TC (Time Code) is helpful, but you must not become dependent on it. Look for a visual cue to tell you when to start. You already have the rhythm and emotion of the line in your mind from hearing the original sound.

Rehearse at performance volume while listening to a dialogue loop several times with sound coming over the speakers. Sometimes an actor will rehearse mentally, in a low voice or whisper, and then when they go to record the loop, they are way off sync. If you do not rehearse at your performance level, your rhythm and emphasis are different. The next step is to test without sound until the actor and the director are happy with the match and interpretation and the line is ready to be recorded. A good sound engineer has saved many a day by recording the final rehearsal loops, which are often the most convincing.

ADR is a great way for voice actors to learn to work as a team. Whether replacing lines verbatim, adding ambience, recording for animation or substituting dialogue into a new language version of a film, ADR requires concentration but it is fun and creative as you work in an ever-changing lineup of projects.

Anime

The term *anime* refers to a style of animation that originated in Japan, but has become associated with dialogue replacement voiceover work that requires matching lip-sync and timing to the original foreign-language animation. Anime is among the most challenging types of voiceover work.

DEBBIE MUNRO (Vancouver, Canada)
www.debsvoice.com

Debbie's specialty is character voice work for animation and anime. She has voiced roles in *Dragon Ball*, *Dragon Ball GT*, *Benjamin Bluemchen*, *Neverwinter's Night*, and *Max and Buddy*, among others. She runs her own production company where she writes, voices, and produces a variety of audio and video projects. Debbie is also one of Canada's top VO coaches.

Inside Anime

Anime is an art in itself. Fans take anime very seriously. Many watch the original versions (in Japanese or whatever the language) because they get the true intent of the author.

Have you ever watched English dubbed anime's like *Dragon Ball* or *Pokeman*? It's often hard to watch because the mouth doesn't always match the words. Let me explain why. Imagine learning all there is to learn about creating characters (which is A LOT) then trying to implement that in your session, not memorize your script, watch TV, and match voice flaps THAT AREN'T EVEN IN YOUR LANGUAGE—all at the same time? It's near impossible!

It truly takes a gifted person to match voice flaps, adlib, and die 5000 different ways. Seriously! Three quarters of a session is sound effects: falling, dying, running, etc. It all must be done to PERFECT time, matching beeps, mouth work and body actions. Anime is the hardest of all forms of animation voice work.

On top of it all, it pays less. WHY? I asked too! They say that the original voicer is the one who created the character and we are just mimicking it. Although I respect that very much, it's hard to understand, when the anime dubbing work is far more difficult than any other form of voiceover.

If you want to voice animation, learn to act and how to voice anime style. There are a few amateur dubbing sites out there to help get you started from the comfort of your own home, but with anime dubbing, you will more than likely voice in their studios.

Although anime is more work for less pay, animation in any form is one of the most exciting and fulfilling jobs you'll ever do.

17

The Brave New World of New Media

Understanding and keeping up with trends and new technology is a very important part of working in voiceover. For the past few decades there have been gradual, yet consistent, changes in how digital data is recorded and delivered. Many of these technological advances and how they relate to voiceover work, home studios, and content delivery, are discussed elsewhere in this book. During the past several years, many of these new technologies have found a home in what has become known as social networking and other Internet delivery methods, which in the larger sense are referred to as *new media*.

In order to understand the true potential, and anticipate the future of voiceover work, it is important to have some knowledge of what this New media is, how it works, and how we, as voice actors, can utilize it to our benefit. In this chapter, Joe Klein, an expert on the subject, will discuss New media and provide a glimpse into the future of voiceover work.

JOE KLEIN (Laughlin, NV)
www.newmediacreative.com

Joe Klein is truly a veteran producer, director, writer and voiceover artist. For over four decades, Joe has voiced and produced literally thousands of product jingles, pop records, national radio commercials and corporate presentation soundtracks. During the 1970s and '80s, Joe established a reputation as one of Hollywood's leading voice directors, receiving multiple Clio and International Broadcast awards, and was widely known for the "bigger than life" sound he achieved. He has also voiced scores of national radio spots, television commercials and network promos. In 2005, Joe launched The Podcast Voice Guys, which quickly became a leading provider of voiceover content for podcasters, video bloggers and new media networks. In 2008, the company was renamed New Media Creative to broaden the scope of the creative enterprise.

A Revolution in Media Content & Delivery

So, just what is this new media thing, anyway?

For those in their twenties, New media is more than just a familiar term! It's most likely the only kind of media that the majority of those in this demographic consume. For those more "old skool" folks, however, the term has often evoked looks of bewilderment over the last few years.

If you are a voiceover artist or other media professional involved in the field, it's time you got a handle on the brave new world of new media if you're not already familiar with it, as new media is already pervasive, and is definitely where all media is heading.

A SHORT HISTORY OF NEW MEDIA

The term new media refers primarily to newer forms of media content, which, beginning early in the new millennium, were originally being created by alternative and renegade content creators. Falling under the new media genre now is a rapidly growing new breed of alternative content being produced by mainstream media outlets as well, including repurposed versions of the content that appears on traditional media channels.

New media is an outgrowth of a more functional Internet which, during the first decade of new century, came to be known as *Web 2.0*. Web 2.0 utilizes the latest technologies along with high-speed broadband Internet connections to offer content in new formats and genres not previously available via traditional media delivery channels or even on the Internet before high-speed connections became widespread.

The best examples of early new media content were, at first, online audio "podcasts" (so named back in 2004 as a hybrid term combing the name of the hugely popular Apple iPod® name with the word broadcast). Originally referred to as "audio blogs," podcasts had a relatively humble debut, but then exploded in 2005. The number of podcasts online increased exponentially over several years, with hundreds of thousands of podcasts available by 2007. Differentiating podcasts from traditional media was that podcasts were played on media players installed on computers and on small portable media players that became available in the late '90s. Podcasts were commonly encoded into the MP3 file format that became popular for music, since this format significantly shrunk the file size of an audio track.

Podcasts were, at first, created and voiced by amateur hobbyists or "wanna-be" performers. They were downloaded directly off the Internet and stored in computers or in portable media players, where they were available to play whenever the listener desired. A technology called "RSS" (real simple syndication), originally developed for text blogs, was modified for audio blogs and allowed people to "subscribe" to podcasts and automatically download each new episode of a podcast series as it was posted online, in the same way that a DVR records new episodes of television shows.

Portable media players played a major role in the success of podcasts, as people could listen to shows whenever and wherever they wanted. Commutes, workouts, exercise routines and long plane flights were, (and still are) perfect times to listen to prerecorded podcasts. The Apple iPod®, introduced late in 2001, was largely responsible for making the media player a mainstream device. Originally used to listen to personal music libraries, the number of iPods in use grew rapidly in the first few years after its debut. The iPod was very well suited to downloading and playing podcasts, which were flourishing as the "hot new thing" by 2005. In that same year, the popular iTunes music store, which was the primary service used to manage iPod content, began offering podcasts along with music.

By 2006, the podcast world was on fire. That same year heralded the launch of the video site *YouTube*. With the phenomenal success and growth of YouTube (and several similar video networks) further bolstered by the introduction of the video iPod late in 2006, video podcasts started to appear, and these, along with endless other genres and categories of videos, would soon become a major component of the new media mix. By the time of publication of this book in 2010, video content is now dominating the new media landscape, with many formerly audio-only programs now including video (even if that is nothing more than a static shot of the show's host). In fact, online videos became so popular so quickly that the term *viral video* was coined soon after YouTube arrived on the scene, to denote a video that was posted online and spread around the world very quickly, like a virus.

YouTube and other video sites also ushered in the era of "streaming" media content. Up until that time, most podcasts and other audio and video content was downloaded and played at a later time. But, by 2006, broadband Internet connections had been widely adopted and the technology to "stream" audio and video content live had vastly improved. "Streaming" media is not exactly "live" in the sense of terrestrial, cable and satellite-delivered media, but it is close to it. The picture and sound is instantly encoded to digital data as it occurs and is then uploaded to the Internet as thousand of "packets" of digital data on a constant basis. On the receiving end, specialized streaming media players (the most common of which is Adobe's Flash player, decode the packets of data, "cache" them into the memory of the receiving device and then play the video or audio instantly. All this happens in a manner of seconds, so the streaming media is, in fact, "delayed" slightly, but it is still pretty much "live" and, actually, not that much more delayed than a satellite television broadcast. The primary difference between streaming media and downloaded media is that streaming media is not stored on the receiving device but, rather, decoded and played "on the fly."

Audio and video production technology continues to improve, becoming more affordable and easier to use. Broadcast-quality, high definition video cameras, for example, once costing thousands of dollars now cost just a few hundred and are as small as the digital still cameras that first became popular over a decade ago.

In fact, digital video cameras are now everywhere and even incorporated into most cell phones. Whether we're being shot by our friends or observed by the ever-present surveillance lens, more and more of us are ending up "on camera" more than at any other time in history, whether we like it or not. Reality television shows like *America's Funniest Home Videos* and *Cops* paved the way for this video revolution well over 20 years ago (with both shows still airing). Scores of other reality TV shows have followed and flourished. As of this writing, dozens more are currently produced as online programs only, bypassing traditional media networks to reach ever-growing audiences.

Digital photography and video have evolved from a shutterbug's hobby to a worldwide obsession. Now, anyone with a digital camera can become a citizen journalist, news photographer or member of the voracious worldwide paparazzi community. The desire of the masses to capture, share and consume video media is at a fever pitch. It's safe to say that, driven by the presence, power and reach of the Internet, the passion for pictures has reached a level nearing madness, and digital video has become a major component in the new media world.

Audio books, which, in the past, were only available on cassettes and CDs, are now distributed primarily online, where they are downloaded in the same manner as a podcast. The online distribution methodology now being employed to deliver audio books now qualify this genre as yet another one to fall under the new media moniker.

A couple of years back, printed books themselves morphed into what can be considered a form of new media, depending on who you ask. Recently, a new breed of electronic book (e-book) readers hit the market. The most popular is a device called the Kindle, distributed by the online retailer Amazon. These readers have, themselves, become quite popular as a new way to read the printed word. They use an "e-paper" technology to simulate the appearance of a printed page, and connect directly to the Internet to download and display nearly all the popular works of the day.

Further fueling the fire of the red-hot new media scene has been the breathtaking evolution of the new breed of "smart" cell phones, or smart phones. While feature laden phones such as the BlackBerry have been around for several years, Apple's iPhone® turned the smart phone world on its ear in 2007, thrusting the device into the mainstream much like the iPod did for portable media players over half a decade earlier.

Newer, smaller computers, known as *netbooks* are becoming more widely used and yet another new breed of device, called *tablets* (which have been in the development stage for several years) is poised to take off just as this book goes to press. The introduction of Apple's latest "gee whiz" product, the iPad®, in the spring of 2010 takes media consumption to yet another level, by combining the functionality of a portable media player, book reader, magazine viewer and Internet browser into an a single device that is larger than an iPhone or iPod, but can still be carried around far easier than a laptop or netbook computer, and sports a very long battery life.

With desktop computers, laptops, netbooks and tablet devices, new media can now be consumed by the masses in an even more universally-available manner than traditional radio and television was in past decades.

We've discussed how the primary channel of distribution of new media content is the Internet. For all intents and purposes, the Internet is available virtually everywhere now, in both wired and wireless modes, making it the most powerful "transmitter" of data—and media—the world has ever known. The biggest ramification of this is that media is no longer limited to "local" transmission and markets. Anybody can broadcast anything to practically everywhere on the planet!

It's hardly surprising that network television, cable and satellite TV providers now include a mix of new media content into their programming. But even more interesting is that traditional mainstream media companies are increasingly distributing their own content online, using the Internet, and the latest new media delivery channels, with growing numbers now watching mainstream media on computers of all kinds and smart phones.

There are now several popular mainstream counterparts to YouTube. One of the most popular television sites is called Hulu, which was launched a few years back as a joint venture between NBC Universal and FOX. As of this writing, there are several more television distribution sites and nearly every major broadcast, cable and satellite network is making a growing number of their shows available for viewing online as live, "on demand," streaming videos in full (or near) broadcast quality.

For the huge legions of home-brewed, renegade new media producers, several services that allow high quality, live streaming of video content are now online. One of the most popular is a service called Ustream, once again, a new technology, not even available a few years ago. Suffice it to say that online video has moved forward by leaps and bounds over the recent years and the improvements show no sign of slowing down anytime soon.

While only a few years ago, new media was limited to "renegade" content creators, citizen journalists and amateur producers, everyone now seems to have jumped on the new media bandwagon. Quite literally, every type of business and industry is now actively employing new media, not just for delivering programming, but for promotion, training, marketing, and a growing number of innovative alternative advertising campaigns.

The world of new media is vast, and just keeps expanding. The line between new media and mainstream content gets thinner and more blurred every day. Soon, new media will *be* the mainstream.

Social Networking & New Media

No discussion of new media would be complete without including another element that has become a companion component to the genre over the last few years. This is the phenomenon known as *social networking*.

Concurrent with the unprecedented expansion of new media (and just when it seemed like we had seen just about everything the Internet made possible), another major technological — and social — evolution emerged online with the birth of social networking.

A social network is a place online where people connect with each other and engage in dialogue and an ongoing interchange of thoughts, ideas, photos, videos, music and just about anything and everything else.

Back in 2005, social networking sites, most notably MySpace and, shortly thereafter, Facebook appeared online. These sites offered a whole new way for people to find each other, meet, network, communicate and promote themselves. Social networks are pretty much outgrowths of long-established online communication methods like chat rooms, message boards, forums and even online dating websites. Social networks take the concept of meeting, communicating, networking and the sharing of thoughts, information and media to a degree never practiced before.

Social network sites offer a place to easily post online anything and everything about oneself. Besides all of the usual profile information of a person or organization, photos, videos, and numerous other ways to share information are offered online. Members of a social network become friends with each other and share thoughts and other information by posting it on each other's pages. Then friends become friends with others, and friends of others and, before long, one can establish a huge network of hundreds, or thousands, of online friends.

These online social networks have become so pervasive over the last few years that the most popular networks and their features have become verbs as well as nouns in every day conversation. Several years ago, we started "Googling" people to find out about them. Now, it's become commonplace to be told that you've been "friended" by someone on Facebook or asked about your MySpace page before anything else.

MySpace gained popularity as more than just a place to meet other people. Since the overwhelming majority of its users are young, MySpace quickly established itself as a very effective place to market the artistic endeavors of young people, specifically musicians, bands and other creative artists. A band without a page on MySpace became a rarity and not having a page was simply unheard of, with good reason, as thousands of groups have successfully used the social network to launch successful careers.

Following in the footsteps of MySpace, Facebook has emerged over the last few years as the true dominant player in the social network space, with hundreds of millions of users. People of all ages and backgrounds are now using Facebook to find long lost friends, make new ones, find mates, share information or media, network and market themselves or their businesses.

Other more business-oriented social networks have also emerged, the most popular being LinkedIn, and are also being widely used as practical tools for business networking. Of course there are hundreds of social networks, large and small, now online, with scores more coming and going.

Like new media content, the most popular social networks are now available on the iPhone (and other smart phones) and the iPad, and most have their own applications to access and use the network from anywhere these devices can access a signal. The ever-increasing power and functionality of phones is the biggest reason for the phenomenal growth of social networks, as people now use them to "stay in touch" with their social and business contacts anywhere and everywhere. In fact, smart phones are well on their way to replacing computers and even other portable media devices for many tasks and uses like social networking.

In 2006, social networking merged with blogging (the original "new media" of the text-based world) and instant text messaging in a new service called Twitter. Twitter quickly emerged as yet another very successful social network, and the very nature of its service created yet another new category of online communication, now often referred to as *micro-blogging*.

The Twitter service is unique, in that it only allows messages of up to 140 characters to be posted. Twitter is like a huge public chat room or message board accessible to everyone, and the 140 character limit served to put an interesting spin on the dialogue that takes place there. It allows people to send and receive text messages from dozens, hundreds or thousands of people simultaneously through its own type of social network. Twitter now has over a hundred million subscribers, many of whom post endless tweets about anything and everything they may be doing or thinking at any moment. It's not all meaningless chatter, however. Many use Twitter to inform others about social or business gatherings and use the service to stay abreast of what is happening in their own social or professional circles.

Like YouTube, MySpace and Facebook, Twitter's growth was as exponential as it was phenomenal. After humble beginnings in 2006 as an experiment by a group of employees of a podcasting service, by late in 2008, millions were using Twitter to "tweet" others. Celebrities started to jump on the Twitter bandwagon that same year to communicate with their fans. Currently some celebrities claim to have millions of "followers" and are using Twitter to communicate with their fans, viewers, listeners or readers.

Hundreds of worldwide and local news organizations are now using Twitter to push out news headlines and, in the process, drive traffic to their primary websites and media outlets. Over the last couple of years prior to this writing, thousands of companies began using Twitter to communicate with their customers on a real-time basis, gathering feedback, fielding complaints and dealing with customer concerns in an effort to promote customer relations and good will. Other business and entrepreneurs of all kinds, including voice actors, are also using Twitter to promote their enterprises and make announcements. To help manage the endless flow of dialogue streamed by Twitter, there are now scores of tools available to organize your "tweeps" (fellow Twitter users) and make following, sending and receiving tweets more dynamic and interactive than Twitter's own online interface. You can find them by searching for "Twitter tools" online.

In addition to all the social networks, sites, and services, a growing number of aggregators and support services are springing up. With one of the most popular being Friendfeed, there are now dozens of enterprises with sites that help users navigate, maneuver and manipulate the tangled web of social networks. Like the tools available to enhance the Twitter experience, these services coordinate and consolidate myriad other services out there, by automatically posting one's blog headlines to Twitter, tweets to Facebook, pictures and videos to Flickr and YouTube and so on, for all the social network services one belongs to. Keeping up with these aggregators can be even more of a daunting chore than using the social networks !

While social networking itself doesn't truly fit into the new media category per se, it is most definitely intertwined with it. Social networks now include infinite amounts of new media flowing within their structures. They are ripe with links to viral video and other sites that contain new media content. The long and short of it is that social networking relies heavily on new media, and new media's growth and success is due, in a large part, to social networks.

The Present & Future of New media

The new media and social networking landscapes are thriving and continue to grow unabated. The World Wide Web is a vast ocean overflowing with new media content, and never before has there been such a quantity and diversity of available media content.

The current state of new media, as it is with all things tech, is very fluid and dynamic, with changes occurring on a daily basis. There is now an unstoppable metamorphosis and merging of new and traditional media. Advancing technologies and shifting socio-economic trends continue to drive an ongoing evolution of the new media genre, resulting in one new phenomenon after another, keeping new media, and technology in general, in a constant state of progress and flux.

Every week, new services and companies arrive on the Web 2.0 scene. In fact, the term Web 2.0 is being replaced by Web 3.0, which is a moniker for a yet far-more powerful and mobile interactive web that allows for anyone to live a totally interactive online life anywhere and at any time.

A key factor that continues to drive the growth and vitality of new media content is the interactivity that now exists, enabled by a Web 2.0 and 3.0 Internet. Unlike radio, television, magazines and newspapers, online content offers real-time consumer interaction, allowing listeners and viewers to interact with, manipulate and even customize media content, which was simply not possible in decades past. For years now, it's been common for blogs and other sites (called wikis) to allow content to be commented on or edited by visitors of the site on an ongoing basis. This interactivity gives today's online content a fluidity and life of its own, maintaining or, in some cases, even increasing its relevance with time. This level of interaction adds significant value to the content and engages the consumer to a far greater degree than traditional media could or did.

In the middle of the last century, there were three television networks and a few local television stations competing for audiences in each broadcast market. Today, there are hundreds of channels of cable and satellite television programming and countless thousands of websites offering online video programming. Podcasts and other forms of new media content have been added to the mix over the last decade, resulting in a dazzling array of content that is now available 24 hours a day. The content available is massive, and unprecedented.

Creators of content now face a degree of competition for the eyes, ears, hearts and minds of consumers that is much greater than in the past. Attaining a measurable audience for content today is challenging, and building a "significant market share" is a much higher mountain to climb than at any other time in the history of media.

Still, the power of the Internet makes reaching the masses an attainable goal for all. The right viral video or live camera feed posted online can attract millions of viewers within days, or hours, once the word spreads!

Besides the battle for audiences, there is another aspect of the new media phenomenon that warrants discussion. This is the role that demographics play in the consumption of new media. For younger people, the Web 2.0 and 3.0 world is commonplace and part of their everyday lives. For the older demographic who grew up in a far less interactive world, the concept of new media and the dizzying array of content choices available today can be cumbersome and daunting. How this "digital divide" will ultimately play out is yet to be seen as the baby boomer generation continues to age and younger generations mature. But it does appear that the aging boomer and senior demographics have been embracing new technologies at a steadily rising pace, as the technologies become more widespread, intuitive, and easier to understand.

The future promises to be filled with massive quantities of media, available in new formats and consumed on new devices that will surely be awe-inspiring. There will be more and more "specialized" content of all kinds, targeting specific niches, interests and demographics.

The sheer quantity of content and number of producers creating it will keep rising. An ever-increasing percentage of new media will be created and produced by a new crop of young, talented performers, podcasters, pundits and videographers. While there will always be a demand for professionally-produced media, the mix will continue to swing toward more user-generated and reality-based content, due to the much lower cost to produce these forms of content and the constantly improving production tools available at such a relatively low cost.

Even with all the changes in media content, one thing will remain constant. Voiceover artists and voice actors will always be in demand!

The Voice Actor's Place in New Media

The ongoing advancement and expansion of new media is unquestioned. Likewise, the migration to new media outlets of current media applications such as commercials, training video, film, television programming, and many other genres of entertainment and business is opening doors to many new opportunities for the observant voice actor.

In a nutshell, new media is a boon to voice actors, announcers and voiceover artists of all kinds. As more and more new media producers are, finally, moving toward "professional quality" production, there is an increasing amount of voiceover work up for grabs. Online commercials and promos, promotional videos, training videos, marketing tools, and instructional and informational media of every shape and form are being produced at a constantly growing rate. Even short form media elements, such as fully produced show opens, closes and promos are being widely used in new media content.

The sheer quantity of content now being produced. presents a wealth of opportunities for producers, performers and others involved in the creative process. But, with the opportunities come an unprecedented level of competition to achieve artistic success.

Besides competition, the biggest challenge to today's voiceover artist is, and will continue to be, keeping up to speed with the market and all the technologies related to it. As mentioned previously, staying on top of the vast array of products and services available to produce and market audio and video products and services, including voiceovers, can be a numbing and never-ending task. So, staying organized, learning to be an effective multitasker, and utilizing one's time efficiently are very important skills to master these days, in addition to those needed to perform as a voice actor.

Another pressure is on pricing. Intense competition has made pricing an important issue to deal with. Those paying for voiceovers, like other employers, want more for less, so it is more important than ever for a voiceover artist to be more than just an able performer. Being skilled in the art of pricing and negotiation are now vital. If you don't feel comfortable dealing with these areas, seek out an agent or representation of some kind. But, keep in mind that a growing number of voice actors bite the bullet and negotiate for themselves these days, so paying for the services of a representative may add a cost that hinders your ability to compete. It's probably better to check out the blogs and newsletters of other voiceover artists and professionals. Tips on pricing and negotiation are often topics of conversation and dialogue.

Lastly, it is now more important than ever for today's announcers and voice actors to take a direct and active role in marketing and promoting their own services and "brand." To compete effectively, it is simply a given that the latest techniques must be used. Again, extensive use of the Internet and participation in social networks are, increasingly, the keys to success in any profession or business enterprise, and voiceover is hardly an exception. Educating oneself as to where the work is and how to get a piece of it is the key to achieving success!

As mentioned above, a good place to start is by following the blogs and newsletters of other voiceover artists and voiceover related sites and communities. To find them, perform a search on Google or other search engine in the blog category for "voiceovers." As you visit these places online, you will most learn about where the bulk of the work is these days and get ideas about how to market your own voiceover-related endeavors.

You'll also find links to others with your interests and to seemingly endless sources for news and related information about the field. When going online, "one link leads to another."

Establishing an account on a popular social networking site like Facebook is the first step to making your presence known and finding others to communicate with. Make sure your profile is complete, up to date and includes your skills, accomplishments and history. Become "friends" on Facebook with other voiceover artists and others in related fields, such as producers, engineers, agents and recording studios. Do the same on at least one professional social network such as LinkedIn. Learn as much as you can about using social networks and how to enhance your own visibility on them. The more you interact and participate, the more you will be seen, followed and contacted.

Establish an account on Twitter and start to "follow" other voiceover artists, groups and entities. There are several to choose from. Do a search for the words "voiceover" or "voiceovers" and the site will return results of "tweets" that contain those words. You can then start following other people and organizations that post voiceover related items on Twitter, which will usually include links to sites and pages with articles about the subject. As you follow these other Twitter users, its a good chance that they will, in turn, begin to follow you as well. Post your own tweets on Twitter from time to time that you feel may be relevant to others in the voiceover community and start to communicate with other voiceover professionals via Twitter "direct messages" or messages on Facebook.

I would also recommend joining a few online "groups" of voiceover professionals. To find them, as you did with the blog search, do a search for "voiceovers" in the Google or Yahoo groups categories.

Remember, it's all about networking in the new online world. See, and be seen. Follow, and be followed. Communicate, and people will soon start to communicate back to you. These days, dialogue leads to success!

I am constantly asked how one can navigate the rapid sea of change constantly occurring in these areas. To be sure, staying on top of the latest developments and innovations in new media and social networking is very much a full-time job in itself. This chapter has offered only a basic and very abbreviated history and overview of new media and social networking. By the time you are reading this, it is a certainty that much of the information offered here will be dated or superseded, and hundreds of new products and services will have come — and gone. That is why I opted to encourage online searches rather than including links to sites. The best way to stay current is to go online, search for and then gather the latest information.

Here are a few suggestions that I hope will stay current, however. To keep abreast of what's going on, there are thousands of blogs online that cover the latest developments and trends in new media, social networking and related fields. As of this writing, a few of the better, and most popular tech blogs are TechCrunch, Gizmodo, Engadget, Technorati and Slashdot. My vote for the best blog covering the world of social networking is Mashable. All of the blogs above also do a good job of covering new media news and views, Of course, all of these blogs have Twitter feeds as well, so you can follow them post by post on Twitter, which, in effect, acts as headline service for the blogs.

Besides blogs, a good source to stay on top of all things tech and the online world is to regularly listen to podcasts or view some of the thousands of podcasts or online video programs out there. I highly recommend a collection of weekly programs offered up by the TWiT Network, which, as of this writing, can be found at **www.twit.tv**. TWiT offers several shows about tech in general, the Internet, and several more specialized subjects, such as Google, Windows, Macs, digital photography and much more. You can watch them live as they're being recorded or download them at any time. Another popular online new media network to check out is Revision 3. Both Revision 3 and TWiT are great sources for information about tech, new media and other current topics.

However you choose to stay informed, try not to be intimated by the endless heaps of new information flowing out. There is definitely way too much information for most of us to consume, so find a few blogs that you like best, and just follow them as best as you can and at your own pace.

As it has always been in the art of voice acting, the new media voice actor's job will continue to be one of communicating a compelling message through the creation and presentation of interesting characters and reads. And, although the fundamental performing skills remain the same, the business model, technology, and marketing methodologies are ever-changing and do require constant monitoring in order to adapt to trends as they evolve.

Keeping the vast amount of content and information that is available today in perspective should be a priority for all. In this brave, new world, maintaining the proper balance is critical to the future of new media and all those involved in it. Evolving technologies should continue to educate, empower, and stimulate creativity, rather than confuse, intimidate, and hinder it.

There is a new breed of creative new media producers offering new opportunities for voiceover talent to help them deliver their content. As voice actors, it is our job to evolve along with the new media marketplace.

18

Your Voiceover Demo

Your Professional Calling Card

Your voiceover demo is your best first—and sometimes your only—opportunity to present your performing skills and abilities to talent agents, producers, and other talent buyers. Many times, you will be booked for an audition or for a session simply based on something the producer hears in your demo. The purpose of a demo is to get you work!

In the world of voice acting, your demo is your calling card. It is your portfolio. It is your audio résumé. It is your letter of introduction. You don't need headshots, or a printed résumé, but you absolutely do need a high-quality, professionally produced demo. It is the single most important thing you *must* have if you are to compete in the world of professional voiceover.

Demo Rule #1:
Don't Produce Your Demo Until You Are Ready

HOW WILL I KNOW?

OK, you've just completed several weeks of a voiceover workshop and you're excited about getting started on your new career path. You received lots of positive feedback and encouragement during the workshop, and now you're ready to produce your voiceover demo, right? Probably not!

The single biggest mistake beginning voice actors make is to produce their demo too soon.

The reality of voiceover is that unless you already have a strong performing background, you are simply not going to be ready for your demo after taking a single workshop—or, for that matter, possibly several workshops or even months of private coaching. You certainly won't be ready for your demo after just reading this book. Mastering your performing skills may take a considerable amount of time. Producing your demo too soon may result in a presentation and performance quality that is likely to be much less than is needed to be successful in this business. Producing your demo too soon is simply a waste of time and money, and can potentially affect your credibility as a performer later on.

Before you even think about having your demo produced, make sure you have acquired both the business skills and good performing habits necessary to compete in this challenging business. Remember that there are a lot of other people trying to do the same thing as you. Anything you can do to improve your abilities and make your performing style just a bit unique will be to your advantage. Study your craft, learn acting skills, and develop a plan to market yourself *before* you do your demo. Take *lots* of classes—you'll learn something new from each coach! Acting and improvisation classes will help you develop your performing skills and voiceover classes, workshops, and conferences will hone those skills for the unique demands of voiceover work. If possible, find a voiceover coach who will work with you one-on-one to polish your technique before you go into the studio.

So, you're probably asking the question: "How will I know when I'm ready to do my demo?" Good question! In her *Demo and Marketing Magic for Voice Actors* workshop and E-book, my coaching partner, Penny Abshire, breaks this question down to four possible answers.

- **The short answer is:** You'll know.
- **The medium answer is:** You'll know when you're ready for your demo after you have the proper training, you've researched and studied other professional voice talent and their demos, know what the marketplace is looking for and feel confident you can deliver at the same level as other professionals doing similar voiceover work.
- **The long answer is:** You'll know you're ready to produce your demo when you can be handed *any* script—cold—and, within two or three takes, you can perform that script at a level comparable to other professional talent. You also feel confident that you can get work with your demo that is enough to cover the costs of its production.
- **The best answer is:** You'll know you're ready to produce your demo when you can stop asking yourself the question "Am I ready to produce my demo?" Which takes us back to: You'll know!

HOW DO I GET TO BE THAT GOOD?

The craft of performing for voiceover is, for most people, an acquired skill. There are some rare individuals who are natural-born performers, but most of the working professionals in this business started out by mastering their fundamental acting skills and moving on from there.

Of course, you certainly have the option to take some short-cuts and produce your demo after only reading a book or taking a single class. I wouldn't recommend it, but you can do it. And you may get lucky.

Remember this: Your performance, as heard on your demo, will be compared to every other demo a producer listens to—and most producers listen to a *lot* of voiceover demos. After a few hundred demos, it's not hard to separate the great talent from the "good," and the "good" from the rank beginners. Most producers will know within about the first five seconds of listening to a demo. You want your demo to present you as one of the great talent and keep them listening a lot longer than that first five seconds!

To do that, you must become an expert at communicating with drama and emotion before you have your demo produced. Here are some things that will help you get where you want to be quickly.

- **Study acting.** Acting is the key to an effective performance. Learn how to act so your performance sounds natural and believable, and learn how to use your voice and body to express drama and emotional tension.

- **Do your exercises.** Set up a daily regimen for doing your voice exercises. Get into the habit of keeping your voice in top condition. Your voice is the "tool" of you trade—take care of it.

- **Take classes and workshops.** Have an attitude of always learning. You will learn something new from each class and workshop you take or repeat. The voiceover business is constantly adapting and new trends become popular each year. You need to be ready to adapt as new trends develop. Voiceover workshops will give you a foundation upon which you can develop your craft and business.

- **Read other books about voiceover.** Every author on this subject presents his or her material in a slightly different manner. You may also learn new techniques or get some fresh ideas from reading a variety of books on the subject. If you learn only one thing from reading a book, it was well-worth its purchase price.

- **Practice your skills and techniques.** When you are working on a piece of copy, rehearse your performance with an attitude of continually perfecting it. Have a solid understanding of the techniques you are using and polish your performance in rehearsal.

- **Get personalized coaching.** Every performer is unique and it takes personalized coaching to truly discover and refine that uniqueness. As valuable as they may be, you won't get this from a voiceover workshop. Seek out a qualified private coach to help you become the best you can be before you produce your demo.

A WORD OF CAUTION

The tendency of many people who enter the world of voiceover is that they want to fast-track their training and rush into promoting themselves as "professional" voice talent. This inevitably means producing a demo before they are ready. There are several companies who take advantage of this natural tendency for instant gratification. They will teach a very encouraging introductory class for a minimal fee, then up-sell to a very expensive three-, four-, or five-day workshop, often held at a recording studio in a different city. As icing on the cake, they will include, or offer for an additional fee, a complete demo session as part of the course.

Consider this: Would you be ready to perform at Carnegie Hall after only a week of piano lessons? Would you be ready to perform major surgery after only a single year of medical training? Of course not! It would be foolish to even consider such things. Professional competency in any craft is only achieved after consistent study and considerable experience. It never happens overnight. Yet a beginning voice actor will often jump at the opportunity to spend several thousand dollars to take a fast-track voiceover course so they can get their demo produced long before they have acquired either the essential business or performing skills. Although the information and basic training may be of some value, the production of a demo after only a few days of training inevitably results in a demo that is both costly and unmarketable.

Before you spend the money for a demo, I strongly encourage you to study both the business and craft of voiceover to the point where there is no doubt in your mind that you are ready to get that demo produced.

Demo Rule #2:
Your Demo Must Accurately Represent Your Abilities

Your voiceover demo must be "great!" It cannot be merely "good." To create a great demo, you must make the effort to develop and hone your performing skills. Since your demo may directly result in bookings, it is extremely important that you be able to match the level of your demo performance when under the pressure of a session. It is quite easy for a studio to create a highly produced, yet misrepresentative demo that gives the impression of an extremely talented and polished performer. If the performer's actual abilities are less that what is depicted on the demo, the shortcomings will be quickly revealed during a session.

Your demo should be professionally produced by someone who knows what they are doing. Even if you've assembled a top-of-the-line, state-of-the-art home studio, without practical knowledge of the business and extensive production experience, don't even think you can put a demo together at home and expect it to sound professional. Even with my many years of experience as an audio producer and sound designer, I still want the assistance of a good director when I'm in the booth. It is extremely difficult for one person to deal with both the engineering and performing aspects of producing a demo at the same time. You need to be focused on your performance and not dealing with any equipment.

You need a director to listen to your performance objectively, help you stay focused, and help get you in touch with the character in the copy. In today's world of voiceover, it is essential that you develop self-directing skills because you may find yourself recording many projects in the privacy of your home studio. However, performing effectively without a director, or by directing yourself, is very challenging and it's the last thing you want to do when producing your demo. Although many professional voice actors believe they don't need a director—and can actually perform quite well without one—all voice actors do, in fact, need a director to bring out their best work. The top professionals will tell you that they perform much better when they have a good director to guide them through their performance.

When you go to the studio to produce your demo, you should consider the session to be just like a real commercial recording session. You need to be able to get to your best performance, in three or four takes. If you need more than six or seven takes to get the right delivery, you may not be ready. Realistically, anything after about the fifth or sixth take should be aimed at fine-tuning your delivery. Be careful that you don't spend a lot of time just getting into the groove of the performance for a script. If you do that, your end result may not be an accurate reflection of your abilities. The same can be true of over-rehearsing your copy. Too much rehearsal can result in setting a groove for your performance that may be difficult to break out of. Your copy should be rehearsed just enough so you are comfortable with it, yet can be directed into alternative deliveries.

I know of one voice actor who was booked through an agent based solely on the demo. The demo sounded great and had logos of major television networks and other advertisers on its cover. The impression was that this performer had done a lot of work and was highly skilled.

The performance was recorded during an ISDN session in San Diego with the performer in a New York studio. A few minutes into the session, it became apparent that the voice actor could not take direction and would not be able to perform to the caliber of the demo. The producer gracefully ended the session and a different voice actor was hired to complete the session the following day. The producer refused to pay the talent agency's commission because she felt the talent agency had misrepresented the voice actor's abilities. The original performer was never told that the session was unsatisfactory and actually sent a nice thank-you note to the producer. The performer did get paid, but the recording was never used.

It turned out that the agent had never worked directly with the performer. They had been promoting the person based solely on a highly produced demo. A few days later, it was learned that the performer was actually attempting to memorize the copy due to a problem with dyslexia that made it difficult to perform the lines live, as written.

The agent later apologized to the producer and mentioned that this performer had done some excellent self-marketing, had extensive stage experience, and presented a very professional image during the interview. There were only two problems: The first was that the performer had created a demo that clearly exceeded actual abilities. A secondary problem was that the agent signed this performer without first testing those abilities with a cold-reading. Bottom line: You *must* be able to perform to the level of your demo when booked for a real session.

Versatility Is Your Selling Tool

There are two schools of thought for demos. One suggests that a demo should reveal the performer's range of versatility through a variety of examples showing different emotions, attitudes, and characters. This type of demo reveals a wide range of vocal styles, placements, and character voices, and is commonly used to market for animation and video games.

The other approach to a voiceover demo focuses on the performer being the most real and natural person he or she can be while demonstrating a range of emotion within the natural delivery of their own voice. This is the most common approach to commercial, narration, IVR, and telephone messaging demos.

Good voice actors can do dozens of voices, emotions, attitudes and characters because they are able to find a place in their body from which to center the character and place the voice—even if the character and voice originate from their natural personality. Other good voice actors have developed a highly defined performance style that is at the center of everything they do. These two basic types of performers are the *character* and *celebrity* voice actors discussed in detail in Chapter 10.

Regardless of which approach works best for you, it is the range and variety of performance in a demo that represent a voice actor's abilities. The essence of who you are needs to be present in every track of your demo. You need to capitalize on your strong points and present them in the best possible manner in your demo. The range of attitudes, emotions, and characters you can express during a voiceover performance is your own, unique, *vocal versatility*. Your strongest, most dynamic, and most marketable voice is called your *money voice*. This is the voice that will get you the work and may eventually become your trademark or signature voice. Your other voices are icing on the cake but are necessary to clearly show your range and versatility as a voice actor.

Demo Basics

DEMO FORMATS

One interesting thing about technology is that it is constantly changing. What was in favor yesterday may be out of favor tomorrow. Reel-to-reel and cassette demos are history! The audio CD, which was the standard for voiceover demos for many years, has now given way to demos that exist as electronic data files in a variety of digital audio formats including Flash audio files. Digital audio files of voiceover demos have several major advantages over the audio CD: there is no expense for packaging; the demo can be posted on a website, which makes it immediately available for listening and/or downloading worldwide; files can easily be copied and cataloged on a computer in folders that identify a performer's style, performance genre, or in whatever way works best for a producer or agent. They can be emailed, renamed, edited, uploaded, and assembled onto a compilation audio CD. For auditioning purposes, it's very fast and efficient to simply open a folder on a computer and click on successive demo files until the right voice is found. Isn't technology wonderful? It's anybody's guess what the future may hold for voiceover demo distribution formats.

Even with the advances in technology, there are still some producers and agents who prefer holding a CD in their hands. Although it may not be as efficient as working with electronic files, there is something to be said for reading the label of a demo CD and studying the performer's visual marketing image while listening to their demo.

For the purpose of marketing voiceover talent in today's highly technological world, it is an absolute must that your voiceover demo exist as an electronic file stored in an easily accessible location on your computer. From this electronic file, and with the proper computer software, you can create "one-off" audio CDs that can be packaged and mailed to clients and prospects if or when needed. Even in this age of electronic files, it's still not a bad idea to have at least a few CDs on hand for those times when they are requested.

Types of Voiceover Demos

Before producing your voiceover demo, it is important that you study current demos of professionals in the area of voiceover you are interested in. As with any business, you need to know what your competition is. One of the best places to study demos is **www.voicebank.net**. Here, you will find voiceover demos from nearly every major talent agent in the U.S. Click on "Demos and Clients," choose a category, select an agent, then start clicking on demos. You'll be amazed at what you will learn as you begin listening to some of the top voiceover professionals in the country.

Through your study of the craft and business of voiceover, you will be discovering the delivery style, characters, and performance techniques that work best for you. You will also be making choices as to the types of voiceover work that you are best suited for. Very few voice talent are able to effectively perform in more than a few categories. Animation and character voice actors will rarely work in narration, but many will do commercials. Audio book narrators may not find telephony voice work to be at all satisfying, but they may do industrial narration or documentary voice work. Find the areas of voiceover work you enjoy the best and are most suited for, and focus on mastering your craft in those areas.

There are no hard-and-fast rules for producing a voiceover demo. Your objective is to effectively reach the talent buyers who hire voice talent in the area of your demo. Not too long ago, it was acceptable to produce a single demo that featured multiple voiceover genres, such as commercial, narration, and character all wrapped up in a two-minute demo.

Today, producers want to hear a specific demo for a specific type of work—and they want to hear it quickly. Your demo should feature examples of what you can do that meets, or exceeds, the expectations of those who are looking for voice talent in that area. In other words, your commercial demo should consist of only your commercial delivery style, and your character demo should feature your character voices. There may be some crossover, such as a character voice in a commercial, but you should avoid the temptation to combine multiple demo types into one demo.

Ideally, your demo should be compiled from actual projects you have worked on. However, if you are just starting out, this is not possible. You will need to create your own copy and design a demo that will catch the listener's attention and hold it. Even working professionals will sometimes write original copy to create a demo that really puts their voice in the spotlight. This can be a challenging task, but it can pay off big.

There are several schools of thought as to the type of voiceover demo to produce when just getting started. One is to produce a commercial demo first, because this type of demo can present performing styles that apply to almost every other voiceover niche. Another is to produce a narration demo first, primarily because this niche represents the largest percentage of voiceover work. Yet another approach is to produce a demo based on a determination of the specific niche area that the voice actor wants to work in. All three are viable approaches, but the first two will be far more likely to result in bookings for an entry-level voice actor. Whether a commercial or narration demo is produced first will largely be a decision based on the performer's skills and consultation with their coach.

Voiceover demos are generally produced on a bell curve with the performer's *money voice* at the beginning and end. In-between are a variety of performance attitudes and styles that reflect the performer's range and abilities. Having the same voice at the beginning and end can provide a reference point for the listener and give him or her an opportunity to categorize the performer's vocal age and personality type.

There are two basic structures for demos. The first is a *compilation demo* with examples of a variety of performing styles. The second is a *concept demo* that combines the demo clips into a logical sequence or story where each clip leads into the next. In both formats, as the demo progresses, there will be changes in attitude, pacing, energy, and character. Concept demos are relatively rare as they are challenging to write, must be thought out very carefully, and fall into one of two categories: extremely good and unbelievably stupid!

COMMERCIAL DEMOS

Even though radio and TV commercials are only about 10 percent of the business, a well-produced *commercial demo* can demonstrate a performer's abilities for nearly every other type of voice work. Every segment should reveal some aspect of the real you through your delivery and the characters you create. There is little demand for accents, dialects, and wacky characters in radio and TV commercials. Unless you are marketing your natural accent, those voices should be saved for other demos.

Your demo may contain as many as 12 to 15 segments, each of which demonstrates a different emotion, attitude, level of energy, personality trait, or delivery style at a variety of tempos. The copy chosen must be typical of what is commonly heard on radio and TV commercials. This type of demo should begin with several very short "clips" of about :03 to :05, followed by somewhat longer elements, each fully produced to sound like a real-world commercial. The trend over the past few years, especially with agents, is for a commercial demo no longer than one minute. However, you might want a slightly longer version of no more than 1:15 to 1:30 for your website or direct marketing.

INDUSTRIAL NARRATION DEMOS

Corporate and *industrial demos* tend to contain copy that is somewhat longer than the copy in a commercial or character demo. The longer length of copy allows the producer time to more accurately assess your reading and delivery skills for this type of storytelling. It also gives them an opportunity to hear how you handle complex words, concepts, and sentences. As with the other types of demos, your money voice and strongest material should lead the demo, followed by a variety of styles, range, and versatility. Industrial demos offer a good opportunity to use various microphone techniques, a range of delivery speeds, and storytelling techniques to good advantage.

Where the average length of an individual segment for a commercial demo might be 6 to 10 seconds, a segment on an industrial demo might run 15 to 20 seconds, or even slightly longer. You'll need more time to complete the descriptive text for a procedure or technical discussion.

A typical full-length industrial demo will run about 1:45 to 2:00 and will include five to seven segments. As with a commercial demo, your agent will ask for a one-minute edited version for their marketing purposes.

CHARACTER AND ANIMATION DEMOS (INCLUDING GAMES)

Character and animation demos are designed to feature your talents primarily for animation and video game work. They also demonstrate your ability to create marketable voices for believable, "real" characters.

Character voice work for animation is probably the single toughest area of voiceover to break into, so both your performing abilities and your character/animation demo must be of extremely high quality.

At its essence, a character or animation demo features voices that are recognized as "real" people, but which are actually voices you create that are different from your real voice. For animation, the characters are often exaggerated or quirky in some way, while video-game characters are usually "real people." Each clip features a different attitude, vocal characterization, or personality. Not all voice talent have the ability to create voices that sound completely different from their normal voice. If you don't have this ability, a character voice demo may not be for you.

Producing animation demos is a specialty area of production. Each segment of an animation demo must sound like it came from an actual show and similar voices should be separated from each other. Most studios have extensive music and sound effects libraries, but not all studios have the proper music and sound effects needed for an animation demo. Call around to find a studio that can do this sort of work and ask to listen to several animation demos they have produced. There are several studios in Los Angeles that specialize in animation demos.

A typical animation or character voice demo will be about the same length as a commercial demo, around 1:00 maximum, and may include 10 to 15 individual, fully produced elements. Again, if you have an agent, be sure to ask what length they prefer.

AUDIO BOOK DEMOS

The format for an audio book demo is far different from all other voiceover demos. An audio book producer wants to hear how you tell a story over an extended period of time.

There are a few important differences between this type of demo and other voiceover demos. Audio book demos are the only type of demo that should include a slate of your name. The slate should be spoken by someone else who simply says your name followed by the word "reader." After this brief introduction, you begin by giving the title of the book and start reading the story. Your choice of material should include a variety of emotions, attitudes, and characters. You'll need to find something unique for each character in the book you are reading for each segment. This can be a challenge for characters of the opposite sex. Usually a shift in energy or attitude will reveal a character far more effectively than a change of pitch. Change-ups of tempo, rhythm, and the use of other techniques can also help to differentiate the characters as you tell the story.

Practice for your audio book demo by recording yourself reading out loud, finding the drama, emotion, and attitudes for each scene and character of the story. You must develop the skill to be consistent with your delivery style for a very long period of time. Your audio book demo should reveal that consistency.

A typical audio book demo will run from 5 to 12 minutes and may consist of as few as 3 to 5 fairly long segments. Ideally, you should be able to perform each segment of your audio book demo as a continuous reading with a minimum of stops and starts. If you find you need to stop frequently, have difficulty reading the text, or need to go back for pick-ups, you may not be ready for audio book work.

VOICE IMAGING, BRANDING, SIGNATURE VOICE DEMOS

Unlike other types of voiceover work, those voice actors who work in the area of station imaging need special knowledge of the broadcasting industry. If you don't have a radio background or thoroughly understand the purpose of imaging and how this aspect of the voiceover business works, imaging may not be for you. An imaging demo features a single, specific, often "edgy" delivery style throughout the demo. Most imaging voice talent have a separate demo that features a different, specific attitude for each radio format. Also, unlike other types of voiceover work, imaging often presents a detached delivery—more typical of an announcer, rather than the conversational delivery style for commercials, character, and narration.

An imaging demo should be no longer than one minute in length. Although any studio capable of producing a commercial demo should be able to handle an imaging demo, you would be wise to find a producer who knows and produces imaging work. As with animation demos, imaging requires special music and effects that may not be available at all studios.

PROMO AND TRAILER DEMOS

A promo and trailer demo focuses on television programs and films. This is the only demo in which two genres are commonly combined, but many voice talent will market these separately. A *promo* demo can include examples of both TV promos and movie trailers. Television promos are essentially commercials that promote a specific television program instead of a retail product. The program being promoted could be a local show, a movie, news program, news feature segment, or other station programming. A *trailer* promotes a movie. To properly produce promo and trailer elements for your demo, you'll need to find suitable *sound bites*, or excerpts from television shows or movies that you will wrap around your voice work. As with a commercial demo, the length will be about one minute to no more than 1:30.

Most movie trailer work is done in Los Angeles, and there are a handful of voice talent who are consistently hired for this type of work. That doesn't mean you can't break in to trailer VO work, but you'll need to find the companies that produce trailers, you'll need an agent to represent you for trailer work, and you'll need a killer trailer demo. Television promo work is usually booked directly by a television station's Promotions Department and can be an entree to trailer work.

TELEPHONY DEMOS

Sometimes referred to as a *message-on-hold demo* (MOH) or *IVR* (*Interactive Voice Response*), this type of demo is pretty basic. It usually consists of one or two examples of outgoing messages, one or two on-hold messages for different types of businesses, and perhaps even an example or two of a concatenation project or interactive voice responses. Examples should include appropriate background music. The idea of an MOH demo is to demonstrate what you sound like delivering information over the phone.

One might think that because a telephone connection has a reduced frequency response (about 8 kHz), an MOH demo should be equalized so it sounds as though it is being heard over the phone. Although this is certainly an option, I would not recommend it primarily because the reduced frequency response of your demo will not accurately reveal the subtlety of your performance. Even though the nuance and detail of your delivery may be lost during an actual phone message, you want your demo to show you at your absolute best. I'd recommend producing your MOH demo at the highest possible quality. As with most other demos, this one will also be in the one to one-and-a-half minute range.

Producing Your Demo

PREPARATION IS THE KEY TO A GREAT DEMO

I've said this before, but this is so important that I'll say it again: do not produce your demo until you are ready, and even when you are ready, have your marketing plan in place before you spend the money on producing your demo. Spending some time on a critical self-evaluation to determine your performing strengths and weaknesses, your vocal style, the market you want to work in, and researching the talent buyers in that market is homework that will save you time and money.

From this point on, I'll assume you know who you will be marketing to, you've got a marketing plan in place, you have a functional home studio, you've mastered your performing skills, and you've gathered a number of scripts.

Rehearse your copy with a stopwatch—you do have a stopwatch, don't you? An analog 60-second sweep stopwatch is best because it is easy to start, stop, and reset. Some newer cell phones have a built-in digital stop watch that can work well, but many hand-held digital stopwatches can be cumbersome to use. It may take some searching, but you can find analog stopwatches online starting at about $50. Time yourself with each rehearsal, recording yourself if possible, and do a complete analysis for each script. Make notes on your scripts about the character, attitude, and emotional hooks, as well as ideas for music style and sound effects if appropriate. Consider mic placement for each script. Mark off what you believe to be the strongest :10 to :15 of each script and consider a possible sequence for the demo. Be prepared to record the entire script during your demo production session, and be flexible enough to understand that everything you rehearse will probably change. The work you do at this stage will pay off later.

Ideally, you should find a director who can assist you with the production of your demo. Hiring a director is like having a second set of ears. It allows you to focus on your performance so you will not have to worry about the technical details of the session. Many recording studios have engineers experienced in directing voiceover.

Above all, when you are in the studio recording your demo, have fun and enjoy the experience! I encourage you to stay through as much of the process of producing the demo as you possibly can. This is your primary tool for establishing yourself in the business, and you're paying a lot of money for it. It is not only your right, but your responsibility to make certain your demo is produced to the highest possible standards and completely meets your needs as a marketing tool. Your input will be important for your producer or the engineer to create an effective demo. You take a great risk if you simply go in and record your tracks, then leave the production up to the engineer or your demo producer. By observing the production process for your demo, you will learn a lot about what really goes on behind the scenes in a recording studio.

HOW LONG SHOULD MY DEMO BE?

The question of your demo length was partially discussed in the section covering the various demo types. We'll take things a bit further here.

There are two answers to the question of length. The first answer is: if you have an agent, ask her what length is preferred by their agency for the type of demo you are producing. The second answer is: if you don't have an agent, you'll want your demo to conform to the current conventional length for the type of demo you will be producing. As with other trends in voiceover, the length of a voice demo has changed over the years. The general standard lengths for various types of demos has been discussed earlier, but preferences in your market may also be a determining factor for the length of your demo. The best way to discover the appropriate demo length is to listen, and time, other voiceover demos. Visit websites for talent agents in your area or listen to demos at **www.voicebank.net**.

In major markets like Los Angeles, Chicago, and New York, you'll find the average length for a commercial demo to be around 1 minute to no longer than 1:30. The trend has been moving toward shorter demos, and some Los Angeles agents are requesting demos as short at :30. Other markets may prefer longer demos between 1:30 and 2 minutes. It will be rare to find commercial demos longer than 1:30 in today's voiceover world, although demos up to 2 or 3 minutes were the standard length only a few short years ago.

WHAT ABOUT PRODUCT NAMES?

Your demo is an advertisement for you. The clips in your demo do not need to mention any product names, but should demonstrate your ability to communicate emotionally with a variety of styles and attitudes. There are actually two schools of thought on this. Some agents and producers believe that including product names lends credibility to the performer (especially if the spot is one that the performer actually worked on) and that they give a good opportunity for the producer to hear how the performer "sells" the client, or puts a spin on the product name. Other producers feel that the most important aspect of a demo is the performer's talent and ability to communicate the message or tell the story, and that product names can actually become distracting or that the use of product names may be misrepresentation if you didn't do the original spot. The reality is that most talent buyers, agents, and producers don't really care if you actually did the spot or not. They want to hear what you can do with the words. They know that most voiceover demos aren't the real thing—and it doesn't matter.

If you choose to include product names in your demo, I'd suggest including only one or two and let your delivery and performing abilities shine for the rest of the copy. It may be a good idea to include a product name in your first demo just so you can demonstrate how you can give value and importance to the client or product. If you do choose to include product names, just don't overdo it. You can also change the client or product name to avoid any concerns of misrepresenting that you actually did a spot. As you acquire copies of projects you have worked on, you should include a few product names from actual spots in your updated demos. Unless you have actually worked for the clients you mention, you should never include their logos on your CD cover, on your website, or in your marketing, and when you do, it should always be only after obtaining proper permission. To do otherwise may be viewed as misrepresentation.

WHERE DO I GET THE COPY FOR MY DEMO?

There are many approaches to obtaining copy for a demo. Be creative!

Some demo studios provide the copy and handle all the production. This is fine if you don't mind taking the chance of other people in your market having the same copy on their demos. The only real advantage of having the studio provide the copy is that your demo session becomes more like a real recording session; that is, you won't have the opportunity to see the copy in advance. The downside is that your session may take considerably longer because you will be working the copy cold and relatively unrehearsed. You might also feel rushed when you are "on the clock" to get through all the copy necessary, which could easily affect the quality of your performance. However, the major problem with the studio providing the copy is that you can easily end up using copy that is not right for your performing style. The purpose of your demo is to present your talent in the best possible manner. Performing copy that is not right for you can only work against you, no matter how well the demo is produced.

A better approach to finding copy for your demo is to listen to radio and TV commercials and browse through magazines. By listening to radio and television commercials, you can find copy that matches your style. Record commercial breaks and transcribe the ads that fit your abilities, putting each script on a separate piece of paper. Transcribe the entire commercial even though you may end up using only a small portion if that script makes it into the demo. Having the entire script in front of you will help you discover the emotional content of the commercial and the target audience.

You can also find copy by rewriting magazine ads—especially from women's magazines. There are also a variety of resources for copy on the Internet. However, here's an important point to keep in mind: the scripts you'll find on an online script database are there for the purpose of providing rehearsal copy, and may not be the best choice for your demo. The best copy for your demo will be copy that is uniquely yours and presents your performing abilities at their best.

Magazines are a great resource for a potential demo. Look for ads that include a lot of copy. Look for ads that target specific audiences: men, women, young, older adults, and so on. Look for products or services that will allow you to perform the copy in a variety of styles: serious, humorous, hard-sell, soft-sell, dynamic, emotional, and so on. Look for key phrases and sentences that have emotional content—these will be your keys to an effective performance. Most important, select only those ads that might actually be potential radio or TV commercials. The tracks in your demo must sound like real-world radio or television spots.

Technical, news, travel, and women's magazines often have ads that can be easily adapted for voiceover. Most print ads are written for the eye, designed to be read, and include a lot of text that may not be appropriate for voiceover. However, if you think about it, you'll realize that the people who write these print ads are often the same people who write national-quality radio and television ads. Since print copy is written to be read silently, you will usually need to rewrite the copy so it can be used for voiceover. You don't need to completely rewrite a print ad; just take the strongest sections and rework them so they make sense for voiceover. If you have a talent for rewriting copy, here's an opportunity to let your creative juices flow. If this is not one of your strong points, your demo producer or one of your coaches may be able to help.

If you have some writing experience, you can even write customized material for your demo. But be aware that if you write your own copy, it must sound as though it was written by a professional.

Obtain as much copy as you can and narrow the scripts down to about 30 to 40 different ads from radio, TV, magazines, and technical journals. Include a variety of styles that will reveal your full range of capabilities: slow, fast, dynamic, emotional, character, and so on. Also make sure each script you choose is appropriate for the type of demo you will be producing.

This may seem like a lot of copy—and it is—but by starting with that many possibilities, it will be easier to determine the copy that best fits your style. By the time you start recording, those 30 or 40 scripts will be whittled down to about 12 to 15.

Be prepared to perform the entire script at your demo session. The reason for this is that you may actually end up with an extremely effective delivery on a segment of the script that you may not have expected. If you only rehearse portions of your scripts, you might overlook an opportunity for a perfect transitional element, or an especially emotional performance. Your director can help create the best performance for each script and you may end up actually recording only a small portion of the copy.

Thoroughly woodshed and do a character analysis for each piece of copy, making notes on the scripts. Practice your performance for each script just enough to become familiar with it. Be careful not to get yourself locked into any specific attitude or character. Keep in mind that your session engineer might direct you into a performance completely different from what you had decided on. If that happens, you need to be able to adapt to the direction. If you can't, or if you find yourself getting stuck in the same delivery for each take, then you are not ready to have your demo produced.

As you prepare your demo copy, make notes about music style or sound effects, but don't worry about finding them. The engineer will handle that at your session. Your job is to focus on finding suitable copy that you can perform effectively. You have the luxury of being able to prepare and rehearse. Take advantage of it! You will not have this luxury in a real-life studio session. Take at least three clean copies of each script with you to the studio: one copy for yourself, one for the engineer, and the third for your director.

WHAT ABOUT DIALOGUE, MUSIC, AND SOUND EFFECTS?

The purpose of a demo is to feature *your* voice-acting performance. Including other voiceover performers should be done judiciously. If you include a dialogue spot, make sure that yours is the featured performance and that the other voice is of the opposite gender. This may seem obvious, but you'd be surprised at the number of demos with two voices that are hard to tell apart. Also be certain that the other performer knows how to act. I've heard far too many demos that include a dialogue spot where the second player showed little or no acting ability, or worse, the other performer showed superior acting ability. And don't be tempted to do both voices yourself. Producers want to hear how you work with other voice talent, not how clever you can be performing multiple voices.

The use of music and sound effects is essential to creating a demo that sounds like it contains real-world spots. If used, music must be appropriate for the mood and energy of the message. It is an infringement of copyright to use music from store-bought CDs or downloads, or even arrangements of popular tunes that you perform yourself. Only suitable music from licensed *music libraries* should be used in your demo. Sound effects should only be used where appropriate, and although sound effects are not copyrighted, they must be of high-quality. This aspect of your demo requires a knowledge of audio production, editing, and postproduction that is acquired only through many years of experience. You might have the best recording equipment and software, but if you don't know how to produce and mix professional audio, you would be best to leave the production to an experienced demo producer. Attempting to do this yourself will generally result in a demo that is substandard and leaves a bad impression with the talent buyers who hear it. And, as a voice actor who is attempting to present the image of a professional, that is something you cannot afford.

WHAT IS THE PRODUCTION PROCESS?

I can't speak for other demo producers, but when my partner and I produce a demo through our VoiceActing Academy, we follow a nine-step process:

1. We begin with an in-person or telephone consultation during which we identify the type of demo to be produced, determine the voice actor's primary delivery style, and discuss the development of their marketing plan. The task of finding copy is given to the voice talent. We don't provide scripts, although we will occasionally write original copy or help rewrite ads selected by the voice talent. This can take from a few days to several weeks. We ask for two to three times the number of scripts that will actually be used.

2. About a week before the session we'll have a second consultation where we'll review all copy with talent to sort out the best scripts and eliminate duplicates. The resulting scripts will still number about twice what will end up in the demo. The voice talent rehearses those scripts, but not too much.

3. On the day of the demo session the first thing we'll do is go through the scripts again. We look for variety, and we'll ask the voice talent to perform a portion of each script. This time, we eliminate scripts until we have what we believe are the strongest scripts for our client. The rest we'll keep handy in case we need them.

4. Next, we'll record the dry voice tracks for each script. Occasionally, we'll determine a script isn't right, so we'll drop that one and move on to the next. Each script will be recorded numerous times, striving for the "perfect" delivery. The session is handled no differently than any other commercial session. It's just that we're recording about a dozen commercials in a fairly short period of time. It may take up to two hours or more to record the 8 to 12 voice tracks for a demo.

5. As each voice track is recorded, I'm thinking of the production value that will bring each clip to life. Music and sound effects are critical. I

actually start thinking about the "sound" of each clip during the pre-session meetings, but it's after the recording where we really get down to business. The music and SFX search can be a time-consuming process because for a powerful demo, each clip needs to have music with the proper energy, mood, and dynamics. It takes a skilled audio producer/engineer to do this quickly and efficiently.

6. When all the individual demo clips are produced and mixed, the next step is to sequence them into the most effective order. We'll always start the demo with what we consider to be the talent's strongest voice. The *money voice*! This is the voice we want the talent buyer to remember. From there, we'll place the clips in a sequence that provides variety, interest, and changes in mood. The idea is to create a sequence that will keep the talent buyer listening. If everything on a demo sounds the same, the buyer will stop listening after a few seconds. And we only have about five seconds to grab their attention.

7. The master sequencing will result in the full-length demo. For most demos, a talent agent will want a one-minute cut-down version. So, the next step is to edit a one-minute version from the full-length demo. A second or two over or under is OK.

8. The final step is to burn an audio CD and render both versions of the demo out to MP3 files, which are then emailed to the talent and put on a data CD-ROM for future use.

9. I'll send my client home with their demo in hand asking them to call me if they hear any problems or if their agent needs something changed. It doesn't happen often, but since demo production can move fairly quickly, it's possible to miss an edit or for a music or sound effect to need adjusting in the mix. And occasionally an agent will request a different sequence.

This process is the same regardless of the type of demo we're producing. The amount of time it will take to produce a demo will vary depending on several factors: the type of demo being produced, how quickly the talent is able to get to the best take; how long it takes to find the proper music and SFX; how long it takes to do the postproduction for each clip; and how many changes or rerecords are needed.

I insist that our demo clients sit in on the entire demo production process. I want their input at each stage of the production, and I want to know that they are happy with each segment. I also want them available in case I need to record a pick-up or if we discover a clip that just isn't working and we decide to replace it with a different script.

HOW MUCH WILL MY DEMO COST?

Production of your first demo will very likely be the single most expensive part of breaking into the business of voiceover, possibly second only to what you spend on training. No matter what your level of experience may be, you should seriously consider hiring a demo producer who knows the business and what talent buyers are looking for. The additional ears and professional direction can save you a lot of money.

The cost of producing a voice-acting demo will depend on the type of demo, the market you're in, and your demo producer. It can vary widely from market to market. To a certain extent, it will also depend on your performing abilities. For recording studios, time is money, and the faster you can record a high-quality performance (fewer takes), the sooner your demo will be completed and the less it will cost.

There are two basic ways studios will price out demo production. One is to book the studio on an hourly rate and add on other related costs. The other, and more common approach, is to book the demo production as a package. You can often save some money with a package deal, but if your demo is produced quickly, it may cost you more than if you had gone with hourly rates. It's a choice you'll may need to make, and if the studio you'll be working with works only one way it won't be a problem.

If you're in a major market (L.A., N.Y., Chicago), you can expect your demo to cost in the neighborhood of anywhere from $1,500 to $4,000. In other markets, you can expect to pay anywhere from $600 to $1,500, or more, for the production of your voice demo. The actual price you pay will depend a lot on the studio and producer you are working with. Studios that are known for producing excellent demos will give you a great product, but it will cost more. You definitely "get what you pay for" when it comes to demo production. However, beware of the *demo mills* that will offer to include production of your demo as part of a short—but very expensive— course. The demos produced by these operations will rarely get any work.

Although actual session fees vary, and may be somewhat lower or higher, the following gives you an example of how the cost for a typical demo session might break down at a studio rate of $125 per hour.

PRODUCTION ELEMENTS	TIME AND FEES	SUBTOTALS
Studio time (voice recording)	2 hours @ $125/hour	$250.00
Postproduction (editing, music)	4 hours @ $125/hour	500.00
Track sequencing and/or dubbing	1.5 hours @ $125/hour	187.50
Music license (for music used)	1 blanket license	400.00
Outside producer/director	1 flat fee	400.00
Materials (CD), including tax		50.00
Total cost (not including duplication)	7.5 hours in studio	$1,787.50

The actual time it takes to record your copy may vary considerably, and some of these items may not be required, thus affecting the price. The cost of studio time varies greatly from city to city and depends on the complexity of your session. Some studios charge a fee for the music used in your demo while other studios will provide the music at no charge. In most states, the cost of studio time and music license fees are exempt from sales tax, but the materials and recording media are not. Check with your studio to find out what portions of the session or materials will have sales tax applied. If you've set up your voiceover business properly, the entire cost of producing your demo can be deducted as a business expense.

WHAT DO I NEED TO KNOW ABOUT STUDIOS?

As you prepare for your demo, you will be wearing your producer hat. In that role, you will have already prepared your copy and directed yourself in your performance during practice and rehearsals. Some of your other duties as producer will be to make all the arrangements for studio time, hiring a producer, printing, CD duplication, and distribution of your demo.

Most larger cities have at least several recording studios and radio stations. In this age of easily accessible high technology, many small towns have studios capable of recording a high-quality demo. You will find commercial recording studios advertised in the telephone book and on the Internet. However, there may also be many excellent home-based project studios in your community that are not advertised anywhere. Even though recording services and studios may be plentiful in your area, this does not mean that all studios are able to produce a marketable demo.

The majority of commercial recording studios are designed for music sessions. The engineers at these studios are usually very competent at recording music, but may know very little about producing commercials or directing voiceover talent. Home-based project studios are most often designed to handle the recording needs of musicians and composers, but may not be suitable for, or capable of, recording quality voiceover work. Larger recording studios and production houses and even some radio stations are expanding their production capabilities to include a much wider range of services, including voice recording and commercial production. Even if you have a state-of-the-art home studio, you should hire a professional demo producer to produce your demo.

After you have selected your scripts, rehearsed them, and are confident that your performing skills are up to par, it's time to start calling the studios in your area to schedule your session. Also check with your coaches and with other voice talent you've connected with through social networking sites and voiceover discussion boards.

When you book a recording studio, you may be assigned an engineer who is not interested in demo production, and may not be skilled at directing voiceover talent. If you are producing the demo on your own, you need to be prepared for this. As the producer of your demo, you need to be ready to guide your engineer through the process and have a good idea of what you want in your demo, including the selection of music and sound effects, and the final sequencing of clips. If your engineer is not capable of directing you, and you haven't hired a demo producer, you'll need to rely on your self-direction skills to get you through. This can be a real challenge because you should be focusing on your performance—not on the details of directing.

HOW DO I BOOK A DEMO STUDIO AND PRODUCER?

The following pages contain some tips and questions to ask as you call around looking for a studio to hire to produce your demo, as well as some important basic information about recording studios.

- **Find a studio that records radio and TV commercials:** If the studio is primarily a music studio, they may not be capable of handling your needs for a voiceover demo. Look for a studio that is experienced in producing commercials or demos.

- **Does the studio have an engineer who knows how to direct voiceover talent?** Unless you have hired a director, you *will need* an engineer who can direct you as you perform your copy. Many studios have engineers who know how to record the human voice, but don't know the first thing about directing talent for an effective voice-acting performance. When you enter the studio, you need to take off your producer hat and become the performer. Even if you hire a director, you need to find a studio that has an engineer who knows how to produce and direct for voiceover.

- **Does the studio have any experience producing voiceover demos?** This should be one of your first questions. You may also have this question answered when you find out if the studio has an engineer who knows how to work with voiceover talent. Even if a studio does a lot of radio commercials, it does not mean that they also produce voiceover demos. Unlike a :60 radio commercial that is a continuous script, your demo will consist of anywhere from 8 to 15 very short clips. The sequencing of these clips will play an important role in how your demo is perceived by the final listener. If the studio has produced demos in the past, ask to hear what they have done for others or for the names of other voice performers for whom they have produced demos. Then follow with phone calls.

- **Ask to listen to other demos produced by the studio or producer:** You'll be investing a considerable amount of money in your demo, so it's important that your producer or studio is willing to let you hear previous work they've done and provide you names of other talent you can talk to. If they are reluctant to release any information, find a different studio or producer.

- **Does the studio have session time that will coincide with your availability?** If you can't book the studio at a time when you are available, you need to find another studio. Many recording studios offer evening or weekend studio time, and may either offer a discount or charge an extra fee for those sessions. You may be able to get a reduced fee for late-night sessions, but you may not be able to get an engineer experienced with voiceover.

- **What is the studio's hourly rate for voice recording?** Many studios have a sliding scale of prices depending on the requirements of the project. Other studios book at a flat rate, regardless of the session. Shop the studios in your area to find the best price for your demo production. Find out if there are any price changes between the voice session and the production session. Find a studio that will give you a flat hourly rate for your entire project. Some studios will give a block discount for sessions booking a large amount of time. A demo session probably won't fit this category, but it couldn't hurt to ask. If you're working with a studio that charges a package price for demo production, their hourly rate won't matter.

- **Does the studio use analog or digital equipment?** In today's digital recording age, this question is usually irrelevant. The difference between digital and analog production in a recording studio is in the areas of recording, editing and postproduction. Analog quality in a recording studio, although rare these days, is extremely high and should not be a consideration for your demo recording if that is all that's available; however, analog production may take some additional time since it usually involves multitrack recording. Digital workstations are the norm and reduce the production and editing time considerably because the audio is recorded and edited within a computer. Today, most studios use some form of digital recording equipment as their primary means of recording.

- **Does the studio have access to music and sound effects libraries?** Your demo will need music and possibly sound effects to underscore your performance. If you are producing an animation demo or an imaging demo, you'll need some very specific types of music and sound effects. Many recording studios do not have any CDs of music that can be used in a demo, even though their primary business may be recording music. Find a production studio that has one or more music libraries that can be used to underscore your spots, and that are appropriate for your type of demo. A *music library* is a collection of music created by a company that produces CDs of music specifically designed for use in commercial, TV, and film production. As you were preparing for your session, you made some notes on music and sound effect ideas. Discuss your ideas with your engineer at the beginning of your session.

It is not a good idea to use music from your personal music collection for your demo. Even though your demo is meant for limited distribution and will not be for public sale, the possibility of copyright infringement for unauthorized use of the music does exist. Also, the use of familiar or popular music may create a distraction if it is not used wisely. If you have a specific sound in mind for some of your demo tracks, you might want to take in some examples from your personal collection, but keep in mind that they should not be used in the final demo.

- **Does the studio have any additional charges for music or sound effects used?** Some studios charge a fee for any music used in your demo, while other studios include the music as part of a package price. If there is a music use fee charged by the studio, make sure it is a *blanket license* rather than a per use or *laser-drop* license. A blanket license covers all music used in a project and is considerably less expensive than several laser-drop licenses. Usually, there is no charge for sound effects. If you provide your own music, there will be no charge, but you take the risk of any copyright infringement issues that might arise from its use.

- **What other fees will the studio charge for materials, including sales tax?** What does the studio charge for CD one-offs, digital media, and any other materials used in the production of your demo? Does the studio have any additional charges for archiving (backing up and storing) your demo project? What portions of the demo production will have sales tax applied, if any? All of these items will affect the total cost of your demo.

- **How much time does the studio estimate it will take to produce your demo?** You should plan on at least six to eight hours for the completion of your demo, although you may be able to have it completed in much less time. The studio's experience in producing demos will be a factor here, as well as your performing abilities. If the studio has experience producing demos, ask for an estimate of production time and an average cost breakdown. Some demo producers will record your tracks on one day and produce your demo over the next several days.

- **What will you take with you when your demo is completed?** In most cases, you can expect to leave the studio with at least least one audio CD of your demo. You should also make sure you receive high quality MP3 files of all versions of your demo. These may be delivered on a data CD-ROM, emailed to you, posted on an FTP website, or transferred to your own digital storage media.

- **How will your demo be backed up?** A backup is different from the master of your session. The *master* is the final version of your demo in a form that will later be presented to an agent or talent buyer. A *backup* is a copy of all the elements of the project, not necessarily in any special order or structure. Even though the backup will contain everything from your session, often including outtakes, it may be in a format that cannot be used anywhere except the recording studio where your demo was produced.

It's not necessary for you to keep a copy of your entire project. You most likely would not be able to open the files or know what to do with them even if you could. But you should make sure the studio will keep a back-up copy of your session on file for future reference. If they don't keep archives, you'll definitely want to get your entire session backed up on data CD-ROM.

Your Demo Recording Session

If you have practiced and mastered your voice acting skills, have your marketing plan in place, and are prepared and ready to work, your demo session can be a lot of fun, and an educational experience. If, on the other hand, you go to your demo session unprepared or without having mastered the necessary skills, your session can be very uncomfortable.

Because you are the executive producer as well as the performer, even if you have hired a producer, you are the one person responsible for making sure your demo is well-produced and that it will be a useful tool for marketing your talents. This means you will have the opportunity to supervise the entire process of your demo production. You won't often have this chance when you are doing real-world sessions. If a producer insists that you leave during the postproduction process, I'd suggest finding a different producer.

Keep in mind, though, that when you are in front of the mic, you need to be focused on your performance, not on other aspects of your demo. This is where hiring a director or having an engineer who knows how to direct becomes important. Track sequencing, music, sound effects, duplication, packaging, and distribution can all be left for later.

ARRIVE ON TIME AND PREPARED

In recording studios, time is money. If a session is scheduled to start at 10:00 AM and you don't arrive until 10:10, that's at least 10 minutes of wasted time and money—probably more, because it takes a certain amount of time for the engineer to prepare the studio. Recording studios usually bill for their time whether you are there at the scheduled time or not. The lesson here is to *be on time* for your session!

If you live your life in a constant mode of running late, you might want to set your clocks ahead, or do whatever is necessary to make sure you arrive at your session on time, or preferably a bit early. Arriving late for real world sessions will get you a bad reputation in a hurry, and no doubt will cause you to lose work. Arriving late for your demo session will put you under unnecessary stress, costing you valuable time and money.

The same goes for being prepared. In real-world sessions, there is little more you need to do than to show up at the studio at the appointed time, ready to perform. However, for your demo session, you are also the producer, and you must be prepared with rehearsed copy and ideas to discuss with the engineer handling your session. If you hire someone to produce or direct your demo, you need to make sure that the two of you take the time to rehearse your copy to find the strongest material and that you both have a good idea of the results you want from your demo.

Here are some other tips to make your demo session a productive and pleasurable experience:

- Arrive at your session a bit early.
- Eat a light meal or snack before your session.
- Arrive in good voice, fully warmed up and ready to perform.
- Have a bottle of water with you.
- Rehearse your copy *before* arriving at the studio.
- Make a note of which scripts you think are your strongest.
- Plan in advance for a possible sequence of scripts.
- Plan ahead for music and sound effects.
- Be ready to accept new scripts that the engineer might have available.

WORKING WITH YOUR ENGINEER/PRODUCER

Aside from your producer, if you hire one, your engineer will be one of the most important people you work with during your demo session, possibly even working as your director. In any case, it is important that you and your engineer work together as a team on your project. Remain flexible and open to your engineer's suggestions. If you are careful in booking the studio, you will probably have an engineer who knows much more about voiceover work than you do. You can learn a lot from a good engineer and he or she may even become a good contact for work later on.

KEEPING YOUR DEMO CURRENT

Your demo will be useful for at least six months to a year, although you may actually use your first demo somewhat longer. As you begin doing paid sessions, you will want to get copies of your work and update your demo occasionally, perhaps every six months to a year. Your agent may request an updated demo or a cut-down version for their house demo CD or website. Each time you update or change your demo, you will need to book a new session unless you have mastered your editing skills. Fortunately, digital technology makes it easy to update your demo as often as necessary, especially if you or the studio you are working with has your original demo project archived. If you are updating your demo, you will probably not need to spend as much time recording new tracks or in postproduction. And you will most likely not incur any additional music license fees, especially if you are simply inserting some of your recent work.

Plan ahead by budgeting for the studio time and have a good idea of the tracks you want to include. Send your updated demo to people you have worked for. A new demo is a good opportunity to stay in touch with past clients and to inquire about upcoming projects.

Demo Tips, Tricks, Dos & Don'ts

Whether you are producing your first voiceover demo, adding a new niche to your marketing plan, or updating an existing demo, keep in mind that you want your demo to feature your best performing abilities. Every script you select should be chosen for the purpose of demonstrating a different aspect of your performing abilities. Plan the content of your demo carefully to include copy you perform well and that is appropriate for the type of market you want to reach. A demo of commercial copy is not appropriate for a producer of corporate projects, and a demo of character and animation voices is not appropriate for a commercial producer.

If you intend to be seeking agent representation with your demo, be prepared to produce a commercial voiceover demo first. A well-produced commercial demo will demonstrate every performing skill that is used in other areas of voiceover such as narration, character, promo, and so on. If you intend to focus your marketing on a niche area of voiceover work, you should consider a second demo designed to focus on only that area.

Producers often listen to a demo with their finger poised, ready to move to the next demo. This is somewhat due to time constraints, but is largely due to the fact that dozens of poorly produced demos arrive in email and cross the desks of producers every day. Remember that producers make their decisions about a voice within the first 5 to 10 seconds of listening to a demo—some even quicker than that. If you are going to make it past that crucial first 5 seconds, your demo performance must be well-presented and highly skilled.

A good demo keeps the producer listening and it has entertainment value with a new surprise, emotional hook, acting technique, vocal variation, or character twist happening about every 5 to 10 seconds. A good demo does not give the listener an opportunity to turn it off.

Here are some things to keep in mind while preparing for your demo:

- Don't do a demo until you are ready. Make sure you have done your homework and have mastered voice-acting skills.

- A single workshop or class will not qualify you to produce your demo. A single workshop can only give you the fundamentals of working in voiceover. Don't rush into producing your demo after one class—even (and especially) if your instructor recommends it.

- Do your homework: research your market and know your niche before even thinking about going into the studio.

- Have your marketing and business plan in place *before* you spend any money to have your demo produced. You should know in advance who you will be contacting to promote your voiceover talent.

- It's OK to recreate other commercials in your style and mention a few product names. Your demo does *not* need to be a collection of work you have actually done.

- Your first demo should focus on the niche you will be marketing to primarily. You will need a commercial demo when you get an agent.

- When using magazine ads for demo copy, choose only those products or services that might actually advertise on radio or TV, and, more important, that fit your delivery style.

- Don't be afraid to write your own copy or adapt commercials or print ads. Writing your own copy can help to make your demo unique and more effective.

- Each voiceover niche must have its own demo.

- Include a wide range of variety in style, attitude, and character, even if your demo is focusing on your voice style, rather than character. Keep the listener guessing as to what will happen next.

- Limit each "clip" to only a single, concise statement or few brief sentences. You only have a short time to catch and hold the listener's attention. The content of each clip must be compelling.

- Focus on what you currently do best. Then revise and prepare new demos as your abilities grow and you acquire copies of projects you have worked on (usually updating once or twice a year).

- Keep your demo short—no longer than 1:30 for a commercial demo with a 1-minute cut-down version for talent agency compilation CDs and websites.

- Your strongest delivery style should be at the beginning. The producer is going to make his or her decision in 5 seconds or less.

- Have your demo professionally produced. Don't think you can put a demo together at home and expect it to sound professional. A poor-quality demo—in either performance or audio quality—is a waste of your time and money.

- Work one-on-one with a coach or demo producer to hone your performance and make certain the copy you have chosen is the best for your delivery style and the type of demo you are producing.

- Keep your ego out of your demo. Your demo is not about you, it's about what you can do with your voice.

- You do *not* need a demo to get voiceover work, and you *will* need a demo eventually.
- Beware of any coach, business, or workshop that promises to produce your demo after taking their course. If you feel even the slightest "red flag," trust your instincts and seek other training or a different producer.

You will be far better off with a solid 1 minute of fast-moving, well-produced, demo than with a slow-moving 1:30 lacking variety and range. If you have the appropriate performing skills, you can produce demos for virtually every niche area of the business. The actual length may vary depending on the type of demo and your market, so you would best be advised to consult with your local talent agents to find out what they are asking for in your market. As you gain a reputation and become more versatile, you may be able to justify multiple demos, or a compilation of demos on a CD or your website.

An Alphabetical Review of Voiceover Demo Categories

The primary categories of voiceover demos are discussed earlier in this chapter. This list is a breakdown of the many subcategories of voiceover demos. Although any of these might be worthy of a demo by itself, most often several of these related subcategories will be included within the context of a more general demo category.

- Accents and dialects (ethnic)
- Audio book
- Celebrity
- Character—animation and toys
- Character—announcer and tags
- Character—celebrity sound alike
- Character—real people
- Character—sound effects
- Character—video game
- Commercial—radio
- Commercial—television
- Commercial—web
- Dialogue—multiple voices
- Documentary
- E-Learning (online training)
- Foreign language
- Imaging (radio)
- Industrial—training
- Industrial—video kiosk
- Industrial—web learning
- In-store messaging
- Jingles (singing)
- Narration—corporate marketing
- Narration—medical
- Political
- Promo (television)
- Specialty
- Spokesperson
- Talking toys & games
- Telephony—IVR (Interactive Voice Response)
- Telephony—message-on-hold
- Telephony—phone prompts
- Trailer (film)
- Youth (children or children soundalike)

19

Your Demo Is Done, Now What?

You've spent a good deal of time studying your craft, and you have made an investment in producing a high-quality, marketable demo. Congratulations... you're in business. As you begin making contacts for voice work, you will be speaking to and meeting professionals who may have been in this business for many years. These people have seen it all, and have little time to waste on an amateur trying to break into the business. Your first impression needs to be memorable and professional.

Present Yourself as a Professional

CREATING YOUR BRAND

Presenting yourself as a professional is important when you submit your demo to agents and talent buyers. A coordinated *brand* shows that you mean business, and take your career seriously. Your brand is the visual and/or auditory representation of who you are and what you do. It sets you apart from your competition in the mind of your clients.

Creating an identity, or brand, for your business is not always an easy thing to do, and it is something you might not want to tackle yourself. Fortunately, there are quite a few talented graphic design artists in the business who you can hire to assist you. Even if you hire someone to help develop your brand and design your graphic image, you still need to provide some input. You might even want your graphic designer to hear your demo to get a better idea of what you do. Graphic designers can get their inspiration from just about anything, so be as thorough as possible when presenting your ideas.

Your graphic look should reflect your individual personality. It should be consistent in all printed materials and carry through to your website.

Your visual image is an important part of your marketing campaign. It can help set you apart from the crowd and ultimately work toward establishing you as a "brand name" in the world of voiceover.

WHAT MAKES YOU UNIQUE?

There are two elements of your marketing that can set you apart from others who do what you do. The first is a *UPS*, or *unique positioning statement*. This is a short one- or two-sentence statement that clearly defines what you do, for whom you do it, and your unique solution to an urgent need. Writing your UPS can be a challenge because it requires you to fully understand the value of what you do and the critical needs of your clients. This may require some research on your part and a great deal of thought as you hone and refine your statement. It's not as easy as it looks. There are many excellent books and Internet resources that discuss this aspect of business development, and that can help you create your UPS.

The second element is a *USP*, or *unique selling proposition*. A USP is a refinement of the UPS into a short statement that communicates the specific benefit of a purchase. It says: "use our services and you will get this specific benefit." The essence of the USP can then be crafted into a *slogan* that can be anything from a single word to a short phrase.

Here's the unique positioning statement for our coaching and training services website **www.voiceacting.com**:

> *We teach powerfully effective communication and performing skills that we've developed over more than three decades of stage, television, recording studio, and advertising experience. We work with people who want to break into the business of voiceover and with business professionals who want to improve relationships with their customers, increase sales, improve their communication skills, create more effective advertising, or become better presenters and performers.*

This positioning statement is intended to give us a clearly defined focus on what we do, who we do it for, and the results that can be expected from using our services. It positions us as expert performance coaches and as a business that understands business communication. An orchestra conductor is our logo, representing the process of combining several core elements of communication to achieve effective results. For marketing this aspect of our business, we refine the positioning statement to a single phrase that is more concise, yet conveys the story we want to tell. At first, we successfully marketed our performance coaching under the VoiceActing.com banner graphic and used the USP:

> *We make you sound great!*

Our original VoiceActing.com graphic, which is still used, looks like this:

VoiceActing.com
we make you sound great!

As we began to produce the VoiceOver International Creative Experience (VOICE) **(www.voice-international.com)**, the world's only conference for voiceover talent, we realized that, although our positioning statement still conveyed what we did, our USP was limiting and no longer accurately reflected the way we were perceived by the voiceover community and our clients. We spent a considerable amount of time brainstorming and testing various ideas and slogans. As a result, we expanded the VoiceActing.com brand by creating the VoiceActing Academy. A new graphic was designed, retaining the orchestra conductor, and a new USP was created that we believe more accurately positions who we are and what we do. Here's the new VoiceActing Academy graphic with our updated USP:

VoiceActing Academy™
Changing lives one voice at a time

You can create your own UPS, USP, and slogan by taking a close, hard look at what you do, who you do it for, and what makes you different. Only by close examination will you be able to discover what makes you unique from other voice talent in your area. When you discover what that is, write it out in a sentence that describes it clearly and concisely. It should describe who your primary customer is and what they gain from using your services. Use our examples above to get started. This process can take up to several hours or several days and will usually result in numerous variations and possible statements. Once this creative exercise is complete, you'll have a much clearer picture of your role in the world of voiceover. With this understanding in mind, you can now begin to explore various ways of refining the essence of your work into a concise USP and slogan.
Here are a few examples:

- *A Penny for your $pots. She just makes cents!* (Penny Abshire)
- *Changing lives one voice at a time* (VoiceActing Academy)
- *Orchestrate your message!* (James R. Alburger)
- *My Voice, Your Way!* (Debbie Munro)
- *Aural gratification guaranteed* (Lani Minella)
- *Guaranteed to round up more business* (Bob Jump)

As you work on developing your UPS, USP, and slogan, be creative and let your imagination run wild. Come up with as many ideas as you can and narrow them down to a few that work for you. Pick the best one and use it everywhere. Your slogan and logo, if you use them, should be included in every piece of print material, as an email signature, and on your website.

Building your business as a voice talent can be a daunting task that can be made easier when you understand that you don't have to do everything at once. Take things one step at a time. As you complete one aspect of your business development, begin working on the next. Approach your business development from an organized and structured foundation, much like you have done with your performance craft.

There are dozens of excellent books available that can help you develop your USP and business identity. One of my favorites is a small book by Mark LeBlanc titled *Grow Your Business*. This little book provides the tools and processes to give you the focus to create a powerful defining statement for your business. With that in hand, the world is yours! Mark's website is **www.smallbusinesssuccess.com**.

Setting Up Shop

Today's voiceover world revolves around the voice actor's home studio and office. If you are going to be in this business, you will need to dedicate an area of your home for your studio and office. Of course, your office can be put together over a period of time, and you may already have much of it in place. Your home office will, most likely, be run from the same computer that is your home studio.

The purpose of setting up a formal office area is so that you can really keep yourself in a mindset of handling your voiceover work as a business. The recordkeeping and organizational aspects of a business become increasingly important as you begin doing sessions and generating income. If or when you join a union, you will want to keep track of your session work and your union paperwork. There are also certain tax advantages to setting up a formal business and you would be wise to consult a tax advisor or accountant on this matter.

As with any business, it is important for you, as a voice actor, to stay in touch with your clients and prospects. You might want to consider some of these methods for that all-important client communication:

- **An answering machine, voicemail box, or service** This is essential for taking calls when you are not otherwise available. Be sure to check for new messages frequently, especially when you get an agent. There are some interesting virtual phone number messaging services available on the Internet for free or for a minimal monthly charge. When you sign up for one of these services, you get a special local phone number and mailbox extension that gives you access to voicemail, email, and fax—all in one place. This type of service can even "read" your email to you over the phone, and everything is accessible by both phone and over the Internet. An Internet search for "phone messaging service" will reveal many options.

- **A cell phone**—A cell phone can be one of, if not the, most useful tools you own when you are on the road. It's become an absolute necessity for staying in touch with clients and your agent. Please remember to turn your phone off when you are in the booth. Better yet, don't even take it into the booth.

- **Business cards, letterhead and envelopes**—You will be making many contacts as you develop your voice-acting business. As a professional, you should consider each contact as potential work. Your first impression leaves a lasting memory. Even though we live in a largely electronic world, you should consider professionally prepared and printed business cards and stationery as essential ingredients to presenting a professional image.

A business card is an absolute necessity as a voice actor. As part of your personal networking, you will want to let everyone you meet know what you do. Your business card is the first and best introduction to you and your talent, followed closely by your demo. Always carry a supply of business cards with you and hand them out every chance you get.

The two most important things on your business card are your name and a telephone number where you can be reached. The most common problem with business cards is that the telephone number is too small to read easily. The second most common problem is too much information on the card.

The purpose of a business card is to be a reminder of who you are and how you can be contacted. Include only the most important information about yourself on your card. If you are using a slogan or

Figure 19-1: Business card dimensions with sample layout.

logo, those should be on the card as well. Keep the design clean and simple for best results (see Figure 19-1).

- **Thank-you notes**—A frequently overlooked, yet very important, business practice is the thank-you note. A brief note of thanks is often all it takes to leave a good feeling with a producer or client. These little notes can easily be prepared in advance, help generate positive memories of your work, and provide a gentle reminder that you are available.

- **Newsletters and Postcards**—Some voice performers send out a brief printed newsletter on a regular basis to clients and producers. Newsletters can take the form of anything from a simple postcard to a brief letter (mailed in an envelope or simply folded and stapled). Content usually includes a brief description of recent projects and clients and any other interesting information. Of course, your graphic identity should be a part of the newsletter. The purpose of the newsletter or postcard is to keep your name in front of the talent buyer. Keep your copy short, concise, interesting, and to the point. Too much information will result in the mailing being thrown away without even being read. If you have an email address for your contacts, you might compose a brief update on your activities on a monthly basis. The idea is to keep your name in front of the people who book voiceover talent.

- **Blogs and social networking websites** □ *Blogs* (short for web log) have become a popular, and highly efficient, method for voice actors to communicate with their clients and friends in the voiceover community. Blog subscribers receive almost immediate notification when a new post is added that announces a new client or other news. Even if you don't have a website, you can set up a blog to start establishing your brand and serve as a resource for your prospects and clients. An Internet search for "blog" will bring up numerous blog sites, most of which are free.

  There are literally hundreds of social networking sites. Two of the best known are Twitter and Facebook. These sites are similar to blogs in that you can post news and other information, but they are different in that they are accessible to a much broader viewer base.

  Perhaps the best social networking site for voiceover talent, that serves double-duty as one of the best resources for learning the inside "secrets" of the business is **www.voiceoveruniverse.com**. This isn't an online audition site, a discussion board, or a way to stay in touch with clients. Instead, VOU is a gathering place for voiceover performers around the world, where they can share what they know and learn from some of the best in the business. Membership is free, but is limited to voiceover talent only.

Print Materials

Even if your marketing will primarily be through your website and email, there are several marketing items you will want to consider having professionally designed and printed, including business cards, envelopes, and stationery. For best results, take your layout to an experienced printer. However, if you are on an extremely tight budget and possess the necessary computer skills, you can use a laser printer or high-quality, color ink-jet printer to create some of your own print materials.

Consult with a printing service or paper supply company about paper stock and ink colors. These people are in the business of making printed materials look good and may be able to offer some valuable suggestions. If you do your own printing, choose paper stock that reflects your branding. You can purchase specialized papers and even sheets of preformatted business cards, mailing labels, and CD labels, ready to be loaded into your printer. However, be aware that specialty papers or perforated paper for business cards may present a less-than-professional image.

Avery is considered by many to be the standard for blank, preformatted labels. Its website, **www.avery.com**, has a free version of the Avery label printing software that allows for designing and printing on most of its label paper stock. You can find comparable labels for almost every Avery label design at Label Blank, **www.labelblankcorporation.com**.

If you're printing your own CD labels, you should know that you can save a lot of money by purchasing a box of 100 sheets of Label Blank or Avery labels, rather than the smaller packages of 25 sheets that you'll find at most stationary and computer stores.

PHOTOS

One of the nice things about voice acting is that your physical appearance is far less important than your ability to act. Your job as a voice actor is to market your talents and skills as a vocal performer. Unless you also intend to market yourself for on-camera work, it is usually not a good idea to include a photo of yourself in your promotion materials.

No matter how good your demo might be, a photograph is going to give the talent buyer a face to go with your voice. You can easily be pigeon-holed or stereotyped as a result of a photo on your demo or website. Many agents and producers will associate a face to a name before they associate a voice to a name. Although not intentional, this can be a real disservice for the voice actor. My recommendation if you are just starting in this business is to keep your image clean and simple without photos, and let your voice do the selling. Later on, as your image, branding, and credibility become established you might consider adding a photo to your marketing materials.

Of course, there are exceptions. If you are also marketing your talents as a model, an on-camera performer, or if you do live theater, a photo is a must. As a multifaceted performer, a photo can actually work to your benefit because it will tend to associate your versatility with your name in the mind of the talent buyer. Of course, you should also build separate websites or have dedicated pages to feature the different types of work you do.

If you use a photo as part of your packaging, hire a professional photographer who understands performance headshots to take the picture and make sure the photo reflects your money-voice personality. Your photo is an important part of your branding and must be of the highest quality. Storefront portrait studios will rarely be able to provide effective marketing photos. Interview photographers recommended by your agent and businesses you work with, but be wary of any agent who requires that you use a specific photographer or service provider.

CD LABELS

The era of the audio CD as a distribution medium for voiceover demos is rapidly going the way of the audio cassette. CDs will, however, most likely remain an option for voiceover demos for some time to come. If you choose to create audio CD demos, you'll want the cover, label, and tray card designs to reflect your professional image and branding. There are several options for labeling and packaging your CD demo. Perhaps the most common form of CD packaging is the familiar plastic jewel case. Alternative packaging options include a clamshell case, basic paper or cardboard sleeve, a slim-line jewel case, and a DVD case.

Since the purpose of your CD demo is to present your performing abilities in a professional manner so talent buyers will hire you, it is important that they be able to locate your demo quickly when they are in search of the perfect voice. For those producers who use audio CD demos, the standard jewel case will provide easy storage and visible access to your demo. The slim-line jewel case has no spine, or edge, labeling, and can disappear when placed on a shelf with other demos. The clamshell case and paper and cardboard sleeves all provide no space for anything other than the CD, and are not recommended. Although a DVD case provides for extra information on the cover and an insert inside, it's not recommended for voiceover demos because it will not fit on the same shelf with other audio CDs. It can be argued that the larger size of the DVD case will make the demo stand out from the crowd, but the reality is that it will usually be stored someplace away from other CDs and can be easily forgotten.

The standard CD jewel case gives you the best possible presentation of your demo with two areas that can hold labeling. The label for the front clear door is called an *insert*. The label for the back of the jewel case (which has folded portions for the two edge labels) is called a *tray card*. For the "do -it-yourselfer," there are many computer software programs available that include templates for printing both the insert and tray card as well as the round label for the CD. You can find these programs at most office supply stores and many computer retailers, or download the free Avery label software from **www.avery.com**. For the most professional results you will probably need to find a graphic designer to do the layout for your CD labels for printing by a full-service print shop.

If you are competent with graphic design software, like Adobe Photoshop or Microsoft Publisher, you may be able to design your own CD label and packaging. You'll find lots of resources on the Internet for CD label, insert, and tray card design templates to use with your design software. Search for "CD template." If you do create your own designs, keep in mind that they must have the appearance of being professionally designed. It's simply not worth trying to save a few dollars if the final product screams "amateur."

Your jewel case labels should contain the essential information about your demo: your name, a catchy slogan (if you have one), your logo (if you have one), your agent's name (if you have one), and a contact phone number (yours or your agent's). This information should also be included on both the *insert* and *tray card*. The back should also include the CD contents, especially if there is more than one demo on the CD. You might also include a short bio of yourself or perhaps a brief client list for added credibility. A website and email address might be other items to include, depending on how you are marketing your demo. For example, if you are represented by an agent, it would not be a good idea to put your personal website or email info on your demo CD.

Regardless of the outer packaging, the CD itself will need some sort of label. Labeling for the CD comes in two basic forms: a paper label and imprinting on the CD. If you are duplicating your own CDs you will be using paper labels. Some CD duplicators will also use paper labels for short run CD duplication. However, other duplicators will use the more professional-looking process of imprinting your label design directly on the CD. Before placing your duplication order, check around for the best pricing and labeling format.

Getting Your Demo Out There: To Duplicate or Not to Duplicate

One of the things you should determine when designing your business and marketing plan is how you will distribute your demo. Within the space of a few short years, demo distribution has moved from reel-to-reel tape to audio cassette to audio CD to electronic data files. In today's voiceover world, your primary distribution will be via an MP3 audio file on your website, sent as an attachment in an email, accessible through an online audition site, or sent from your agent.

When the CD was king, it was often necessary to spend a lot of money for a duplication run of anywhere from 100 to 1,000 or more CDs plus the print material that went with them. In just the past few years, that has all changed! Although the trend today is electronic files as the primary means for demo distribution, the audio CD continues to be found on agent and talent buyer shelves.

It probably isn't a bad idea to have at least a few audio CDs available for those prospects or agents who might request them. Although you can certainly take the high-end route described in the previous section, burning your own CDs directly on your computer is a common and cost-effective method for creating a small supply to keep on hand. Assuming your demo was produced professionally, there is no difference in quality between burning your own CDs or using a duplication service. The advantage of using a CD duplication service is that the final CD will have a more professional look with the label printed directly on the CD. For CD duplication services, perform an Internet search for "CD duplication" in your city. By the way, the term *burning a CD* comes from the fact that during the recording process, the CD laser literally burns tiny pits in the CD media.

CDs can generally be duplicated in any quantity you need and on relatively short notice. Burning a single CD duplicate on your computer is called a *one-off,* and is a perfect copy of the original digital information. Multiple one-offs are made during the process of CD duplication. CD replication is an entirely different process for producing hundreds or thousands of perfect copies. You don't need to worry about CD replication until you are doing mass marketing of your demo CD on a national level.

Because electronic files are so popular as a delivery media for voiceover demos, I wouldn't suggest burning any more than about 25 copies of your demo to start with. These would be reserved for distribution to people you contact who specifically request an audio CD of your demo. When you get an agent, he or she will let you know how many copies the agency needs to keep on hand—if any.

If you've never burned a CD on your computer, it would be to your benefit to learn how to use software designed for that purpose. You'll find a variety of software manufacturers at your local computer store or with an Internet search for "CD burning software." Most software for creating CDs will also burn DVDs. For your voiceover demo, you'll want to use commonly available media for audio CD or CD-ROM. DVD media won't play in an audio CD player.

More important than the CD is the MP3 electronic file. The MP3 file originated as the audio portion of a more complex digital video format. The Motion Picture Engineering Group (MPEG) is the organization responsible for developing the standards for digital recording. The original Mpeg 1 video format had several component parts, one of which was known as "audio layer 3." As digital audio recording became popular, this part of the video format was modified to be a standalone format for digital audio. Thus was born MPEG Layer 3 digital audio, or MP3 for short.

An MP3 file is actually a compressed (or compacted) form of a larger .wav (WAVeform audio format) or .aiff (audio information file format) sound recording. Imagine an inflated balloon. The rubber balloon represents the primary digital information and all the air inside represents the duplicated data present in the file. An MP3 file can be seen as a deflated balloon. The primary data is the same, but all the duplicated data has been removed, thus reducing (or compressing) the size of the file. An MP3 file created at the CD sample rate of 44.1 kHz will be roughly 1/10th the size of the original uncompressed raw audio file.

Most audio recording software can easily create MP3 files, and knowing how to do this is essential if you are going to submit auditions or your demo as electronic files.

Your Website

Believe it or not, there was once a time when the Internet did not exist, and no one knew what a website was. Hard to believe, but it's true!

For voiceover talent today, a website is an absolute necessity. It's often the first stop a prospect makes to learn who you are and what you can do. A website is your 24/7/365 brochure, available to be visited by anyone, any time, anywhere in the world. If you don't already have a website, I'd suggest you seriously consider learning about the Internet and getting online! A website is an invaluable tool in marketing your voiceover talent.

The mere thought of building your own website can be daunting. But, it's really not all that difficult if you have some basic computer skills. Many website hosting companies offer website templates as part of their hosting service. With these, you design your website online. It can take some time to add all the bells and whistles, but you can often have a simple site ready to go in just a few short hours. There are also several software programs that will allow you to design your own site on your computer. These will offer more options and capabilities than the online templates, and allow you to keep a backup of your site on your computer. Their ease-of-use, which is generally reflected by price, ranges from pretty simple, to very complex.

No matter what approach you take for designing and building your website, remember that the purpose of the site is to market you as a voice actor. Flashy animation, glitzy graphics, and clever font styles may look nice, but they will not serve the purpose of branding your voiceover business.

WEBSITE ESSENTIALS

There are many important considerations for putting your website online. Here are just a few:

1. **URL** (Universal Resource Locator), also known as the *domain name*: This is the name of your website. You want yours to be simple, short, and descriptive. Your domain name must be

registered before you can set up an account with a website hosting company. Most hosting companies can help you with registering your URL. Check for availability and register or host through **www.magicinet.com** or one of many other registrars.

2. **Site design:** Your website should be designed to reflect who you are and what you do. Carry your branding through to your website to keep your visual image consistent.

3. **Email:** Using an email address such as "you@yourdomain.com" only makes good marketing sense. Most websites allow for email to be viewed through the website's webmail account, or for email to be forwarded to another address. Avoid common .aol, .hotmail, .msn, .yahoo, and similar email addresses that shout "amateur."

4. **Tell your story:** Your website is the perfect place to let prospects know who you are, who you've worked for, and what you can do for them. A one-page site, if well-designed, will present you as a professional. A very good example of a single-page site is **www.pennyabshire.com**. You'll notice that everything you need to know about Penny as a professional voice talent is easily accessible on that one page.

5. **Post your demos:** Your demos are your primary marketing tool, and you should absolutely post them on your website. MP3 is the recommended format as it is a fairly small file size and will download quickly.

   Another way to post your demos is to convert them to streaming audio files. This is a more complex process that is often best handled by your website designer. Real-media and Flash are just two popular methods for streaming audio files from a website. The major difference between MP3 and streaming audio is that MP3 files must first be downloaded before playing, which can take some time, while streaming audio will play almost instantly. The downside of streaming audio is that the files cannot be downloaded for future reference. If you are planning to use Flash audio to stream your demo, you might also consider including separate, downloadable MP3 files as well.

Building your own website can be an educational and fun experience—if you have the time and inclination. If you'd rather put your time and energy into developing your performing skills and voiceover business, you might want to consider hiring a web designer to build your site. There are a handful of web designers who specialize in creating sites for voiceover talent. Before hiring any web designer, take a look at their work and ask for referrals. If you expect that you might need to update your site on a regular basis, you may want to consider building your own.

20

How to Work in the Business of Voiceover

Promoting and Marketing Your Voiceover Talent

Getting voiceover work is a numbers game: The more you hustle, the more contacts you will make. The more contacts you have, the more you will work. The more work you do, the better known you will become. The better known you become, the more people who want to hire you, and you get more work. It's not quite that simple, but you get the idea—it truly is a numbers game. This chapter will help get you started on the right track.

If you're just getting started in voiceover, you'll be doing all the work; making the calls, sending the auditions, recording the sessions, handling the billing, and doing the follow-ups. When you are just getting started, this can seem overwhelming, but if you are organized and know what you're doing you can reach whatever level of success you desire.

But be prepared… it will take some time. Voiceover is not an "overnight success" kind of business. Achieving any degree of success will take an organized, concerted effort on your part, supported with knowledge of your market and competitive performing skills.

Before embarking on an all-out promotion campaign for yourself, do your homework and get organized. If you have an agent, they may be able to recommend specific businesses for you to contact, or they may ask that you let the agency handle all your promotion—or they may be of no help at all. If you agree to let the agent do all the work, set a time limit for your representation contract. During that time, you can see how many auditions you are called for and get a sense of how well you and your agent work together. Working with your agent is the best way to have an organized and consistent promotion campaign.

If you do not have an agent, and are not planning to get one in the immediate future, you are on your own. If you expect to get any auditions or any work, you must devise your own promotion and marketing campaign and do all the legwork. You'll need to find the names and contact information for your prospects, make the calls, send out the letters, compose the emails, work the social networking sites, and design your marketing campaign. This can be a time-consuming process, but you can make it go a bit easier if you take it in stages. As you create your promotion campaign, keep your long-term objectives in mind and continue honing your performing and engineering skills.

There are many good books on marketing and advertising from which you can gain a tremendous amount of information. You can also learn a great deal by taking an adult education or college extension advertising and marketing course. The Small Business Administration (**www.sba.gov**) offers a variety of classes, services, and business tools that you may find helpful in organizing and running your business. Through these and other resources, you will not only learn some excellent ways to promote yourself, but you will also learn what goes into creating the marketing and promotional copy that you work with as a voiceover performer.

When you promote and market yourself, you are acting as your own agent and ad agency. These roles are simply additional aspects of your business and you must become familiar with them if you are to be successful.

Finding and Working with an Agent

A common question is "Do I need an agent to do voiceover work?" The short answer is "No, you don't need an agent" but this isn't really the right question to be asking. A better question is "Will a talent agent help me in my voiceover career." For most voiceover talent, the answer is "Yes." If you work without an agent, you are limiting yourself to only those voiceover jobs you can find for yourself, and you will be responsible for negotiating your fees and collecting payments. One major advantage of having an agent to represent you is that you will gain access to auditions and clients that you might never have met if you were not represented. Your agent will also handle fee negotiations and collect payments. Having a talent agent working for you is definitely to your advantage; however, this does not mean that you *must* have an agent to be successful.

There is a belief among beginning voice actors that landing an agent means they can just sit back and watch the work roll in. Sorry, but it doesn't work that way! The truth is that your agent is only one part of your larger marketing plan. According to Gabrielle Nistico of **www.vocareer.com**, although an agent will create and distribute marketing materials designed to reach industry professionals, those materials are generally intended to promote the agency, and not an individual talent. It is the voice actor's individual marketing efforts that ultimately promote their unique skills and abilities to the voiceover marketplace.

It is important for you to network constantly and let your talents be known. Networking with other voiceover performers keeps you up on current trends, and, if you are nonunion, you may get a better idea of the fees other performers are earning. Joining or networking with professional associations like Media Communications Association International (**www.mca-i.org**) will help to keep you connected with local producers, production companies, and others who may ultimately utilize your services. Always keep a few demos and business cards with you and be ready to pitch yourself when the opportunity arises. It's a subtlety, but maintaining an attitude of professionalism communicates credibility and integrity.

GETTING THE GIGS

You will probably get your first few voiceover jobs through friends, networking, or some other contact you make yourself. As you begin working, your skills will improve, producers will begin to know about you, and your talents will become more valuable. When you reach the point where you are confident with your abilities, have developed a repeat client base, and are ready for more work, it's time to find an agent. Remember, most working pros have an agent. To present yourself with a professional image, you should too. So, how do you go about finding yourself an agent?

The first thing to understand is that an agent works for you! Some beginners in this business think it's the other way around. Most agents are very selective about who they represent, and may even give the false impression that the performer is working for them. Not true! It is their job to get you work by sending you out on auditions and connecting you with producers who will hire you. Once on the audition, it becomes your job to perform to the best of your ability. Your agent may help to market you by making your demo available to casting directors, advertising agencies, and production companies. Once a job is booked, the agent negotiates your fee.

As you begin your search, you will find that no two agents are alike. Some handle the paperwork for the union, while others want the client or performer to handle this task. Talent agents in a large market, like Los Angeles, run their businesses completely different from a talent agent in a smaller market in the Midwest. And working long-distance with an Internet agent is different from working with an agent in your home town. As with much of the voiceover business, there are no hard-and-fast rules. The most important thing is that you are comfortable with your agent, and that your agent is comfortable with you.

SEARCHING FOR AN AGENT

There are a few things you need to know about talent agents before you start seeking representation:

- A talent agent is not in the business of nurturing you or grooming you to be a professional voice actor. They expect you to have your performing skills in place and ready to go.

- A talent agent may not be interested in you unless you have a track record and an existing client list. Agents are in business to get you booked as often as possible, and at the highest fee possible. They only get paid when you work.

- You may have a great track record, and an incredible demo, but you may be rejected simply because the agency already has other voice talent with the same or similar delivery style as yours. Being rejected for representation is not a personal attack on you or your abilities.

One way to find an agent in your area is to contact your local AFTRA office. Even if you are not a member, they will be able to provide you with a list of all franchised agents in your area. Many agents work exclusively with union talent, although some work with both union and nonunion talent.

Be prepared for rejection. Most agents and producers in Los Angeles will not even open or listen to an unsolicited demo, although this policy is different in other cities. You will have much better success finding an agent and finding work if you spend some time on the phone first. It may take a little research on your part, but the time you spend talking with agents and producers on the phone will pay off later on. Don't expect to get results on the first call. Marketing your talent is an ongoing process and results often come weeks or even years later. You should also know that in major markets like Los Angeles, New York, and Chicago, many talent agents will not even be willing to speak with you unless you are referred by one of the talent they currently represent.

You can start your search for an agent with a simple Internet search for "Talent Agent Your City." Talent agents for major markets in the U.S. are also posted at **www.voicebank.net**. Another way to find an agent is to go to a theatrical bookstore. Samuel French, Inc. is among the best. See their website, **www.samuelfrench.com** for store locations in Los Angeles, New York, and Toronto. A keyword search on the site for "agencies" will bring up several resources.

Yet another incredibly valuable resource for locating talent agents in Los Angeles and New York is *The Voiceover Resource Guide*. This is a small booklet—and a website—that includes agents, demo producers, recording studios, union rates, and lots of other information for voiceover talent. It's available at most of the studios in L.A. and New York, and online at **www.voiceoverresourceguide.com**.

While on your search for an agent, you can also call recording studios, TV stations, and production companies in your area. Ask for the production manager. Let this person know you are available for voiceover work, and ask if they book freelance talent or prefer working with an agent. Ask for the names of the talent agencies he or she works with. Let them know your union status. If the company is a union shop (an AFTRA or SAG *signatory*) and you are nonunion, they will not be able to hire you, but may be able to give you some good leads. Don't forget to let companies you contact know that you have a demo you can send to them. Follow up all phone contacts with a thank-you letter.

Many talent agents specialize in certain types of performers, such as modeling, on-camera, voiceover, music recording, theatrical, and so on. You can call the agent's office to find out if they represent voiceover talent and if they are accepting new performers. Keep this initial call brief and to the point, but be sure to get the name of someone to send your demo to if the agency expresses any interest.

Proper phone etiquette is important when calling an agent. Agents are busy people and will appreciate your call more if you are prepared and know what you want. Here's an example of an ineffective call to an agent:

AGENT: Hello, Marvelous Talent Agency.
ACTOR: Hi, uh, is there somebody there I could talk to about doing voiceover?
AGENT: Who's calling?
ACTOR: Oh, yeah. My name is David Dumdum, and I'd like to talk to someone about doing voiceovers.
AGENT: This is a talent agency. We don't do voiceovers, we represent talent.
ACTOR: That's what I mean, I want to talk to somebody about representing me.

This kind of call not only takes a long time to get anywhere, but the so-called actor is not at all clear about what he wants to discuss. Even if this performer had a decent demo, the chances of getting representation are poor simply because of a nonbusinesslike and very unprofessional presentation. Here's a much better way to approach a call:

AGENT: Hello, Marvelous Talent Agency.
ACTOR: Hi, this is Steven Swell. I'd like to know if your agency represents voiceover talent.
AGENT: Yes we do.
ACTOR: Great! I'd like to speak to someone about the possibility of representation. Are you taking on any new performers?
AGENT: We are always interested in looking at new voice talent. If you'd like to send us a copy of your demo, we'll give it a listen and let you know.
ACTOR: That's terrific. I'll get a copy to you in today's mail. Who should I send it to?

This performer gets to the point of his call quickly and effectively. He is polite, businesslike, and keeps an upbeat, professional attitude throughout the call. Even though he didn't connect with an agent, he did get a name and there is now a clear process for getting his demo into the agency.

Narrow down the prospective agents in your area. You can immediately eliminate those who represent only models, print, or on-camera talent. The Los Angeles area has more than 250 franchised agents, and only a handful represent voiceover talent, so in a larger market, you must be very specific in targeting potential agents before sending out your demo and résumé. Smaller markets can have zero to several talent agents, depending on the market size. Representation by a small talent agency in a small market can be an excellent way to break into the business of voiceover. There are also a growing number of talent agents who represent voice talent nationwide, or even worldwide, through the Internet.

Before contacting any agent, prepare a brief and to-the-point cover letter to accompany your demo. This is not a résumé. This is a business letter intended to introduce you to the agency and should be no more than a few short paragraphs. Simply state that you are a voiceover talent and that you are interested in discussing the possibilities of representation by the agency. Depending on your initial contact with the agency, your demo and letter may be sent by postal mail or email.

When sending through the post office, each letter should be an original, and should be addressed to the person whose name you were given during your research. The envelope address should be printed by a computer, not by hand. This gives your envelope a professional appearance. Include any relevant performing experience in your letter. Any reputable agent will require a demo from any talent they are considering, although a résumé is not generally used or necessary when marketing for voiceover work.

Here's an example of a good cover letter that is short, to the point, gives a professional appearance, provides some important information, and suggests the performer's potential value to the agency. Notice that this example requests action from the agency to arrange an interview.

Dear Mr. Agent:

Thank you for your interest in my demo. As I mentioned on the phone, I am a voice actor seeking representation. I have been booking myself as a freelance performer for the past few years and have had several successful commercials on the air.

Additional information about me and my background is included on my website at www.jamesalburger.com. For your convenience, I've enclosed a copy of my current demo.

I believe I can be a valuable asset to your agency, and I look forward to hearing from you so that we can arrange for a meeting to further discuss representation by your agency.

Sincerely,

Once you've been asked to send your introductory letter and demo, *do not* call to see if your demo was received. It will often do you no good, and may even irritate some agencies. And don't expect to get your demo back! Talent agents know you send out demos to other agents in the area. If they hear something they like, they will call you. If you are good, and they're interested, they will call quickly, simply because they won't want to miss out on representing an excellent performer.

Don't be surprised if you don't get a call. There may be many reasons for an agent not accepting you or not getting back to you quickly. Don't expect or ask for a critique of your demo. If an agent is kind enough to critique it for you, use that information to learn how to improve your skills and create a better demo. You may need to produce two, three, or more demos before landing that first agent.

Sooner or later you will find a talent agent who is interested in talking to you. But be aware, the agent's interest does not mean you have representation. It only means that he or she is interested in learning more about you and your talent, and to determine if you will be a good fit with their talent agency. When you are selected for representation, expect your agent to request changes in your demo. Your agent knows their clients and the best way to market you to them.

INTERVIEWING AN AGENT

It may take some time, but when you do get a positive response, you may be asked to set up an appointment to meet with the agent. This can be quite exciting. What will you wear? How should you act? What will you say? If you are handling your correspondence via email, you may be asked to call the agent, or the entire process may be done over the Internet.

Remember that although it may appear as though the agent is interviewing you, the reality is that you are interviewing the agent. Handle this interview just as you would an interview for a new job. Dress nicely, and present yourself in a businesslike manner. Be careful to wear clothes that do not make noise. A good agent will probably ask you to read a script as part of the interview. Enter the office with confidence. Play the part of the successful performer. Create your character for the interview just as you would for a script, and act as if you are a seasoned pro and already represented. Your chances of signing with an agent will be much better if your first impression is one of a skilled and professional performer.

Interview all your prospective agents as thoroughly as possible. Listen carefully, and don't be afraid to ask questions at any time. What types of work have they booked in the last month? What is the average scale they get for their performers? What is their commission? Is their commission added to the performer's fee, or taken out? How many voiceover performers do they represent? How long have they been in business? You can even ask whom they represent and for a list of some performers you can contact.

During your meetings or phone calls with agents, you may talk about everything except your voiceover work. They will want you to be comfortable so that they can get a sense of you as a person, and you will want to get to know them a bit. You need to decide if you like them and have confidence that the agency will be able to get you work. They need to determine if you can work with them as a team.

Take your time. Don't rush to sign up with the first agent who offers to represent you. Also, if any agent gives you the impression that you are working for him or her, you might want to consider eliminating that person from your list. The agent works for you—not the other way around. If an agent requires a fee of *any* amount before they will represent you, they may be operating illegally and you should end the conversation and leave. By law a talent agent is only entitled to a commission based on the work they obtain for you. When an agent directly charges you a fee to be posted on their website, or for headshots, or for anything else, they may be *double-dipping,* and that's illegal, or at the very least potentially unethical.

When you sign with a talent agency, it will normally be a contract for one year. Some agencies request a multiyear agreement, but this can cause problems if your agent doesn't promote you, and you don't get work. If you don't have a good working relationship with your agent, they can literally put your career on hold by simply not sending you auditions. If you are uncertain about the relationship, you may ask for a six-month trial, but if you show a lack of confidence with the agent, it might be best to simply seek representation elsewhere. Even with a good relationship, it is generally a good idea to renegotiate with your talent agent every year.

A large agency may have many people in the office and represent a large talent pool. A small agency may have only one or two people handling the entire business. It is easy to become a small fish in a big pond if you sign with a large agency. On the other hand, most large talent agencies sign only voiceover performers with years of experience and a solid track record. Your first agent most likely will work for a smaller agency that can give you more attention and help guide your career.

WORKING WITH YOUR AGENT

Once signed, you should keep your agent up to date on your work. Let him or her know how your auditions and sessions go, and keep the agent current with an updated demo as needed. Calling your talent agent once a week should be adequate, unless he or she requests you call more or less frequently. Your agent can also be a very good indicator of the areas you are weak in, and may recommend classes and training if necessary. The key to working with an agent is to stay in touch and ask for advice. They generally know the business far better than you.

One good question to ask your agent is how you should handle work you obtain on your own. Some agents will allow you to handle your own personal bookings without paying a commission. However, it may be advisable when someone directly approaches you for work, that you refer the company or person to your agent, especially if you are a union member. As a professional performer, your job is to perform. Your agent's job is to represent you and negotiate for the highest fee. Although it is generally wise to let your agent handle the negotiations, there may be some situations where it might be best for you to handle the money talk yourself. This is something only you and your agent can work out, but if you have a good relationship with your agent, and the situation warrants, you may have a better chance of landing the job.

I know one voice actor who auditioned for a CD-ROM game and noticed that the other voice actors who said they had an agent were being passed over for callbacks. With this in mind, he called his agent to discuss the situation. Their mutual decision was that the voice actor would avoid any mention of representation until after he was booked. He handled the negotiations himself and actually managed to get a higher fee than most of the other voice actors booked for the project. Even if you are an accomplished negotiator, your agent is your representative.

As a career grows, it is common for performers to change agents several times. A word of warning, however: Changing agents can be traumatic. You are likely to have a case of the "guilts" when leaving an agent, especially if the person has done a lot to help promote you and develop your career. When this time comes, it is important to remember the reasons why you must change agents. You may have reached a level of skill that is beyond your agent's ability to market effectively, or you may simply be moving to a new part of the country. On the other hand, you might be changing agents because your current agent is simply not getting you the kinds of jobs you need.

A Business Plan for Voice Actor You, Inc.

You have probably heard the phrase: "If you fail to plan, you plan to fail." This is as true in voiceover as it is for any other business. You need to have a vision of where you want to be and have some sort of plan as to how you will get there. If either of these is missing, chances are you will not be as successful as you hope to be as quickly as you would like to be. Things will get in your way from time to time, and you will be distracted by just living your life. However, if you have a plan, you will be prepared to work around those obstacles when they jump in front of you.

As an independent professional, you need to look at what you do as a business. With that in mind, my coaching and business partner, Penny Abshire, has adapted a simple business plan that you can use to develop focus on the business side of voice acting. You will wear many "hats" as you operate your business. You are the CEO, CFO, Sales Manager, Marketing Director, Director of Education, and finally, a performer. It is critical to your success that you understand what you are doing for each of your duties and that you have a direction in which you are moving. The "Business Plan for Voice Actor You, Inc." on the following pages is something to which you should really give some serious attention. Don't just skim through this and forget about it. Copy these pages and read through the questions. Set it aside for a few hours to think about how you will plan your career, market yourself, sell your services, learn new skills, and protect your future. Some of the questions will be fairly easy to answer, while others may take a great deal of time and thought.

The time you spend preparing your plan will be time well spent. Refer to your plan on a regular basis and review it about every six months, or at least once a year. Things do change, and your goals and objectives may change. This is intended to be a guide to keep you on track for your career.

YOUR DEMO

This is your audio brochure and your product (at least at this point in time). Your demo is what your potential employers (your customers) will use to judge your talent as it applies to their projects. Your demo is your primary marketing tool. You will need a high-quality demo to market your talent and sell your services. Chapter 18, "Your Voiceover Demo," covers this subject in detail.

MAKING CONTACTS WITH PROSPECTIVE CLIENTS

Sales calls are an art form all their own. This primer will give you some basic ideas, but you should also consider some additional study on the subject of sales and marketing.

You will need to spend a fair amount of time on the phone, contacting potential clients. Know what you want to discuss before making any calls. Know your niche, what you do best, who you are marketing to, and be specific about the type or types of voiceover work you are promoting. If you are trying to get into animation voiceover, it's not appropriate to call ad agencies or discuss your expertise with telephone messaging.

Before calling, do some research on your prospects to learn how they use voice talent. When you call, let your professionalism speak for itself and show your prospect that you understand their needs and how your voice work can be of benefit to them. Be careful not to be in a rush to sell your services. You'll be much more successful if you engage your prospect in a conversation to let them get to know you, and for you to gather additional information about how you can help them. Have some prepared notes to look at so that you don't forget anything important during your call, and be prepared to answer any questions that might arise during the conversation.

Needless to say, your stationery should be printed, your demo should be produced and ready to mail before you begin making calls, and you should have a system in place for cataloging prospects and following up.

Remember, you need to talk to someone who is directly responsible for hiring voiceover performers. If you do not have a contact name already, tell the receptionist the purpose of your call. She will most likely direct you to the person you need to speak to, or refer you to someone who might know to whom you should speak. If you can't get connected right away, get a name to ask for when you call back. If you get voicemail, leave a clear and concise

Business Plan for Voice Actor You, Inc.

This simple business plan is designed to help you focus on your business and propel you in the direction you want to go. Give each question some serious thought before answering and review what you've written at least once or twice a year.

1. As **Chief Executive Officer**, what is your vision or plan for a career as a voice actor; it should be specifically designed to ensure your growth, profitability, and financial gain?

_____

_____

_____

   What change(s) must take place to bring this plan to fruition?

_____

_____

2. What strategic alliances are you forming to ensure the achievement of the vision or plan of VOICE ACTOR YOU, INC.?

   a) With whom are you aligning?

_____

   b) How will this be beneficial?

_____

3. As V.P. of ☐uality Control, what are you specifically doing to ensure and/or improve the quality of the service provided by VOICE ACTOR YOU, INC.?

_____

_____

_____

4. As **Chief Financial Officer**, what plans must be made to accommodate the financial and marketing continuity of VOICE ACTOR YOU, INC.?
   Current strategy: Anticipated cost:

_____

_____

_____

   a) Alternative sources of revenue?

_____

_____

   b) Probability of primary revenue continuation over next 5 years?
      Excellent____   Very good _____   Fair____   Poor____

   c) Back-up strategy:

_____

_____

5. As **V.P. of Marketing**, what steps are you taking to seek new or additional target markets for your services?

_____

_____

_____

   a) Local markets?

_____

_____

_____

   b) Other markets?

_____

_____

_____

6. As **V.P. of Promotions**, what steps are you taking to complete the following:
   a) Seek representation?

_____

_____

   b) Collect materials and prepare for demo?

_____

_____

   c) Demo production?

_____

_____

   d) Graphic design (logo, USP, business cards, stationery/thank-you cards, etc.)?
      Slogan _____
      Design _____
      Printing _____
   e) On-going promotion (blog, etc.) _____

7. As **V.P. of Sales**, what is the projected revenue for year-end?
      $_____
   a) Is that enough to cover company expenses? ___yes ___no
   b) What about expected revenue growth for next year?
      $_____

8. As **V.P. of Education**, what is the training plan *specifically designed* to ensure the services offered by VOICE ACTOR YOU, INC. are equal to, or exceed, industry standards?

_____

_____

_____

_____

What is the time-line for implementation of the training program?

By _____ I will be enrolled in _____ Completion date: _____
By _____ I will be enrolled in _____ Completion date: _____
By _____ I will be enrolled in _____ Completion date: _____

By _____ I will read _____ Completion date: _____
By _____ I will read _____ Completion date: _____
By _____ I will read _____ Completion date: _____

By _____ I will study and/or research _____
_____ Completion date: _____
By _____ I will study and/or research _____
_____ Completion date: _____
By _____ I will study and/or research _____
_____ Completion date: _____
By _____ I will study and/or research _____
_____ Completion date: _____

9. As **V.P. of Human Resources**, what needs to be done to protect the mental, physical, and spiritual health of the primary employee (*you*)?

_____

_____

a) Vacation allotment, family leave, and general mental health maintenance?

_____

_____

b) Maintaining connection with corporate stockholders? (*family*)

_____

_____

c) Your spiritual health?

_____

_____

10. As **Director of Maintenance**, what adjustments should be made to improve the visual appearance and physical health of the primary employee (*you*), the product, or service?

a) What do you plan to do?

_____

_____

_____

_____

b) When will you get started□-*specifically*?

_____

11. As **Chief Benefits Officer**, what financial planning is in place to ensure your future financial security (*i.e., retirement*)?

a) What do you plan to do?

_____

_____

_____

_____

b) When will you get started?

_____

_____

_____

_____

12. As **Accounting Department Head**, what steps are you taking to maintain accurate invoicing, recordkeeping, and IRS accountability?

_____

_____

_____

_____

_____

_____

_____

message that includes your phone number at the beginning and end. It's a very good idea to write out your message so you know what to say when you are forwarded to voicemail. Keep it conversational so you don't ramble or sound like you are reading as you leave your message.

It may take a few follow-up calls before you connect with someone. If you already know how the company you are calling uses voiceover, your conversation should be of an introductory nature. If you don't know, your call should focus on how voiceover work might be used to benefit their business. Either way, the call should be more about them than you. You probably will find some companies that have not even considered hiring an outside professional for their voiceover needs. Undoubtedly, you will also find many that are not interested in what you have to offer. Remember, this is a numbers game, so don't let yourself get discouraged.

Offer to send a copy of your demo to those who are interested. Follow up by mailing your demo with a letter of introduction. It is amazing how many people never follow up a lead by sending out their promo kit. You will never get any work if you don't follow up.

FOLLOW-UP

You will need the following basic items for follow-up:

- A cover letter on a professional-looking letterhead
- Business cards
- Labels and envelopes to hold your print materials and demo
- A voiceover client list detailing any session work you have done
- A website you can refer prospects to for additional information about you and your services, and where they can listen to your demo
- Your demo as an audio CD for mailing and as an MP3 file for posting on your website and emailing.

First impressions are important, and the more professional you look in print, and sound on the phone, the more your prospect is likely to consider you for work. If you use color in your logo or graphics design, you should consider using special paper designed for color ink-jet or color laser printing. With appropriate computer software and a good printer, you can design a simple form letter that can be adapted to your needs.

You will need several different versions of your letter of introduction, depending on whether you are following up from a phone call, or if the follow-up is from a personal meeting.

Keep your letter to no more than three or four short paragraphs in a formal business style. Personalize the heading as you would for any business letter. Thank the person you spoke to for his or her interest, and for the time spent talking to you. Remind them of who you are and what you spoke about. Let the company know how you can help them and how they can contact you. Also, mention in the letter that you are enclosing your demo. Be sure to include your website and email address in your letter. The following is an example of a typical follow-up letter:

Dear Mr. Client:

Thank you for taking the time to speak with me yesterday, and for your interest in my voiceover work.

As I mentioned during our conversation, I am available to help your company as a voiceover performer for in-house training, marketing presentations, and radio or television commercial advertising. I am enclosing a list of some recent projects I have voiced and a copy of my demo, which runs approximately one and a half minutes. You can learn more about me and listen to more of my work at www.voiceacting.com. This will give you a good idea of the types of voiceover work I do that can be of benefit to you.

Should you be in need of my services, please feel free to call me anytime at the phone number above or send an email to info@voiceacting.com. I look forward to working with you soon.

Sincerely,

Unlike the other performing arts, in the world of voiceover, a résumé is not a requirement. Most talent buyers are more interested in what you can do for them now, rather than what you have done in the past, but a list of clients may be helpful. If you have an agent, include the agent's name and phone number in the letter. In larger markets your agent's number should be the only contact reference. In smaller markets you may want to include your own number as well as your agent's. (*Note*: Your agent's name and phone number should be on your demo, but mention it in the letter as well.)

You do not need to mention your union status or fees. Your union status should have been established during your phone call, if that was an issue, and it should be noted on your demo CD label or in your email signature. Your fees are something to be negotiated either by your agent, or by you, at the time you are booked. If it comes up in a conversation, just tell the person that your agent handles that, or that you cannot quote a rate until you know what you will be doing. If they insist, quote the current AFTRA scale for the type of work they are asking about. At least that way you will be quoting a rate based in industry standards. If you are booking yourself as nonunion, freelance talent, mention that your fees are negotiable.

During your initial call, you should have set up a timeframe for any follow-ups. After sending your thank-you letter and demo, call your contact at the scheduled time to confirm that the package was received. This helps to maintain your professional image and serves to keep your name on their mind. Don't ask if the person has listened to your demo. That's not the purpose of your call. If they bring it up, fine, but you should not mention it.

Before completing your follow-up call, ask if there are any projects coming up in the near future that might take advantage of your talents. If so, and if the company is considering other voiceover talent, be sure to make yourself available for an audition. Phrase your conversation in such a way that it seems like you are offering to help them. This puts you in a position of offering something of greater value to your potential employer, rather than just being someone asking for work.

Once you have established a list of possible employers, you will want to stay in touch with them. Consider sending out a brief note or postcard every six months or so and on holidays. The purpose here is to keep your name in front of the people who book talent.

Perhaps the only rule for follow-up is to be consistent and persistent. Maintain a professional image, keep your name in front of your prospects, and you will get more work. Here are some ideas for follow-up reminders:

- Thank-you card (after session, meeting, or conversation)
- Holiday and seasonal cards
- Birthdays and anniversaries (if you know them)
- Current projects you have done
- Generic reminder postcard
- Semiannual one-page newsletter updating your activities
- Special announcement about upcoming projects

REACHING THE PEOPLE WHO BOOK TALENT

Many large companies have in-house production units, while others hire outside production houses and work with agents. There will usually be someone who is in charge of coordinating promotion and advertising that may require the use of voiceover performers.

One problem in reaching people who use voice talent is figuring out which companies are likely to need your services. Some possibilities are:

- **Watch local TV and listen to the radio.** Look for local advertisers who are doing commercials with voiceover talent.

- **Call advertisers and ask who coordinates their radio and TV advertising.** Radio stations frequently use station staff for local commercials, and will not charge their advertisers a talent fee. You need to convince these advertisers why they should pay you to do voiceover work when the radio station does it for free. This can be a real challenge! When talking directly to radio advertisers, you need to put yourself in a class above the radio DJ. Some advertisers like the celebrity tie-in by using station talent. Others may simply prefer to spend as little as possible on advertising. You can get work from these people, but you may need to educate them so they understand the value of using you instead of doing it themselves or using a DJ for their commercials. You may find that they have other uses for voiceover talent for which you would be far more qualified than a DJ.

- **Contact your local chamber of commerce.** Get a list of the largest companies in your area. Many of them will use voiceover performers and some will do in-house production.

- **Check the local newspapers.** Call advertisers that you think might be likely prospects.
- **Use resource directories.** Many cities have a resource directory or a service bureau that can provide you with specific information about businesses in the area. Or, your chamber of commerce may be able to provide this information.

When you contact a nonbroadcast business that has a production unit, start by asking to talk to the creative, promotion, or marketing department. You should talk to a producer or director. Don't ask for advertising or sales, or you may be connected to a sales rep. If you ask for the production department, you may end up talking to someone running an assembly line.

Television stations can be a good source for bookings. They use voiceover for all sorts of projects, many of which are never aired. At a TV station, the production department handles most audio and video production. Some TV stations may even have separate production units for commercials, station promotion, and sales and marketing projects. Start by asking to talk to the production manager, an executive producer, or someone in creative services. You may end up talking to someone in the promotion department, because a promotion producer frequently uses more voiceover talent than anyone else at the station.

Recording studios usually will not be a good source for work, simply because most recording studios specialize in music recording. Usually, those that produce a lot of commercials work with performers hired by an ad agency or client. Some studios do a limited amount of producing and writing, and may book their voiceover talent from a pool of performers they work with regularly. In most cities, there are at least one or two studios that specialize in producing radio commercials. Use good judgment when sending your demo to recording studios. You might be wasting your time, but then, you never know from where your next job might appear. Some studios will recommend voice talent when asked.

Of course, contacting advertising agencies directly is another good way to reach the person who books talent. At an ad agency, the person you want to reach is the in-house agency producer (AP). Some ad agencies may have several in-house producers, and some agencies have account executives (AE) who work double duty as producers. If there is any doubt, ask to speak to the person who books or approves voiceover talent.

There are no hard-and-fast rules here. As you call around, you just need to try to find the correct contact person. Once you connect, use the basic marketing techniques described in this chapter to promote yourself.

As you can see, the amount of marketing research and legwork can be daunting. If you take it in small steps, and in an organized manner, you will, in time, develop a consistent and effective marketing plan.

In Chapter 22, "Shedding Light on the Dark Side of Voiceover," Robert Sciglimpaglia, a voice actor and practicing attorney, provides a detailed look at the legal aspects of working in voiceover.

21

Setting Your Fees and Managing Your Business

Setting Your Talent Fee

The first, and most important thing you need to know about setting and negotiating your fee is that *you have value*. You can do something your client cannot. You have something to offer that is of value to your client, and your client needs what you have to offer. There is something about *you* that the producer believes is right for his or her project. It could be the way you interpret the copy; it could be a quality in your voice; it could be anything. You are the chosen one! Congratulations! If your client didn't want what you have to offer, they would be talking to someone else. You've got the job! All that's necessary now is to work out the details.

Because you have value, you should be fairly compensated for your work. At first glance you may think that a client's proposed budget for voice talent is very reasonable. But be careful not to rush into accepting voiceover work simply based on what the client is offering without first doing a little research. When the script arrives, you may be unpleasantly surprised by the amount of work you really need to do, and as a result, how low your compensation really is. The details of your work need to be clearly defined before you agree to the job.

The second thing you need to know about setting or negotiating your talent fee is that *your time is valuable*. You've made a considerable investment of time and energy to get to a place where you are ready to market yourself as a professional voice actor. You've invested in training and workshops, purchased books, and probably spent more than a few dollars to build a home studio. You've built a business that is intended to produce an income that will recover your investment and result in a profit. For many, the goal is to eventually move into voiceover as a full-time career. If you expect to ever see a return on your investment, you need to give some serious consideration to how you will set your talent fees and how you will work with clients. You need to think like a business person.

279

If you plan to get paid for your voiceover work (and you should), you'll need to learn some negotiating skills. As a voice actor, you are in business for yourself, and fee negotiation is part of doing business. Even if you have representation, you should still work on your negotiating skills if for no other reason than you will be able to discuss your fees and marketing strategies intelligently with your agent. Since part of an agent's job is to handle fee negotiations, the next few sections of this chapter will address setting fees and negotiating techniques for independent voice talent who do not have agent representation.

The primary job of a voice actor is to deliver an effective and believable performance. This can be a challenging task when you are placed in a position where you must multitask by running the computer software, making sure your recording quality is up to standards, finding the proper character and attitude, and delivering a performance that meets the client's needs. With a home studio, you're a one-person-shop, and you do it all! It has taken some time for you to learn how to do all of these things.

An often overlooked consideration when setting fees is the investment of time and money in getting started in the business of voiceover. The cost of books, workshops, and demo production can easily add up to several thousand dollars. Add to that the cost of your home office furnishings; business software; office equipment; supplies; business development and marketing; the cost of your computer; your audio equipment; Internet connection; website hosting; website design services; graphics design; printing; and telephone lines.

It doesn't take much effort to discover the true financial investment you've made in your voiceover business. It is only good business to expect a return on this investment (*ROI*). And, in order to see a return on your investment, you'll need to consider those expenses as you determine your fee structure. There is no easy formula for doing this, but it is something you should consider.

EVALUATE THE JOB REQUIREMENTS

Many experienced producers have a very good understanding of what it takes to record a quality voice track, or produce a complete production. They have been through the production process many times and know what it is like to work with voice talent of all levels of experience. There are many others, however, who have no experience working with voice talent, and have absolutely no idea of what is involved in voice recording and audio production. For a voice talent just getting started, the unfortunate reality is that many first-time clients will be inexperienced and uneducated in the world of audio production, voice recording, and voiceover work in general. As a voice actor, it's not your job to educate your client, but that may be something you'll need to do as part of your negotiations.

Another aspect of voiceover reality is that other producers eager to maximize their profits may be willing to take advantage of beginning voice talent. If you don't know your personal worth, and how to negotiate your fee, it could be a very long time before you begin to see any financial success as a voice actor.

Keep in mind, as you talk to prospective clients, that the fee they offer is not necessarily the fee you will actually receive for professional work. Your potential client may have a price in mind, which is based on completely uneducated and unrealistic expectations. You, on the other hand, may have a fee in mind that is considerably different, based on your knowledge of your investment, your understanding of what it will take to complete their job, and your level of skill. The purpose of a negotiation is to arrive at a level of compensation that is mutually agreeable to both parties. The bottom line in this business is "everything is negotiable." If a prospective client is unwilling to negotiate with you regarding your compensation, you may be wise to reconsider working with that individual.

It is not uncommon for clients booking through Internet audition sites to offer a fee that, at first glance, may appear reasonable, but upon closer examination is little more than minimum wage for a considerable amount of specialized work. Here are two examples of how you can evaluate a potential booking to determine if it will be worth your time and energy:

Example #1: The Trial Transcript

> We have a trial transcript of 2,000 pages double spaced that we need read for an audio book. Contains male and female characters - you would read all parts. Pay is $1,000 +

One thousand dollars—not bad for a few hours of recording time, right? But take a closer look: the project is two thousand pages long. A quick calculation will reveal that this producer is offering only $0.50 per page to record this project! Still, $1,000 is a lot of money! Or is it?

Let's say you estimate that an average completed double-spaced page will take about 1 minute to read. Now triple that because you'll need to edit your recordings and it will take at least an additional 2 minutes of editing time for every minute of completed voice track.

We're now up to 6,000 minutes for recording and editing. Divide 6,000 by 60 minutes and you get 100 hours of work to complete this project. And that's assuming everything goes extremely smoothly.

But wait a minute! You take a look at the script, and you discover the trial had something to do with the biotech Industry and there are lots of technical terms sprinkled throughout the script. Better be safe and add another minute for each page to allow for mistakes and retakes.

Let's be conservative and estimate that it will take about 5 minutes of recording and production time for every minute of completed voice track. We're now at 10,000 minutes—or roughly 166 hours—or about 4 weeks! At their offering fee of $1,000 you'll be making a total income of about $6.00 per hour. The reality is that it will probably take 6 or 7 minutes for each completed minute, so your actual work may be more and your compensation considerably lower. Even if this client is willing to negotiate a higher talent fee, it will most likely not come even close to anything reasonable for the amount of your effort involved. And don't forget that you won't be able to work on anything else while you're recording this epic project. Although 50 cents per page may be a reasonable price for the client, after factoring in your time, it really isn't a very good deal for you. Is your time—and are you—really worth that little?

This example, based on an actual audition request, shows that you need to have a very clear understanding of what your involvement will be in a project before you can realistically discuss price. Unfortunately, there are some voice talent who only look at the offered fee and don't take the time to properly evaluate projects like this.

Before you can provide a realistic estimate, or discuss your talent fee with a prospective client, you need to know as much as possible about what you will be doing. You need to know the going market rate for comparable work, and you need to place a value on your time and performing abilities.

Example #2: The Short Session

Consider this: You've auditioned for, and landed a voiceover job for a 60-second radio commercial for a midsized market. The audition took you 10 minutes to record, edit, and send out. The job will pay $150, and based on the script, you expect it will take you about a half hour to record, edit, and deliver the final project. That works out to $150 for about a half-hour's work, or $300 per hour. Pretty good pay, right? Wrong!

That $150 gig may be the only job that came in that week—or that month. Let's say you spent 10 hours recording and sending out auditions before you got this job, plus another 5 hours on the phone and sending out email. Now, consider what you've spent on phone calls, postage, your website, marketing, training, and everything else that led up to this job.

Just taking into consideration the 15 hours you spent that week, you're looking at a gross income of about $10/hour for that $150 job. But don't forget that the IRS will want part of that income, so you'll actually net something in the neighborhood of $5–$7 dollars/hour for that $150 job.

The point here is that before you can negotiate a reasonable fee for your voiceover work, you must know the value of your time and talent.

THINK LIKE AN ENTREPRENEUR

If you haven't already, start right now thinking of your voiceover work as a business. Your objective as a business owner is to make a profit, and to do that you have to be smart about how you use your time and energy, and how you price your services. As voice talent, we may never be able to change the way producers think. However, we can control the way we think about what we do, and we can control what we charge for our services.

When one of my students asks about what they should charge, I suggest they first do some homework. Find out what the best Union voice talent would be paid for the same work. You've spent a lot of time, money, and energy getting yourself to the point where you can market yourself as a professional voice talent who can compete with the best talent out there. You deserve to be fairly compensated for your work. Even if you're booking your first job, that is no reason for you to undercut your worth. If you have the talent and ability to provide the same quality of work as a veteran voice actor, you should be compensated accordingly.

If you establish yourself as "working for cheap," you may get yourself into a rut that could be difficult to get out of later on. At the very least, it will be extremely difficult raising your fee for a client you've already worked for at a "bargain basement rate." In voiceover work, it's always easier to pull you back than to push you out. The same is true with your fees. You can always lower your fees, but it can be extremely difficult to raise them.

SETTING YOUR FEE

As with many things in this world, perception is a very large factor for determining value. The way you perceive your personal value as a voice actor will affect how you determine your fees.

If you are a member of AFTRA, SAG, or another performing artists union, your talent fees are set by your union. Through a process of collective bargaining, these unions have determined what are considered to be reasonable performance fees for different types of work. These *scale* fees are posted on their websites at **www.aftra.com** and **www.sag.org**. These posted talent fees are not negotiable and are considered as the lowest level of compensation. Signatory producers have agreed to pay the posted minimum fees, or a higher fee that might be negotiated by an agent. This is one advantage of being a union member—you know in advance what your base talent fee will be for any given type of work. Another aspect of being a union member is that you are automatically perceived by the talent buyer as having a certain level of expertise and professionalism.

If you are nonunion, you will need to negotiate your talent fee with your client at the time you are booked. But before you can begin any sort of realistic negotiation, you need to establish a *fee schedule* that outlines your specific fees for specific types of work.

Only you can determine your personal value as a freelance voiceover performer. The process begins by identifying the type or types of voiceover work you are best suited for. Once you've figured out what you do best, the next step is to identify the market price for comparable voiceover work in those areas. It used to be that you could simply make some phone calls in your city to gauge the current talent fees, but no more. The Internet has changed all that, and your market is now the world. When you submit an audition, you may have no idea what city the producer is in or how your recording will be used, so you may have no real information upon which to base your fee—yet most producers want you to provide a quote for the job you are auditioning. This is why it is important to establish your personal value as a voice talent.

You have made a major investment in developing your business and performing skills to get where you are. If your performing skills are at a level where you can effectively compete with other professional voice talent—and the fact that you are getting calls for work proves that you are—then why would you consider yourself any less professional than they are? Why should you accept a talent fee of anything less than other professional voice talent? *Low-balling*, or under pricing, your talent fees may get you the job, but the practice does a disservice not only to the voice talent accepting the fee, but to everyone else in the business as well. It tends to lower the bar, which can only result in lower quality work at cheaper prices. To get, and keep, the best clients, you need to develop a high perceived value, and provide excellent work at fair and competitive prices—not the "cheapest" price. You may be better off starting with a higher fee and negotiating to an acceptable middle-ground.

One way to set your personal talent fees is to use union scale as a starting point, even if you are nonunion. When negotiating with clients you can, of course, mention that you are nonunion and therefore can be somewhat flexible with your talent fee. By starting at union scale, you are telling your client that you are a professional and there is an industry wide value for the work you are being asked to do that needs to be appreciated. Where you go from there is up to you, and it's what the rest of the negotiation process is about. But you've got to start someplace. Here are some considerations as you set your personal talent fees:

- **Your experience and abilities:** How good are you at setting character quickly, finding the right interpretation, seeing the big picture, working as a team player, taking direction, etc? The more skilled you are as a performer, the more likely you will be able to demand a higher fee—especially once you have established a name for yourself and are confident with the work you do.

- **Prior experience and clients:** Have you already done some work for a few satisfied clients? If so, their names may help to establish credibility and thus help to justify a higher fee. Be sure to consider any recent work for inclusion in your demo, but make sure it's good enough in both recording quality and in performance quality.

- **The client's budget:** If you're nonunion freelance voice talent, you'll need to be flexible and decide if you want to work for a minimal fee (which is all that many small or independent producers are willing to pay). Keep in mind that local radio stations will often give away production and voice talent for free just to get an advertiser to buy time on their station, and many independent producers will offer to do the voice work themselves in an effort to save a few bucks.

Your challenge as a voice artist is to offer a service that is superior and more effective for the client than what they can get anywhere else.

- **Can you justify your fee?:** This gets back to your abilities. If you market yourself with professional print materials, a dynamite demo, and an awesome website, you had better be able to meet the level of expectations of your client when they book you for a session. If you give the appearance of an experienced pro, but can't deliver, word will spread fast and it may be a long time before you can overcome a negative image. The challenge in setting your fee is to match the fee to your abilities and the market, and still be within the range of other freelance talent, without creating an impression that you will "work cheap" or that you are "overpriced."

- **Consider your market:** Nonunion talent fees vary greatly from market to market. In order to set an appropriate fee for your services, you'll need to find out what other voice actors are getting paid in your area or for similar work. In your own city you can call the production department of local radio and TV stations, and advertising agencies to ask what they usually pay for nonunion work. However, if the work is out of your city, you may have no other option than to simply decide if the fee offered by a producer is worth your time and energy.

Your training is of less importance than your abilities as an actor. Of course, you must have a great sounding demo, but you need to have the abilities to match. Don't ever think you know all there is to know about working with voiceover copy. Continue taking classes and workshops, read books, and practice your craft daily.

WHAT ABOUT ALL THAT EQUIPMENT?

You do realize, don't you, that if you are recording professional voice tracks on your computer at home, you have a *home studio*? The operative word here is "studio." OK, so your investment in a computer, a microphone, audio equipment, and acoustical improvements may not amount to the hundreds of thousands of dollars a full-blown recording studio would spend—but the simple fact is this: you've got your own studio! Chapter 23, "Your Home Studio," discusses this in more detail.

From a business standpoint, it makes no sense to set a talent fee that does not at least take into consideration the costs of your studio equipment, office supplies, marketing expenses, training, demo production, and so on. Unless you're performing strictly as a hobby, you'll eventually want to recover all those expenses. One way to do this is to create a separate rate for studio time that you charge in addition to your performance fee.

An hourly fee for studio time is standard practice for virtually every recording studio—and, since you own your own studio, it only makes good business sense for you to use a similar pricing structure. If you have production skills and can offer additional production services, this can be a good way to create an additional revenue stream. Separating out your studio rate and talent fee also gives you some additional negotiating leverage because you can always discount one or the other if needed and still have enough income to make some profit.

However, if you are marketing yourself as strictly voice talent and you have limited production and editing skills, you may not want to take this approach. For you, the best way to factor your investment into your talent fee is to simply keep your fee a bit higher. As a nonunion voice talent starting your negotiation at union scale, you'll still have room to adjust your fee if necessary. Of course, your performing skills will need to be at a level where you can justify the higher fee. The important thing to remember is that just because you may be new in this business, it doesn't mean you need to charge unrealistically low talent fees.

BUYOUTS

Projects which are, by nature, limited in their distribution and use, such as industrial training programs, marketing videos, documentaries, and audio books can reasonably justify a *buyout* agreement, meaning that the producer pays you a one-time flat fee for your work, and then has the right to do whatever he wants with that recording for as long as he wants—without ever having to pay you another cent.

Take a look at the way AFTRA handles its talent fees and you'll notice that most categories have a time limit for the use of a performance. If a client wants to reuse a performer's work, they pay the talent a new fee called a *residual*.

As a freelance voice actor, you do *not* need to accept a buyout talent fee for any commercial voiceover work, even though this is the most common type of booking for nonunion talent. If you agree to a buyout fee for a radio commercial, there's a very good chance you may be hearing that commercial for years to come. Or a portion of the radio voice track may be used for a television commercial, an in-store message, a telephone message, on a website, or any number of other uses—and you'll never get paid a dime beyond your original buyout talent fee.

You can certainly negotiate a timeframe for the use of your performance. If your client agrees to this, you'll need the terms clearly stated in your agreement, and you'll want to create a follow-up system to remind your client of the agreement. Enforcing a reuse clause may be difficult as a nonunion voice talent, but if you don't include it in your original contract, potentially, you'll be leaving money on the table; and you'll have no legal recourse if or when your client reuses your work. Whether or not discussion of a reuse fee is appropriate will depend entirely on the needs of your client, your willingness to compromise, and your ability to "read" your client during the course of your negotiation.

THE AGREEMENT

Every booking is a separate business arrangement with unique time constraints, performance requirements, and payment terms, among other specifications. Whenever you exchange your time, energy, or services for money, the only way you will be protected is if all the details of the business arrangement are detailed in a contract. Depending on the type of project and its ultimate use, you may want to negotiate for certain conditions. For example, if the project is to be sold, you might negotiate a clause that includes a residual payment when sales exceed a certain number of units or you may want limitations on how long your voice track can be used in a commercial, or for which other kinds of media it can be used. Everything is negotiable! The goal is to reach an agreement that is mutually acceptable.

The manner in which your performance can be used, and the duration of its use, are most definitely negotiable points that you should consider and discuss with your client. There are no hard and fast rules here, nor are there any specifically worded contracts available. Every agreement is unique and you'll need to come up with the appropriate wording to describe the terms and conditions for the use of your recorded material. You'll also want to make sure you include adequate controls for tracking any restrictions, and possible remedies for any violations of the agreement.

Your agreement is a contract, and if the project justifies it, you may want to seek legal advice to make sure you are protected and receive the compensation you deserve. Of course, if you have representation, your agent will handle the details of any complex negotiation. The specific details that you might include in your agreement will be discussed later in this chapter.

Negotiating Your Fee

The best way to learn how to negotiate is to do it! If you've never done it, the best way to learn how is to study some of the many excellent books on the subject. A search for "negotiating" on **www.amazon.com** will bring up hundreds of books on this subject. Find one that looks good to you, buy it, study it, and begin practicing.

The ultimate purpose of any negotiation is to create an agreement that is acceptable for all parties. For voice talent, this agreement is ideally in the form of a written contract that is signed by both parties prior to the start of any work. It is a written description of the work to be done, the timeframe within which it will be done, who is doing it, the conditions for its use, the responsibilities of each party, the agreed-upon compensation, and the terms of payment, among other details.

Unfortunately, a great deal of voiceover work is booked on only a *verbal agreement*, which is only as good as the paper it's not written on.

When you begin work with only a verbal agreement, you take the chance of not getting paid, or of having serious problems of miscommunication, or worse. Always get your agreement in writing before you begin work.

Ideally, an agreement should be received in the mail, but with tight schedules, deadlines, and the popularity of email, this often isn't practical. A faxed document will work to get things started and in today's electronic age, most courts of law will accept an email agreement as a legal document provided it contains the sender's email address and name.

There are literally dozens of effective negotiating techniques that can be used to maintain high standards and fees for voiceover work. Here are just a few, with only a very brief explanation of how they might be used:

- **Talk about the project:** No matter what you know about a job when you get the call, it isn't enough. One of the first things you should do in any negotiation is to get more information. Ask as many questions as you can, while avoiding any discussion of your fee. When the subject of your fee comes up, divert the discussion by asking more appropriate questions. This requires extremely good listening skills.

- **Get the client to mention the first number:** This can take some skill, but it can often be achieved by simply engaging the client in a conversation and guiding that conversation to a discussion of what they have paid for voice talent for prior work. If your client is comfortable with you, they will often feel safe in talking about what they have paid in the past. At an appropriate point in your conversation you may ask what their budget is for this project and wait for them to answer.

- **Echo... Pause:** This is a technique for maintaining your fee, or perhaps even increasing it. It has been around for a long time and may not work in all situations, especially if the person you are negotiating with figures out what you are doing. If your client says he does not have a budget, it may be necessary for you to provide some education as to what will be involved for you to do the job. In this way you will be creating a perceived value for your work. The idea is to get your client to tell you how much he is willing to pay for the job in question. Whether he says so, or not, he has a number in mind. It's your job to coax it out of him. Let's say he mentions the number $200. You, in a very thoughtful voice, simply repeat the number as though asking a question to verify that you heard it correctly—then stop talking. Be absolutely silent. It may get uncomfortable, but don't speak. At some point the discomfort will be too great and your client will likely come back with something like ". . . well, we might be able to go to $300." At that point, you repeat the Echo and Pause. Usually by about the third time, your client will say something like ". . . $325—that's really all we can afford for this project." You can then use your best acting skills as you say ". . . $325! I can do that for $325."

- **Discount this fee for future work at full fee:** During your conversation, you may find that your client is reluctant to discuss any numbers, or that he truly has only a very limited budget for this project. If that is the case, you can tell him that you base your fees on union scale, but that as nonunion talent, you can be flexible with your rates. If the project appears to be something that might result in future work, you might even offer to discount your talent fee with the understanding that you will be paid your regular fee for future work. When you deliver an outstanding product, your chances of having a new long-term client will be very good. This technique can be a bit risky, so make sure you have your agreement in writing.

The desired outcome of any negotiation is to get paid for your work based on the terms of your negotiation. The challenge today is that the Internet has created an international marketplace. It is common to never meet, or even speak to, your client with everything handled through email. Even with a solid agreement in place, you still have no guarantee that you will be paid when you deliver your voice tracks.

Getting Paid for Your Work

THE DEAL MEMO LETTER

As a freelance voiceover performer, you need to protect yourself from unscrupulous producers (yes, they are out there). The simplest way to protect yourself is to use a written agreement known as a *deal memo*. Even if you are a union member, having a written agreement is a good idea. It protects you and outlines the details of your work. The format for this can be as simple as a brief letter, an invoice, or an email confirmation, to something more formal, such as a multipage contract for services. It's generally a good idea to keep a deal memo as simple as possible. A complicated, legal-sounding document might scare off a potentially valuable employer.

Preparing your deal memo should be the first thing you do when you book a session. A written agreement is your only proof in the event you need to take legal action to collect any money owed to you, or if your performance is used in a manner that you did not agree to. It's a common practice and should be used whenever possible. Make sure you have a signed agreement in hand *before* you begin any work.

The following is an example of a simple deal memo letter. This deal memo includes all the necessary information to confirm the agreement, yet it is presented in a nonthreatening and informal manner. With minor modifications, this letter could be used for either a studio location session or one that you record in your home studio.

Dear Mr. Producer:

Thank you for booking me to be the voice for The Big Store's new radio commercials. As we discussed on the phone today, I will be doing four (4) radio commercials (including up to 6 tags) for $350 per spot as a limited run 90-day buyout for radio only. If you later decide to rerun the commercials, or use my voice for television spots or other purposes, please call me to arrange for a new session or to modify our agreement. As we discussed, my fee for each additional tag after the first 6 will be $75 per tag.

You have also agreed to provide me with a recording of the final commercials. I'll call you next week to arrange to pick up a CD or you can send an MP3 file to my email address.

As we discussed, I will keep your credit card information on file to guarantee the session.

I will arrive at Great Sound Recording Studios, 7356 Hillard Ave. on Tuesday the 5th for a 10:00 AM session.

For your records and tax reporting, I will bring a completed W-9. Please make your check in the amount of $1,400.00 payable to My Name so that I can pick it up after the session. Should you prefer that I charge your credit card, please let me know so I can bring the proper paperwork with me to the session.

I look forward to working with you on the 5th.

Sincerely,

Getting paperwork out of the way before the work begins is a good way to make sure that the terms of your performance are understood by all parties and that the producer doesn't try to change the agreement or add additional production after you have done the work. If you are booked early enough for a session at a recording studio, you might want to fax a copy of the agreement to the producer in advance. But you should still plan on having two copies with you when you arrive for the session—the producer is probably not going to bring his copy. Leave one copy for the producer and make sure you have a signed copy before you leave the studio. If you're recording at your home studio, you can do everything via email and fax. Although an email confirming the details of your work may be considered a legal contract by some courts, it is still a good idea to use your own document and get a written signature.

How to Guarantee You'll Be Paid

Most voiceover work is due and payable upon delivery, but that usually doesn't mean you walk out of the studio with cash in hand or have money in the bank immediately after uploading the files. For many clients, you'll need to send an invoice that states "payable on receipt." Even with that, you may still end up waiting 30 to 90 days before you receive payment. That's just the way some businesses work.

If you don't want to wait to be paid, there are other options available:

- Insist on clients sending a deposit for talent fee and studio time to be paid in advance with the balance to be paid on delivery. If your client doesn't pay as agreed, at least you'll receive a partial payment.

- Ask your client to make payment through an online payment service like **www.paypal.com** or **www.rbsworldpay.com**. You'll need to set up your own account with these services, which can easily handle credit card payments or your clients can set up their own accounts.

- Set up a *merchant account* for your business so you can accept credit cards. Most small businesses can have a merchant account, including individuals operating as a sole proprietor. A merchant account is easy to establish, but it does have a variety of associated monthly and per-transaction fees. If you are only booking occasional work, you would be better off using one of the online payment services.

- Deliver a partial project (75–80%) or deliver a *watermarked* project for approval, and only send a complete, clean copy upon receipt of payment. A *watermark* is a tone, or sound embedded in your audio that effectively makes the recording unusable, but will allow the client to determine if it otherwise meets their needs.

The specific payment arrangements may be different with each client. If you've never worked with a client before, there is no track record upon which to build trust, so it is reasonable to request a deposit or use one of the above techniques for getting paid. It a client is repeat business, it might be reasonable to invoice them with the payment due net 15 days.

Here's how I work with my clients to make sure I get paid: I have a merchant account so I can accept payment with Master Card and Visa credit cards. I have a stated policy on my website that says I require a valid credit card number at the time of booking to guarantee a session. When booking a session, I take my client's credit card information and run a verification to make sure the card is valid for the amount we've agreed upon, but I do not charge the card yet. Instead, I tell my client that their credit card will *not* be charged until they have approved my work. Before I send my voice tracks, I'll call my client to let them know the session is ready to deliver and to ask how they would like to make payment.

Since I already have their credit card number, most clients simply ask me to charge card. As soon as the transaction is processed, their payment is electronically transferred to my bank account. If, for some reason their card is rejected (after it was originally verified), or the payment bounces, I have legal recourse and a contract. I have some additional protection in that I don't deliver a clean copy of the work until their payment clears. I have never had a client question this policy, nor have I ever had a problem with a credit card transaction for payment of services. The peace of mind I have in knowing I will be paid for my work makes the discount fees and other minimal charges for maintaining a merchant account well worth the price.

Booking Confirmation

Agreement for Voice Talent, Recording and/or Creative & Production Services

Please confirm your booking by signing and returning this agreement by Fax or E-mail.
This agreement for services is between _____ (*Actor*) and the Company or individual named below (*Client*). Your signature constitutes agreement to the terms and fees indicated below. All dates, terms and conditions indicated shall apply until changed by a new agreement. Client agrees that the Laws of [Your State] shall apply to this agreement, and Client agrees to submit to the jurisdiction of [Your State] to resolve any dispute that may arise as a result of this Contract. Services provided under this agreement are not to be considered as "work for hire."
We require a signed copy of this agreement and a credit card number, PayPal payment, or company purchase order number before we can begin your session. We accept payment by Visa, MasterCard, check, or PayPal. Payment in full is due upon completion of services unless noted otherwise.

Today's Date:	_____	Contact Name:	_____
Session Date:	_____	Company:	_____
Session Time:	_____	Address:	_____
On-site or Off-site:	_____	City/State/Zip:	_____
Director (if any):	_____	Phone:	_____ Fax: _____
Phone Patch/ISDN:	_____	E-mail address:	_____
Project #:	_____	Project Title:	_____
Project Details:	_____		
Delivery format:	_____	Delivery Method:	_____

> ➢ **Client copy is due 24 hrs. prior to session. Email copy to _____** ◅
Our standard page format is 12pt Arial, double spaced, 1" margins all around

Contracted Services & Rates:

Voiceover Talent: _____
Talent Fee per voice for: Principal: $____ Secondary: $____ Tag: $____ Character: $____ $_____
Page/Project Rate: (includes talent fee, studio time, and editing/production as requested) $_____
Estimated Studio Time: _____   Hourly rate (estimated): $_____
Music: ___ N/A ___ Library ___Custom/Original   License fee (estimated): $_____
Creative/Writing/Production: _____   Rate: $_____
FTP/Matls/Delivery: ___FTP ___CD/DVD ___Other _____   Delivery/Matls. Fee: $_____
PROJECT ESTIMATE: _____   **TOTAL CHARGES:** $_____
DEPOSIT REQUIRED: ___Yes ___No   A deposit is required to guarantee this session: $_____
BALANCE DUE: **This is your estimated balance due. Payable upon delivery of completed project:** $_____
COPY OF WORK: Client agrees that Actor may obtain a copy of the completed work for use in a demo, website, or in other forms of marketing and/or promotion of the Actor's services.
CANCELLATION: Either party may cancel this agreement with at least 24 hours notice by phone or in writing. A minimum $100 session fee plus 50% of above talent fees shall apply to any session cancelled by client with less than 72 hours notice. If cancelled with less than 24 hours notice, the full estimated fee shall apply.
RE-TAKES/DO-OVERS: A minimum $50 session talent fee, plus studio time, shall apply for client changes after sign-off and delivery of original session. Additional talent fees may apply depending on the revisions required.

PREFERRED PAYMENT: ___Credit Card ___PayPal ___Invoice ___Check (PO# required for check & invoice)
A company PO#, PayPal payment, or valid credit card is required to guarantee your session. Deposits may be paid using PayPal or your credit card. Your card will NOT be charged until work is approved and completed project is delivered. Upon completion, your card will only be charged if that is your preferred payment method, or if an invoiced payment is not received within the terms of this agreement. Payment for invoiced work using your PO Number is due upon receipt, Net 15 days. All credit card information is destroyed upon receipt of payment. Your signature below signifies agreement to these terms and conditions.

____ PO or REF #: _____   Payment Terms: _____
____ Credit Card #: _____   Exp: _____ CVC Code: _____
Name on Card PLEASE PRINT: _____
Billing Address for Card: _____
Signature: _____   Date: _____

Please include or verify all credit card information, sign and date, then fax to _____

Figure 21-1:
Example of a Booking Agreement. A form of this type can be used to gather all the details of a session booking and can also serve as a written agreement, or contract, between the client and voice actor. Using this form for both purposes simplifies the communication process and ensures that the details of the booking and payment are mutually understood.

Some clients will prefer to not provide their credit card number. For these clients, I'll request a deposit or payment be sent to my PayPal account. As with a merchant account, PayPal will charge a transaction fee but there are no monthly fees associated with the account. PayPal (and other online payment services) will accept credit cards, but the card number is never revealed to the recipient. These accounts use your email address for payment notifications and associate your online account to your regular bank account. This makes it easy to transfer funds between the accounts and your online history will give you an accurate record of payments received.

Some clients, however, still prefer to pay by company check. If they do, I request a *purchase order* number to guarantee the session. A PO number is a record of transactions that is kept by the company and used to allocate funds for specific purchases. A purchase order number is as good as a contract. I use their PO number as a reference number on my invoice, and my invoice will state "payable upon receipt." If I don't receive payment within a reasonable period of time (as stated in our Deal Memo or Booking Agreement), I'll call my client to follow up on the payment. If it appears that they are delaying payment, I can still charge their credit card, since I don't destroy that information until after I have the money in the bank. Oh, and if your client is in a foreign country, make sure their payment is in U.S. dollars.

CREATE A BOOKING AGREEMENT FORM

I've developed a form that includes a lot of information about the client, the work I'll be doing, the delivery method, my talent fee, my studio charges, and anything else that applies to the project (see Figure 21-1: Example of a Booking Agreement). I'll fill out the form during the booking conversation and either fax or email a copy of our agreement for them to sign and return. A faxed or emailed PDF copy is good to confirm the session, but I'll also ask that they mail an original to me. The signed agreement and either their credit card information, a PO number, or a deposit constitutes a confirmation of the booking. With that in hand, I'll start recording and complete my part of the agreement.

Note that in my example, there is language that places the contract jurisdiction in the state where you are doing the work. This legal detail may become important if you are working for an out-of-state client who refuses to pay you for your work. By having jurisdiction in your state, you will be able to sue without having to hire an attorney in the state or city where your client lives.

OTHER TYPES OF AGREEMENTS

Some larger companies, such as major radio and TV stations, will not accept or sign a performer's deal memo or contract. These, and other reputable businesses, often have their own procedures for paying talent. If you want the work, you may need to accept their terms. However, you can still insist that you have a written agreement in place and even include a clause that places the contract jurisdiction in your state. With contracts, everything is negotiable until the agreement is signed!

You will be asked to provide your social security number and sign their document before you can be paid. If you are not offered a copy, you should request one for your own records in case payment is delayed. You usually will not be paid immediately after your session, but will receive a check in the mail within four to six weeks. If you have representation, this detail will be handled by your agent. However, if you are working freelance, some producers and large companies may take advantage of a 30-day payment agreement by basing the payment terms on 30 working days rather than 30 calendar days. This can result in your payment arriving long after you expected it. Some companies will even take as long as 90 days or more before mailing your payment. In some cases, this may be due to your client awaiting payment from their client before they can pay you. But often, this delay is simply so the company can hold funds in their account as long as possible. If you have not received your payment by the agreed upon time, it is up to you to call your client and gently remind them.

Another common problem with working freelance is that you can do a session today and be called back for changes tomorrow, but unless you are redoing the entire spot, the producer may expect you to do the pick-up session for free. If you don't like working for free, you should consider including this contingency in your deal memo or booking agreement.

When you are called back to fix a problem, the callback session is technically a new recording session. As a union performer, the producer must pay you an additional fee to return to the studio. As a freelance voice actor, it is up to you to negotiate your fee for the second session or provide for this contingency in your original agreement. Unless the problem was your fault, you should be paid for the follow-up session. The producer must be made to understand that you are a professional and that your time is valuable. You are taking time away from other activities to help fix their problem and you are entitled to fair compensation. A good producer knows this and expects to pay you for the additional work.

If you didn't include pick-up sessions in your original agreement, try to find out what needs to be fixed before you begin talking about how much you should be paid for the new session. If you are redoing most of the copy, you might want to ask for a fee equal to what you charged the first time. If the fix is simple, you might ask for one-half the original session fee. If you are exceptionally generous, and expect to get a lot of work from the client, you might offer to do the new session for free. If you do negotiate a fee for the follow-up session, make sure you get it in writing in the form of an invoice, a new deal memo, or a copy of their paperwork.

Union Compensation

The purpose of this book is to give you the tools and information you need to build a business and succeed as a voice actor. As with most major industries, unions play a role in establishing working conditions, benefits, and compensation. Joining a performing union is a personal decision that should be based on complete and accurate information from all points of view. I neither encourage, nor discourage union membership, but I do believe it is important to know how unions may have an impact on your voiceover work, whether you choose to not join, join at Financial Core, or join as a Rule 1 member. Most of the information in this section can also be found on the AFTRA (**www.aftra.com**) and SAG (**www.sag.org**) websites and in their print materials. You can find additional discussions on the pros and cons of union membership on many of the voiceover discussion boards and blogs.

By joining AFTRA or SAG, and working union jobs, you will be assured of reasonable compensation for your talents and protection from unscrupulous producers and advertisers. Your union-sanctioned agent will normally handle negotiations for your work and will sometimes negotiate a fee above scale. Regardless of what you are paid, the agent will only receive 10%, and that amount is usually over and above your fee. With AFTRA the "plus-10" (plus 10%) is automatic. With SAG it must be negotiated, or the 10% agent commission will be taken out of your fee. A performer just starting in the business may make less than scale, but the agent's commission will still be added on top of the performer's fee. The client also contributes to the union's health and retirement (AFTRA) and Pension Welfare Fund (SAG). For many voiceover performers, the health and retirement benefits are the primary advantage of being an AFTRA or SAG member. However, both unions have minimum income requirements before a member can qualify for Health and Retirement benefits.

Residuals were implemented to guarantee that performers are paid for their work as commercials are broadcast. Each airing is considered a separate performance and talent are compensated based on the period of time a commercial is aired. Commercials produced by an AFTRA or SAG signatory have a life span of 8 or 13 weeks. After the original run, if the advertiser reuses the commercial, a new life span begins and the performer's fees, agent commission, and union contributions must be paid again. This happens for every period in which the commercial is used. In radio, residuals begin on the date of the first airing. In television, residuals begin on the date of the recording session, or the "use" date.

If an advertiser is not sure whether the company wants to reuse an existing radio or television commercial, a *holding fee* can be paid. This fee, which is the equivalent of the residual fee, will keep your talents exclusive to that advertiser, and is paid for as long as the spot is held. Once the commercial is reused, residual payments are made just as for the original run. If the advertiser decides the spot has lived its life, your residuals end. At that point, you are free to work for a competing advertiser.

Union recording sessions are divided into several fee categories and specific types of work within each category. For radio and television work, the performer's pay varies depending on the type of work and the market size where the product will be aired. The following is a description of the basic AFTRA performance fee categories. Although some of the details may change from time to time, this will give you an idea of the broad range of work available in the world of voice acting.

- **Session Fee:** The session fee applies to all types of union voiceover work and will vary depending on the type of work you are doing. A session fee is paid for each commercial you record. For radio and TV commercials, an equal amount is paid for each 13-week renewal cycle while in *use* (being rebroadcast) or if the spot is on *hold* (not aired).

   Session fees for dubbing, ADR, and looping are based on a performance of five lines or more, and residuals are paid based on each airing of the TV program.

   Animation voice work is paid for individual programs or segments over 10 minutes in length. Up to three voices may be used per program under one session fee. An additional session fee applies for each additional group of three voices, plus an additional 10% is paid for the third voice in each group of three voices performed.

   For off-camera multimedia, CD-ROM, CDI, and 3DO, a session fee is paid for up to three voices during a four-hour day for any single interactive platform. Additional voices are paid on a sliding scale and there is a one hour/one voice session fee and an eight-hour day for seven or more voices. Voices used online or as a lift to another program are paid 100% of the original session fee.

   Industrial, educational, and other nonbroadcast narrative session fees are based on the time spent in the studio. A day rate applies for sessions that go beyond one day.

- **Wild Spot Fee:** Paid for unlimited use of a spot in as many cities, for any number of airings, and on as many stations as the client desires. The Wild Spot *use rate* is paid based on the number and size of the cities where the spot is airing.

- **Tags:** A *tag* is defined by AFTRA as an incomplete thought or sentence, which signifies a change of name, date, or time. A tag can occur in the body of a radio or television commercial, but is usually found at the end. For radio, each tag is paid a separate fee.

- **Demos:** "Copy tests" for nonair use, paid slightly less than a spot fee. An advertiser might produce a demo for a commercial to be used in market research or for testing an advertising concept. If upgraded for use on radio or TV, the appropriate *use fee* applies.

- **Use Fee:** This fee begins when a commercial airs. Voiceover performers for national TV spots earn an additional fee every time the commercial airs. A standard of 13 weeks is considered a normal *time-buy* that dictates residual payments.

PRODUCT IDENTIFICATION

Radio and television commercials are unique in that they both create an association between the performer and the product. This is most common when an advertiser uses a celebrity spokesperson to promote their product. The viewing audience associates the performer with the product, and the advertiser gains a tremendous amount of credibility.

Product identification can, however, result in some serious conflicts, usually for spots airing in the same market. For example, if you performed the voiceover on a national television commercial for a major furniture store, you may not be able to do voiceover work for a local radio commercial for a competing furniture store. You will need to make sure both spots are not airing in the same market, even though one is for radio and the other is for TV. Conflicts are not a common problem, but they do occur from time to time and usually with union talent. As usual, if you have any questions, the best thing to do is to call your union office.

LIMITED RELEASE PRODUCTIONS

Many projects are never broadcast, such as in-house sales presentations, training tapes, programs intended for commercial sale, and point-of-purchase playback. For most of these projects, performers are paid a one-time-only session *buyout* fee with no residuals. These projects usually have no identification of the performer with the product or service in the mind of the audience, and therefore present little possibility of creating any conflict.

Documenting Your Session

Now that your booking is confirmed with a signed agreement, you're ready to get down to work. You've already got a good idea of what you need to do, you've got a general idea of what will be involved to complete the project and deliver it to your client, and best of all, you are confident that you will be compensated for your work. However, Murphy's Law will inevitably come in to play as some time or another. Anything that can possibly go wrong... will! And it will happen at the least opportune time.

You need to be prepared for Mr. Murphy and one way to do this is to document the time and details of your work through the use of a Work Order or Session Booking Form (see Figure 21-2: Example of a Session Booking Form).

Session Booking Form

Pg: _____ of _____

Client Name: _____ Start Date: _____ Time: _____
Company: _____ Day: _____ Invoice #: _____
Address 1: _____ Terms: _____
Address 2: _____ Confirmed: _____
City/St/Zip: _____ Phone: _____
Project Title: _____ Fax: _____
Project #: _____ Alt/Cell: _____
Ref. PO #: _____ E-Mail: _____

Description: _____

Talent Fee: _____

DATE:	IN:	OUT:	WORK DONE – STUDIO TIME/TALENT FEE:	TIME:	UNIT:	AMOUNT:	OFFICE:

Miscellaneous: **Delivery/Format:**

DATE:	QTY.	ITEM:	AMOUNT:	FORMAT/DATE:	VIA:	AMOUNT:
		ISDN/Phone				
		Outside Studio Fees				
		Travel				
				TRACKING NUMBER		

Comments – File Name – Contracted Talent:

Billing:

	RECEIVABLE:	FEES PAYABLE:	PAYABLE TO – CHECK NUMBER – DATE PAID:
Talent Fee:		/////////	
Studio Time:			
Miscellaneous:			
Shipping:			
Sales Tax:			
Other Income:		/////////	
Other Expense:	/////////		
TOTAL:			**Net:**

PAYMENT: Date Rcvd: _____   Check #: _____   Other: _____

Figure 21-2:
Example of a Session Booking Form or Work Order. This form can be used to document recording time and other expenses of a booked project which can then be summarized in your invoice for services. This form and the Booking Agreement should be kept on file for future reference.

You can call the form whatever you like: Work Order, Session Booking Form, Time Sheet, or a name of your own creation. The purpose of the form remains the same, and that is to document the time and processes that comprise the project you have been hired to complete. Virtually every service business uses some sort of documentation for the work they do. As a voice actor, and a good business person, you should do no less.

Everything you note on your Session Booking Form should be directly related to an aspect of the project you are working on, most of which should be part of your chargeable fees. Of course, you can choose to not charge for certain things or bundle items for a single fee, but the idea is that this form will give you a way to keep track of what you did for any given booking.

On the surface, this may seem like extra work or may even appear as completely unnecessary. However, when you document your sessions you will have a reliable negotiating tool for future bookings. For example, you may have contracted for a specific fee, but during the course of recording the project you discover that it actually requires considerably more time to complete, or there were some things that you neglected to include in your negotiations. By documenting the session you will be in a position to discuss those issues with your client when they book you in the future, and you will be better armed to discuss the realistic production requirements with other clients who might wish to book you for similar projects.

Some of the things to keep track of on your Session Booking Form are:

- Rehearsal time
- Copy editing or creative writing (if applicable)
- Consultation calls with your client
- Time spent to research pronunciation
- Time spent casting other voice talent (if applicable)
- Talent fees for other voice talent (if applicable)
- Studio time used for voice track recording
- Studio time used for editing
- Studio time required for file conversion, burning to a CD, or uploading to an FTP site
- Studio time required for pick-ups and subsequent delivery
- Time spent researching music and sound effects (if applicable)
- Music licensing fees (if applicable)
- Postproduction editing and mixing (if applicable)
- Other related items that may come to mind

As you can see, a Session Booking Form can be a very useful tool that can ultimately help you to identify ways to work more efficiently and even help you increase your revenues through a better understanding of exactly what it takes to do what you do. The form in Figure 21-2 is a simplified design based on the Work Order we use at VoiceActing, LLC. Our session work order has been refined and honed over more than three decades to a point where it perfectly fits with the way we handle our recording sessions.

Voice Acting Expense Report

Use this expense report on a weekly basis to track round-trip mileage for classes, sessions, errands, and other business-related expenses.

DATE	DESCRIPTION	START MILEAGE	END MILEAGE	MEALS	OTHER	TOTAL
TOTALS						

Week of: _____

Entertainment:
NOTE: Attach all receipts to this expense report.

DATE	PERSON(S) ENTERTAINED	BUSINESS PURPOSE	PLACE	TOTAL

Figure 21-3: Sample expense report for documenting business travel and other expenses relating to your voiceover business.

Keeping Records

As an independent businessperson, whether you have an agent or not, you need to keep complete and accurate records of income and business-related expenses well beyond just what you do for a particular session. This is not just for your tax records, but also so you have a way of tracking your career as a professional voiceover performer. Consult a tax advisor as to the best way to set up your record-keeping or refer to some of the many books or computer software on the subject.

You will want to keep records of clients you have worked for, what you did for them, and when you did it. When you get called by a producer you worked for last year, you can avoid undercharging by checking your files to see what your fee was last time. You can also use these records for future promotion and reminder mailings. A simple scheduling book can serve the purpose nicely, or you can even set up a database on your computer. Personal money management computer programs are another excellent way to keep records. Prices range from under $50 to several hundred dollars.

Under the current tax code, just about any expense you have that directly relates to your business can be deducted as a business expense. Even if you work another full-time job, you can still deduct expenses that directly relate to your voiceover business, providing you are operating under standard business guidelines and not doing voiceover as a hobby.

Depending on your situation, you may want to obtain a business license in your city, and eventually may want to incorporate. Setting up a legitimate business entity may have certain tax advantages. A tax advisor can help you with these decisions. In Chapter 22, "Shedding Light on the Dark Side of Voiceover," Robert Sciglimpaglia, a voice actor and practicing attorney, provides a detailed look at the legal aspects of working in voiceover

The following are some of the things you should keep records of:

- **Income**—Keep separate account categories for income from all sources of income received.
- **Expenses**—The costs of doing business.

  Taxes and deductions: Document anything deducted from your pay, including income taxes, social security taxes, Medicare taxes, state disability taxes, union fees, and any other deductions from a paycheck.

  Demo production: Keep track of payments for studio time, costs and materials, duplication, printing, letterhead, business cards, envelopes, postcards, résumés, and CD labels.

  Telephone: Keep track of phone calls made to prospects or your agent, especially any long-distance charges. You might consider a separate phone line to use exclusively for your business. If you have a cell phone or pager, these costs may be deductible as well.

  Website: The costs of registering your URL (domain name), website hosting, and website design are all deductible expenses.

Internet access: The portion of your telephone bill, cable bill, or DSL bill that applies to Internet usage may be a deductible expense.

Transportation: Keep a log book in your car and note the mileage for all travel to and from auditions, sessions, and client meetings. Include parking fees. (See Figure 21-3: Voice Acting Expense Report.)

Other business expenses: Keep track of postage, office supplies, office equipment, computer equipment, and other supplies. The IRS tends to view computers as personal equipment, rather than business equipment, unless the use is well-documented. Identifying your computer as an "audio workstation" may be a more accurate business description of how your computer is used.

Classes, workshops, and books: Classes, workshops, and books may be deductible as expenses for continued education and training in your chosen field.

In-home office: Deducting a portion of your mortgage or rent, and utilities for an in-home office, although legal, may trigger an audit by the IRS. Consult a tax advisor before taking this deduction.

Two excellent resources for software to manage your voiceover business are Performer Track (**www.performertrack.com**) and Pro Talent Performer (**www.protalentsoftware.com**). Both are popular with both on-camera and voiceover actors. An Internet search for "contact management software" will reveal many other options worthy of consideration.

Banking and Your Business

You may want to set up a separate checking account for your voiceover business and perhaps use accounting or money management software on your computer. This can help to keep all the financial aspects of your business in one place and simplify your tax preparation. The bottom line is that, as a professional voice actor, you are in business for yourself whether you work another job or not. As a business person it is important that you keep accurate records of your business-related income and expenses.

As with business management software, there are numerous options for financial management. Quicken (**www.quicken.com**) and QuickBooks (**www.quickbooks.com**) are among the most popular. Quicken is designed for managing personal finances, while QuickBooks has much greater flexibility and is intended for managing business finances.

22

Shedding Light on the Dark Side of Voiceover

Robert J. Sciglimpaglia, Jr. Esq.

INTRODUCTION BY JAMES R. ALBURGER

One of the least understood areas of voiceover lies in the shadows of the legal world. Yes, lots and lots of laws directly apply to our work as voice actors. Understanding how these laws might affect our work is very important. And, as with most laws, ignorance is not an acceptable excuse.

Now, I have extensive experience with legal matters, and I've been told that I can write "legalese" with the best, but I'm not a lawyer. However, I did once play the role of an attorney in a voiceover project. So, as accurate and well-intentioned that any legal advice I provide in this book may be, the fact remains that I cannot be considered an authoritative source.

So, I asked voice actor and attorney, Robert Sciglimpaglia, if he would be willing to discuss the legal side of voiceover and dispel some of the many myths that surround our business. Robert's knowledge of the law and his legal expertise far exceed mine, and the fact that he is a practicing attorney means that the information contained in the pages that follow may be considered as accurate legal advice. However, as with anything legal, a review and confirmation from your own attorney would be wise.

Because this chapter deals with legal issues, the tone may appear to be somewhat negative. Please keep in mind that laws are generally intended to protect an injured party, and will usually provide some sort of penalty to the offending party and compensation to the injured party. On the surface, this can appear negative, but the reality is that laws are what they are.

I'll admit this chapter will not be the easiest to read and it may take a few times through to comprehend all that is here. However, the information contained within these pages is critical if you are going to be successful in voiceover. Take your time with this one, and if your brain starts to hurt, just take a short break. Or, maybe read a good book… out loud, of course.

ABOUT ROBERT J. SCIGLIMPAGLIA

Robert J. Sciglimpaglia, Jr. is a practicing attorney, as well as voiceover artist and on-camera actor in the New York City area. He is the owner of All in One Voice, LLC (**www.allinonevoice.com**), a company that helps voiceover artists and actors with business and legal issues.

Robert has appeared in numerous national voiceover projects, including commercials, promos, and narrations, plus on-camera appearances for television and feature films. For more about Robert, visit his website at **www.robertpaglia.com** and his Internet Movie Database listing at **www.imdb.com**. Although the recommendations and advice in this chapter are based on Robert's knowledge of the law and experience as a practicing attorney, it is highly recommended that you consult your own attorney or qualified tax professional with any questions you might have regarding legal, business, or taxation questions.

Legal and Business Issues in the Voiceover Industry

Due to the growing popularity of the home studio, the voiceover industry is quickly becoming a legitimate and extremely fun way to generate income from one's home. The growth of the home studio has also allowed a diverse cross-section of the population to get involved with the fun. In this writer's experience, many people are now entering this field in retirement, or as a second career.

Fun aside… just like any other home-based business, the "business" side must not be ignored or some serious tax and legal consequences can result. This is especially true for those entering voiceover as a "second career" who may have built up some assets during their prior career.

Like many other businesses, the voiceover industry has its own set of legal and business issues, which this chapter will examine. Of course, this chapter is not meant to be an all-inclusive discussion of every legal and business issue associated with the voiceover industry. Rather, it merely scratches the surface, and some of the issues discussed herein are common to any business regardless of the industry.

Although there are no requirements to actually take an exam and become "licensed" to do voiceover work, as there are for other professionals like attorneys or doctors, there are many jurisdictions that require any business to obtain a license to operate within that jurisdiction. Voiceover, if one is undertaking it as a business and not a hobby, is no different. These licenses are more of a general license where you need to register with either a local or State authority, usually by paying a fee or tax, and filing annual or other regular reports.

In addition, another regulatory compliance that must be investigated by a voice talent starting a voiceover business are the local zoning regulations.

If a voiceover artist plans to record from their home, then they must check to see if businesses are allowed to be run from their home. Some jurisdiction's zoning laws are very strict and do not allow *any* form of business being run out of certain residential zones. If that is the case where the voice talent lives, then they will either have to ask for a variance, or set up a studio somewhere outside of their home by renting space in a mixed or commercial use zone. This is obviously something a voiceover artist will need to know so they will know how much capital they will need to start up their business.

Due to the local nature of business licensing and zoning laws, the author highly recommends a voice talent either research this on their own with their local authorities, usually the Department of Revenue or Taxation Board and local zoning boards, or hire competent legal advisors to assist in this process. Many States also have special departments you can contact that assist small businesses with start-up issues such as these

To Incorporate... or Not

The question about whether or not to incorporate has been subject to much debate in the voiceover community. It is this writer's firm belief however, that there should be no debate about it whatsoever. The question should not be *whether* to incorporate, but *how best* to incorporate.

Just like engaging in any type of business venture, a decision needs to be made about what "form" the business will operate. Will it be a sole proprietorship; a Limited Liability Company (LLC); a C Corporation; or an S Corporation? Normally, voiceover artists don't even consider their business formation and commence operating as what this writer terms "a sole proprietorship by default" as they just jump into the business, willy-nilly, without a plan. They will start sending out their wonderful professionally produced demos as soon as they receive them, set up their websites, MySpace pages, sign up for one of the many online audition sites, commence auditions, and so on, without even giving a second thought as to how their business is structured. This may be because some people getting into voiceover treat it as more of a hobby than a business.

The logic is "until I start to earn some money at this, what's the point?" However, this same logic doesn't prevent many of those voiceover artists from doing business under a "trade" or "stage" name, like "ABC Terrific Voice", or some other catchy phrase. This is what is referred to in the legal world as a "DBA" or "Doing Business As." Nor does the logic prevent voice talent from spending thousands of dollars on home studio equipment, training, production of their demos and marketing efforts. All expenses that the talent believes will pose no problem deducting from their income taxes.

Even after many voiceover artists do start to earn income in the business, they still don't get around to setting up a more formal structure and they only think about such things around tax time, or if they are ever subject to a legal action.

In this writer's opinion, the cost of setting up a formal entity, like an LLC or Corporation, far outweighs the potential cost involved in defending a lawsuit or action by the IRS and, in some cases, will also result in tax savings, depending on the revenues generated from the voiceover business. In fact, an LLC is extremely simple to set up in almost all States and, unless a voiceover artist has employees working for the LLC, it doesn't even require a separate Employer Identification Number (EIN) like a corporation would need.

Further, the IRS makes a very serious distinction between a "hobby" and a "business." If the IRS determines that someone is pursuing voiceover as a hobby, then it will *disallow* any deductions for any voiceover related expenses, like production costs of the demo, home studio costs, workshop or conference expenses and so on. The number one factor that the IRS says you should ask yourself concerning whether an activity is considered a hobby or a business is "Do you run the activity in a businesslike manner?" In this writer's opinion, operating your voiceover business as an LLC or Corporation *certainly* would go a long way in convincing the IRS that your voiceover activity is a bona fide business rather than just a hobby.

One of the many other advantages of setting up an LLC applying specifically to voice talent is that, in most States, it eliminates the need to file a "Trade Name Certificate" or "Fictitious Name Certificate." Most States require individuals who are operating under a name other than their proper legal name as a sole proprietor to file a form with a designated governmental entity. This includes names like "ABC Terrific Voice" or a "stage name" the voice actor uses that is different from their legal name.

Filing an LLC eliminates this requirement in most jurisdictions because the LLC paperwork is filed with the Secretary of State and is deemed to be notice of a "trade name" in that State. This would equally apply to doing business as either an S or C corporation. The failure of a sole proprietor to file such a "trade name" certificate can result in punitive damages, and can even result in criminal proceedings in some jurisdictions.

In short, some form of corporate structure should be the first step in protecting a voiceover artist's personal assets in the unfortunate event that they are sued, and eliminates the problems mentioned above concerning operation as a "DBA." To gain maximum protection, I would recommend operating under a corporation rather than an LLC, but either of these entities is much preferable to operating as a sole proprietor.

As a voiceover artist I agree, in theory, with the statement I often hear from other voiceover artists that "this is a liability-free industry." As an attorney, however, I *know* better than that! The sad fact in the United States these days is that whenever money changes hands or an injury occurs the potential for a lawsuit exists.

Potential Liability Issues
Specific to the Voiceover Industry

CELEBRITY IMPERSONATING

One area of the industry that voiceover artists should be cognizant of is celebrity impersonating. Celebrity impersonating falls under the auspices of the area of law known as "right of publicity" laws. The right of publicity is the right of an individual to commercially exploit their name, voice, signature, photograph or likeness. A handful of States have specific laws concerning the "right of publicity" and some other States that do not have a statute follow the common law rules concerning the right of publicity.

"Right of publicity" laws would allow a celebrity to sue a voice talent who impersonates their voice for commercial purposes. Nevada's statute, however, specifically exempts impersonators from liability for infringement of a celebrity's right of publicity.

Such an exemption does not exist in other State's statutes, however, so a voiceover artist must always be alert when asked to impersonate a celebrity as to how the impersonation will be used. In general, the First Amendment allows certain uses of impersonations, but generally not when those impersonations are meant to generate profits. Such profit-making use most certainly can expose both the voice talent, and the producer to a lawsuit.

This goes for impersonating celebrities who are either alive or deceased, as many Statutes provide a protection to the celebrity for some years after they have died. For instance, in Nevada, the celebrity is protected for 50 years after death, where in Indiana the protection remains for 100 years. For deceased celebrities, their heirs will be the ones deciding who is able to use their loved one's likeness and who cannot.

PRODUCT ENDORSEMENTS

Another potential snake pit for the voice actor is product endorsements. Product Liability laws in the United States are generally designed to protect the consumer from dangerous or defective products. These laws are usually couched in terms of "strict liability" rather than "ordinary negligence." This means that anyone involved with the manufacture, sale, or distribution of a product that causes an injury to the end user can not only be sued by the injured party, but will also be held strictly liable without the need for the injured party to prove that the defendants did anything negligent. The fact that the product was put into commerce and caused an injury, in many jurisdictions, is enough for the injured party to recover.

In addition, there are a variety of consumer protection laws, unfair trade practice statutes and Federal Trade Commission (FTC) regulations and guidelines designed to protect consumers from being ripped off by false and misleading advertisements.

This raises an interesting question concerning whether a voiceover artist is hired to record a commercial that says something like: "This drug is *the* best out there for the prevention and cure of this disease, and I personally guarantee it will work for you" and the drug ends up killing the user, whether the voiceover artist could be held liable for that "guarantee."

If the voiceover artist were a celebrity, then they certainly could be sued under a number of theories, including product liability, but also consumer protection statutes, and Federal Trade Commission (FTC) guidelines against false and misleading advertising. One is reminded of the series of lawsuits against Robin Leach back in 1999 where at least a dozen Attorney's General across the country sued him for endorsing vacation packages in both television and radio ads that turned out to be bogus.

Although one must wonder if such lawsuits would be brought against non celebrity voice talent who are not so "high profile," one of the functions of an Attorney General is to discover collectable assets that could be attached to pay back "victims" of false and misleading ads, or to pay back "victims" of dangerous products, so the possibility certainly exists that such a lawsuit could be brought against a voice actor that has some assets.

LIBEL AND SLANDER, AND INTERFERING WITH A BUSINESS RELATIONSHIP

Although I like to believe our business is filled with nice and great people who are never dishonest (and in my experience, for the most part this is true), as more voiceover artists enter the fold, the competition is bound to increase, and thus, dishonest individuals are bound to sneak in; individuals who will do anything to build their business, including stealing clients and bad-mouthing other talent.

I have witnessed this type of behavior in the legal business in certain areas over the years as more and more attorneys have entered the market. From what I have seen, it tends to happen in large areas where the likelihood of doing business with that particular attorney again is slight. In smaller areas where attorneys have to deal with each other repeatedly, this rarely happens.

I do believe, however, the potential for dishonesty exists as the voiceover field grows and the geographical area of the field expands, and I can see the potential for voiceover artists "bad-mouthing" other talent, who they may never meet, to a producer or client with words like: "that guy/gal is a terrible voiceover artist" or "that guy's nickname is Multi-Take Mike." With the amount of sensitivity and insecurity in this business on the part of advertising agents, and others, it is certainly foreseeable that statements like these could cause a producer to "jump ship" and switch to another voice talent.

I truly wish that this *never* happens in this field, but if it does, this could give rise to defamation lawsuits for libel and slander or lawsuits due to the tortuous interference of a business or contractual relationship.

Contractual Issues and Considerations Relating to the Voiceover Industry

The voiceover industry is a contractual based business. The unions that represent voiceover talent have gone through painstaking processes to ensure that voice talent are protected in the contracts they sign with producers, and that they are fairly compensated for the work they do. On the other hand, the unrepresented nonunion or Financial Core talent are on their own when it comes to negotiating their fees and protecting their interests contractually for nonunion jobs.

All agreements as to what is to be recorded and delivered and the cost of same should be in writing, and equally as important, the usage of the spot being recorded should be *clearly* spelled out. It would be a great idea for a talent to have a standard contract that he/she gets signed for every job, like an attorney has a standard retainer agreement that gets signed at the commencement of each case. Keep in mind that it has been said that the true purpose of a contract is "to keep an honest man honest."

However, due to the often fast-paced nature of the voiceover business, written contracts are not always practical. At a minimum, emails should be exchanged between the client and talent indicating the price, usage and delivery requirements for the final sound files.

It can be as easy as the talent writing a simple email like: "I will FTP the final .wav files to you for this 30-second TV spot by X date, which you can air for a year in the State of X, and you shall pay me the amount of X dollars. Please reply to confirm." Once you get that reply, you have a binding, written contract that is enforceable in a court of law.

Of course, for larger jobs where you are being hired as a "voice" of a company where you will be doing repeated work, a more detailed contract should be signed either prior or simultaneously to work being done.

First, if a contractual term is greater than one year, it *must* be in writing to be enforceable under every States' Statute of Frauds. Second, not only should the contract lay out all the above basic information, but it should also have a provision in there that allows the talent to terminate the arrangement after a certain period of time should things not work out, and it should cover situations such as the sickness or death of the talent.

The reason for this is to prevent the talent from adverse consequences at the hands of the company who has likely hired the talent to associate his or her voice as "branding" of its product or service. As you can see, if a talent just quits once the client has invested a substantial sum of money into such branding associated with the talent's voice, it could potentially be devastating for the client.

A talent's standard contract should also have a clause in it so that the talent can get a copy of the finished spot. The contract should also have a clause that allows the talent to use the finished spot on the talent's demo or website, and if a standard contract is not signed, the talent should again, at a minimum, use email to get specific permission from the client to use that finished spot for marketing purposes.

Although it is customary for a talent to use the final spot for marketing purposes, a talent should never assume it is OK to use the spot without getting the client's permission as the client may not agree, and such would be actionable against the talent as a violation of copyright law.

Another very important point concerning contracts is that, now that the voiceover industry is a global business, even if you have an enforceable written contract, it may not be worth the paper it is written on due to jurisdictional issues. For instance, if you as the talent are in New York, and the client is in California, and you record in New York while the client never steps foot out of California, the New York Courts would have no jurisdiction over the dispute because the individual you are suing resides in California.

Thus, you would have to take a trip out to California, or hire local counsel there in order to collect your money; hardly worth it for the average nonunion voiceover job. As such, it is extremely important to get at least part of the payment sent to you upfront by either PayPal™ or by credit card. If a talent is doing work from around the country, signing up to accept credit card payments will be well worth the expense.

Another specific provision of contract law voiceover artists need to be familiar with is the confidentiality clause, or non-disclosure agreement (NDA). Voiceover artists will often be working on projects that have not been released to the general public. If that is the case, then the producer or owner may request that the talent sign a "non-disclosure agreement" which states that the talent agrees to keep the nature of the work confidential until such time as it is aired, or until the talent receives written permission from their client to release information about the project. The thing to keep in mind here, is such an agreement will prevent the voice talent from displaying the finished spot on their websites, or otherwise, until the piece airs, or until the client gives them permission to display the material.

Non-disclosure agreements or confidentiality clauses are very common in the business, but nonetheless, if a voice talent is uncomfortable with the terms, especially the penalties if they breach the agreement, then they should run that agreement by a competent entertainment attorney, or their agents, to make sure nothing overbearing is being requested.

Intellectual Property Issues: Copyrights and Trademarks

INFRINGEMENT OF COPYRIGHTS AND TRADEMARKS

Voiceover artists have to consider both sides of intellectual property issues, meaning, they have to be concerned with infringing on others copyrights or trademarks, and they also need to protect their own ideas and creations through copyrighting and/or trademarks. The most common area where voiceover talent can run into copyright infringement issues is through their demos. Another common area this author has witnessed concerning infringement are infringement of trademarks in voiceover artist's branding or logos, and unlawful use of trademarked logos on talent's websites.

It is very important that the music used on the demo is properly licensed by the copyright owner. In addition, it is very important for the talent to have proper permission from the producer of a spot to use the spot on their demos. The voice talent does not automatically have the right to use a spot on their demo as that spot has been copyrighted by someone other than the voice talent, unless of course, the voice talent also produced the spot, then they may have copyrighted it.

This author has heard stories of voice talent receiving "cease and desist" letters from companies objecting to either the music on the artists demos, or the use of a particular spot on their demo. Cease and desist letters usually demand a voice talent cease using the copyrighted material immediately or face further legal action, like a lawsuit for infringement.

The cease and desist letter is actually the first step in the legal process and one of the reasons it is employed is to set up a "willful infringement" cause of action. If one uses material they know is copyrighted, then the author and/or copyright owner of the material can sue in Federal Court for statutory damages which go as high as $150,000 for each act of willful infringement. A defense to this willful infringement is that the user was "innocent" in that they did not know the material was copyrighted. That is why the cease and desist letter is sent out because that defense becomes impossible upon receipt of that letter as the receiver of that letter cannot argue they were innocent after they are informed with such a letter.

Similarly, those who produce voiceover demos, commercials or other material that will be broadcast must be cognizant of copyright law. For producers, there are a few things that must be kept in mind. First, using a copyrighted song on a commercial, for example, is an infringement of the copyright owner's rights and that producer could be sued by that owner for willful or negligent infringement, which results in statutory damages ranging from $200 per use up to $150,000 per use for willful infringement.

In addition, organizations like ASCAP and BMI, which regulate royalty payments to artists and publishers, will seek payments on behalf of the owner of the publishing rights to that music for royalties that are supposed to paid each time the song is played over the airwaves, or "displayed publicly." These are actually two separate and distinct rights; one bundle given to the copyright owner, and one bundle given to the publisher, which may or may not be the same.

PROTECTION OF YOUR WORK THROUGH COPYRIGHT

Voice talent also need to understand how to protect their creative works through copyrights. Voice talent are most certainly hired to do most voiceover jobs as a "work for hire," meaning that whoever hires the voice talent is retaining the right to copyright the finished product with the talent's voice on it. This is normal and customary in the business.

However, if a voice talent produces commercials, creates music for commercials, or drafts copy for commercials, that voice talent should do whatever is necessary to retain rights to that spot, meaning, they should specifically state in their contract that their services will not be considered a "work for hire" and that the talent has the right to copyright the spot.

Copyrighting is a very powerful protection as it prevents others from infringing on the work for the entire life of that author, plus 90 years after the authors death. Civil penalties can be as high as $150,000.00 for the willful infringement of a copyrighted work. This provision is probably not going to apply to most commercial spots that are produced as the practical life expectancy of the spot is limited. The provision, however, will prevent another company from stealing the spot, and also allows your heirs to continue to protect your work for you after you have passed away so it can continue to generate income after the author's death.

TRADEMARKS AND SERVICE MARKS

Voiceover artists also need to be aware of trademark law to protect the business names under which they operate. Once a voiceover artist picks a name for their business, like "ABC Terrific Voice," a trademark search should be done to insure that no one else is using the name. A search can easily be done on the United States Patent and Trademark Office (USPTO) at their website: **www.uspto.gov**. A search prior to using the name is wise so as to avoid any legal action for infringement of the trademark by the holder of the trademark.

If available, it would be wise for a voiceover artist to "trademark" the name by registering it with the USPTO. This will help to ensure that the talent will not have to use a different name later on to avoid confusion if someone else starts to use it. However, filing of a trademark can be a very confusing task for the uninitiated so this is something a voice talent would probably want an experienced intellectual property attorney to handle. Technically, since voiceover artists render services, the name would be "servicemarked" rather than being "trademarked." Like copyrights, trademarks also survive the death of the trademark holder and thus your heirs can ensure no one infringes on the use of the name.

This author has also seen several voice talent infringing on famous trademarks through the talent's branding and/or websites. I have seen several instances where voice talent will take a famous logo, for example,

Coca-Cola, and use the logo and name and say something like: Coca-Cola Voiceovers: There Is Nothing Sweeter" or something like this. Not only does the voice talent generally use the name, but also uses the logo and artwork and just replaces the trademarked product name with their own. In this author's opinion, this is complete infringement of the trademark, subjecting the talent to payment of statutory royalties and penalties to the mark's owner, that is Coca-Cola.

I find this both ironic and really bad business. It is ironic because if, for example, this voice talent was asked to do a voiceover audition for Coca-Cola, and Coca-Cola "used" the audition for a commercial or other purpose without the talent's permission, I am sure the talent would be the first to be upset about that, as I have heard this concern many, many times from talent doing online auditions. Yet, these same talent think nothing of using Coca-Cola's intellectual property for their commercial purposes.

It is bad business because if Coca-Cola, for instance, did not do anything about the infringement for a number of years, so that the brand *did* become associated with the voice talent, and then Coca-Cola decides to send a "cease and desist" letter as described above, then where does that leave the talent? They are either going to have to abandon all of that work with their "brand" and "rebrand," or take on a multinational corporation in Federal Court, which has exclusive jurisdiction of United States trademarks.

Likewise, this author has seen numerous examples of voice talent "taking" trademarked logos from websites, for example, Coca-Cola, and putting those logos on their own websites representing clients that the voice talent has done work for. If this is done without permission of the trademark's owner, this is also an infringement that I have personally heard resulted in receipt of "cease and desist" letters. Again, this is the first step in litigation and there is nothing to stop a trademark owner from suing the talent even if the talent agrees to remove the trademark voluntarily.

A Few Words on Business Insurance: Is it Really Necessary?

For voiceover artists that have employees, and/or have a separate studio space outside of their home, the answer to this question is a resounding *yes*! For those who have employees, virtually every State in the Nation requires that you carry worker's compensation insurance to cover injuries to employees sustained while on the job. Most insurance companies sell worker's compensation insurance as part of a business policy. Likewise, if you have a separate studio outside of your home, then you should certainly have business insurance to cover losses from theft and fire, and to protect you in the event that someone is injured while inside your studio.

For voiceover artists who have no employees and work out of a home studio where people occasionally enter their home to use the studio, it would probably be an excellent idea to obtain business insurance, as home insurance policies often contain exceptions from covering "business uses" of your home.

Finally, for voiceover artists who have no employees and work out of a home studio where no one ever enters to use the studio, the business insurance is optional. One advantage for any voice talent to obtain business insurance is that most policies cover "personal and advertising injuries," which will cover claims for certain offenses you commit in the course of business such as libel, slander, disparagement, or copyright infringement in your advertisements, discussed earlier in this chapter.

For solo voice talent with a home studio and no employees, these business policies are likely to be very inexpensive as the risk of loss is very low for insurance purposes.

Generally, if a voice talent has any personal assets, it would be an excellent idea for the talent to work under some form of corporate structure, like an LLC, and to obtain business insurance. This will provide the greatest protection to the talent so that their personal assets will not be at risk. It would take quite a catastrophic loss to exceed these safeguards, and such a loss is very unlikely to occur in this business in this author's opinion.

Tax Tips and Considerations

A question I am asked frequently by voice talent, as well as producers, is "Do I have to charge and collect sales tax for my work?" The answer to this question is very State specific. However, a few generalities exist concerning sales tax. In general, voice talent do not need to charge or collect sales taxes for performing. Producers, or voice talent selling production services or studio time, may have to charge and collect Sales taxes for renting of their studio spaces or other production services. This is controlled by each State, so the producer should check with their local and State Department of Taxation or Revenue to answer this question specifically.

However, another generality, sales taxes only needs to be charged and collected for work performed for clients within the State where the production facility is located, not for work performed outside of the State. For instance, a production facility or studio that is located in New York will only have to charge and collect sales taxes from clients also located in New York.

Are You a Business? Or Just a Hobby?

A voice talent should be sure that they are operating their voiceover business as a business and not a hobby. The quicker the voice talent treats voiceover as a money-making venture, the better off they will be, in this author's opinion, not only for tax purposes, but also in obtaining clientele and increasing income. As discussed in the first section of this chapter, if the IRS determines that an activity is a hobby and not a legitimate business, it will not allow any deductions for any expenses of the "business."

Also, a voice talent, in treating the "business" as a "business", must keep detailed records for all expenses, including receipts. A separate checking account should be set up for the voiceover activities, and a talent should *never* "commingle" fees received from voiceover with personal funds. Any automobile travel should be logged in a book recording the trip, the total mileage, and the business reason for the trip, such as an audition or recording session.

Of course, equipment purchased for the home studio can be deducted. The talent has the option of depreciating the equipment over time, or can take a "Section 179" depreciation deduction, which allows deduction of up to $125,000 in equipment purchases in the year of purchase. $125,000 can certainly build a *really* nice home studio!

Finally, one deduction that voice talent should be particularly cognizant of is the "home office" deduction. This deduction is one that traditionally has raised a red flag for the IRS due to past abuses by taxpayers. In order for one to take the home office deduction, the home office must be used *exclusively* for business.

For instance, if a home studio is set up in the living room where the family also gathers to watch television, then the IRS will generally not allow a taxpayer to deduct the use of the living room as a "home office." If, on the other hand, the studio is located in a segregated area of the house, like a basement or attic, which has a sound booth or acoustic improvements, then a home office deduction will most certainly be allowed as clearly the studio is being used exclusively for the voiceover business.

A SECOND CAREER

Many people out there are taking on voiceover as a second career, or as a "side venture." If this is the case, it is very important that this be done properly or a bulk of the expenses will not be tax deductible.

The IRS does not allow deductions for entering into a "new trade or business." Publication 970 from the IRS states: "Education that is part of a program of study that will qualify you for a new trade or business is not qualifying work-related education. This is true even if you do not plan to enter that trade or business."

However, start up expenses are deductible up to the first $5,000. IRS Publication 535 reads as follows: "Business start-up and organizational costs are generally capital expenditures. However, you can elect to deduct up to $5,000 of business start-up and $5,000 of organizational costs paid or incurred after October 22, 2004. Start-up costs include any amounts paid or incurred in connection with creating an active trade or business or investigating the creation or acquisition of an active trade or business. Organizational costs include the costs of creating a corporation."

This is another very compelling argument as to why to form an LLC or Corporation *prior* to undergoing training and working on your demo, as all of these expenses can be classified as start-up expenses. It is extremely important such expenses are couched in these terms, rather than as "tuition," as "tuition" for voiceover training could certainly be held by the IRS to be a nondeductible expense qualifying one for a "new trade or business," whereas training to prepare one to record a demo could be classified as a start-up cost.

The factors that the IRS says you should weigh to decide if your voiceover activities are a business or hobby are:

- Do you run the activity in a businesslike manner?
- Does the time and effort you put into the activity indicate an intention to make a profit?
- Do you depend on income from the activity?
- If there are losses, are they due to circumstances beyond your control or did they occur in the start-up phase of the business?
- Have you changed methods of operation to improve profitability?
- Do you or your advisors have the knowledge needed to carry on the activity as a successful business?
- Have you made a profit in similar activities in the past?
- Does the activity make a profit in some years?
- Can you expect to make a profit in the future from the appreciation of assets used in the activity?

Final Thoughts

The author of this chapter has done his best to identify many of the legal issues facing the voiceover artist in today's environment. It is not intended, however, to cover every possible situation under all circumstances. Laws may vary from State to State, and many business, tax, and legal issues are unique to the individual and require specialized attention. The author urges the reader to seek out specific advice from legal and tax professionals concerning their specific situation prior to undertaking ANY action.

23

Your Home Studio

A Million Dollar Studio in the Corner of a Room

Since the mid 1990s the business of voiceover has gone through a series of major changes, both in terms of how the business works and in the technology used for recording voice tracks.

Home based recording studios have been around since the first days of sound recording. Early recording equipment was bulky, complicated to operate, and most of all expensive. It used to be that the home studio was relegated to only the serious audiophile or professional musician.

Advances in computer and sound recording technology have changed all that. Today, the home studio is commonplace. With a decent computer, recording software, and some relatively inexpensive equipment, anyone can now have the capability to do what used to require a multimillion dollar facility. And it will fit in a corner of a room at home!

Major advertisers and producers of national commercials and network programs continue to record in traditional recording studios, as do producers of other high-end sound recordings. However, with a properly designed home studio; it is possible to record with a quality comparable to many professional studios. For this reason, many producers, especially those with a low production budget, are moving toward having voiceover talent record voice tracks at home studios. By requiring VO talent to record on their own equipment, producers often expect a faster turnaround while maintaining the same quality they would get from a professional studio, but for the cost of talent fee only. Many times, these expectations are unrealistic, but the fact remains that this is what many producers are demanding, and it seems unlikely that this will change any time soon.

Unfortunately, most voiceover artists are performers and not recording engineers. For many, the technology of a home studio can be intimidating and overwhelming. Still, in today's world of voiceover it is important to at least know studio basics and have functional recording and editing skills.

This chapter will, hopefully, ease your concerns about setting up and operating your own home recording studio.

The Challenge of Recording at Home

At one time, virtually all voiceover work was recorded either at a traditional recording studio, at a radio or television station, or a professional production facility. If you are working in voiceover today, it is expected that you not only have the ability to record in your own personal studio, but that you also have the ability to deliver the same quality that would be delivered by a multimillion dollar recording facility. This expectation by talent buyers creates a serious challenge for the neophyte voice actor.

The monetary aspects of building a home studio are not the issue. The equipment is relatively inexpensive, and the recording software can even be found as a free download. No… the challenge goes well beyond the equipment itself, lying in the complexities of creating a studio-quality environment and with the performer's ability to deliver a recording with proper signal levels and clean edits. It's easy to record audio on your computer! Recording audio that sounds like it came out of a major Los Angeles or New York studio is a different matter entirely. It's not difficult to do this, but it will generally require some serious thought, some research, and, depending on how far you want to go, potentially a lot more money than the cost of your studio equipment.

As a professional voice actor, there are several essential aspects of your home studio that you need to fully understand in order to deliver voice tracks that meet professional standards:

- Basic operation, connection, and functionality of your equipment.
- Operation of your recording software, including an understanding of how to make clean recordings, maintain proper recording levels and handle basic editing processes.
- A basic understanding of room acoustics, how it affects your voiceover recordings, and how to correct acoustic problems.
- An understanding of the principals and applications of various microphone techniques used to produce different results.
- A basic understanding of outboard (external) audio equipment and devices that may be needed for communication with clients or for delivery of voice tracks.
- A functional understanding of computer file structures, software operation, and various delivery options.

If all of this sounds intimidating, or like it's too much to deal with, you might want to rethink your entry into the world of voiceover. Of all the pressures placed on today's voiceover talent, these expectations are completely reasonable and practical. The majority of professional voice actors working from their home studios have developed these skills and deliver high-quality, professional recordings on a daily basis. If you are going to compete with them, you will need to achieve the same results.

Home Studio Equipment

With today's computer systems, it's a relatively simple matter to put together the equipment for high-quality voice recording. Regardless of the equipment you use, you will have a choice between a system that is completely digital or one that is a combination of analog and digital equipment. As with most things, there's more than one way to configure your home studio recording system, and none is better than another. It comes down to personal preference and ease of use.

There are many excellent books on the subject of building a home studio, many of which are available through **www.voiceacting.com** and other online resources. If you need a clear and concise explanation of the component parts of a home studio and how everything works together, my *VoiceActing.com Guide to Professional Home Recording ebook* will answer most of your questions. This downloadable ebook is fully cross-linked internally, and loaded with Internet links to websites with more information about every aspect of a home studio. If you're looking for a book that covers the basics of home studio recording and the production process, Jeffrey P. Fisher and Harlan Hogan's *The Voice Actor's Guide to Home Recording* is an excellent choice. Focal Press, the publisher of this book, has several excellent books on the subject of home recording, including: *Project Studios* by Philip Newell and *Practical Recording Techniques* by Bruce Bartlett. The Focal books (**www.focalpress.com**) are intended more for the home music recordist, but they still contain some valuable information that can be applied to voiceover work.

The entire purpose of your home studio is to record and edit audio on your computer. The first thing you need to do is figure out how to get the sound of your voice onto your computer's hard drive. At first glance this may seem a simple task: you purchase a gamer's headset mic, plug it into your computer's sound card, open the recording software that came with the computer, and start recording. Although that's the basic process, the reality is that it's not quite that simple.

For starters, a computer headset mic is not acceptable for professional voiceover work. These microphones are designed for online gaming and not for recording, and if you use one for voiceover work, you'll be plagued with pops and other noise issues. As for software, the basic audio recorder that comes with Windows is pretty much useless. Mac computers will often come with good audio software, but for a PC, you'll need to look somewhere else.

As I mentioned, there are two general technologies for recording your audio into a computer: analog and digital. The primary difference between the two is that an analog system uses a microphone or an analog mixer that is connected to the computer's sound card, and converts the audio to a digital signal inside the computer. A digital system converts the audio signal to digital information outside the computer either directly at the microphone or by using a *USB* or *FireWire* digital interface. The digital interface may be a standalone device, a digital mixer, or even a digital USB microphone. USB (universal serial bus) and FireWire both refer to high-speed digital connections between a computer and an outboard device.

Of the two, the least expensive, is to connect an analog mixer to the line input of your computer's sound card. Although most computer sound cards are adequate for the job, the fact that the digital conversion process is happening inside the computer, an electrically very noisy place, may result in some unwanted noise in your recordings. If you are using an analog path to your computer, it would be well worth the investment of a few hundred dollars to upgrade your sound card to a high-end digital converter.

If you choose to do the digital conversion outside the computer, you will be completely bypassing your sound card by using the computer's USB port. This can have its advantages and disadvantages. On the plus side is the portability aspect of the USB interface devices, which makes them ideal if you are recording on your laptop, and the relatively low cost of the USB converter. On the downside, if you have numerous devices sharing the USB (like external hard drives, a web cam, etc.), your computer may have difficulty processing the large amount of audio data being transferred, which can result in mysterious glitches or even a system crash.

The equipment and software you purchase for your home studio will depend largely on the kind of work you will be doing. If you're only recording voice tracks or practicing your performing skills, you don't need an elaborate digital mixer with all the bells and whistles.

The following are basic components for a home studio, each with some comments as to their role in the studio. These components are discussed in more detail a bit later in this chapter, and in the Products/Equipment section at **www.voiceacting.com**:

- **Microphone**—You need a microphone that will make your voice sound great. Different mics sound different on different people. It can take some research to find exactly the right mic for you. To get started, don't buy the most expensive mic you can find. There are many excellent mics available for under $100. The mic will connect to a USB interface, a digital mixer, an analog mixer, or directly to the sound card (not recommended). There are even USB mics that connect directly to your computer. A condenser mic is usually recommended over a dynamic mic, due to its superior performance. Condenser microphones require a power source, either by battery or *phantom power* provided by a mixer or USB interface device.

- **Microphone cable**—Professional microphones use a 3-pin XLR type cable that connects the mic to a mixer or USB device. The phantom power for condenser microphones travels through the same wires in the cable that carry the audio signal.

- **USB interface**—The cable from your mic or mixer connects to this device, which converts the analog audio to digital audio. It connects to your computer via a USB cable plugged into a USB port on the computer. A USB device will usually have a recording volume and a volume control for headset monitoring. A good USB interface also provides phantom power for condenser mics. A USB device will replace your computer's sound card as the audio source.

- **Analog audio mixer**—An analog mixer may connect to either the computer's sound card or to the line level inputs on a USB interface. The mixer is used to control the volume for up to several mics or other sources. If connecting directly to the sound card, you'll need some special adaptors, which may be difficult to locate.

- **Digital audio mixer**—As with an analog mixer, a digital mixer is used to control the volume for up to several mics or other sources. Many audio mixers today convert the incoming audio to a digital signal as soon as it enters the mixer. Others will functionally operate as an analog mixer and have multiple analog audio output connectors. However, unlike analog mixers, digital mixers have a built-in USB port that allows the mixer to be connected directly to your computer. If you are using a digital mixer with built-in USB, you do not need an external USB interface.

- **Recording software**—Virtually all recording software will provide for recording and editing of recorded audio. However, all audio software is not the same. Some are more user friendly than others, and price is not a good indicator of ease of use. Software prices vary from free to several hundred dollars. Regardless of the software you choose, you should make sure it will easily save or convert your recordings to MP3 compressed files. When investigating recording software, make sure you have a basic understanding of how you will record and edit before purchasing. Some equipment dealers will recommend ProTools because, they say, it is the standard of the recording industry. That is true. However, ProTools is only the standard for traditional music recording studios—not for voiceover recording. It is very expensive, very difficult to learn, requires proprietary outboard equipment (for a high-end system), and is considerable overkill for recording a single voice track. For anyone just getting started, I recommend Audacity. It's free, very simple to use, and does everything (and more) that a voice actor needs. Audacity is a free download from **www.audacity.sourceforge.net**. To create MP3 files, you'll also need to download the L.A.M.E. encoder plug-in. Audacity is available for both PC and Mac.

- **Microphone stand**—A floor stand with a boom, or a desk stand. Either way, you'll need something to hold your microphone. The most common mic floor stand is a tripod base adjustable stand with a boom attachment. There are also smaller tripod base stands that are short enough to be used as a desk stand.

- **Copy stand** (music stand)—You'll need this to hold your copy so your hands are free for performance and operating the equipment. If you work standing, a collapsible music stand will do the job, but it is usually designed for a performer who is seated and may not have the height adjustment you need. The Manhasset M48 is a much better option: more stable, no knobs to turn, and plenty of height for most uses. I don't recommend any music stands that have knobs to adjust their height or the angle of the paper holder.

- **Headphones**—Conventional stereo headphones or ear buds will work for most voiceover recording. You'll wear your headphones while recording and for other monitoring, if you don't have a set of speakers. You'll want a headset that is comfortable and that reproduces your voice accurately. Some headphones emphasize low frequencies, which will result in a coloration of your voice.

- **Speakers**—Although headphones will serve the purpose for monitoring, and even for editing, they can become uncomfortable if worn for extended periods of time. To start your home studio, a pair of good computer speakers connected to your sound card will do nicely. As you bring in voiceover work, you might want to consider upgrading your speakers to some studio monitors. If you are using a USB device for recording, you may need to adjust your software settings so your computer knows you want to use your sound card speakers for playback monitoring. Basic computer speakers are fairly inexpensive. Studio monitors can run into the hundreds or thousands of dollars for a pair of speakers.

- **Stopwatch**—You'll need a way to time yourself. Stopwatches that beep aren't recommended. They're awkward and they make noise. A better choice is an analog 60-second sweep stopwatch. Prices range from about $50 to several hundred dollars. An Internet search for "analog stopwatch" will bring up numerous sources. Look for one that has a separate reset button. Most of the inexpensive watches require the watch to be stopped first before resetting, but some will allow the watch to be reset while running. Of course, some smart phones, like the iPod® have a digital stopwatch, or you can also use the built-in timing capability of your recording software.

- **Acoustic treatment**—Most home studio areas will need some form of acoustic treatment to reduce echoes and unwanted noise. There are many excellent books on this topic, so it will only be discussed briefly here and in the next section. When designing your home studio, you should be aware that you may need to make some changes to your recording environment for the best recording quality—and that some of those acoustic changes can be expensive.

That's the basic requirements for any home studio. With the exception of a mixer, everything is a necessity. You'll find most, if not all, of it at professional music stores. Computer stores simply don't deal with this type of equipment, and they don't understand home studios. In fact, most music equipment dealers don't understand the needs of a voice actor and will recommend equipment only in terms of what they know about music recording. Although their advice may hold some value, you may end up spending considerably more than necessary.

Understanding Room Acoustics

This section is not intended to give you solutions to your room acoustic issues, but rather will give you a general understanding of typical acoustic problems facing voiceover actors and some ways to solve them. There are literally dozens of books on how to deal with adverse room acoustics, and an Internet search will bring up thousands of other resources.

If you're going to work in voiceover, you will need to have a home recording studio, which means you will need to find a place to record. Your objective should be to record your voice at a quality comparable with that of the best recording studios in the country. OK… that might be a bit lofty a goal, but you *do* need to be able to record with excellent sound quality.

"So," I hear you ask, "how do I find a place in my home or apartment where I can record my voice and have it sound that good?" You've actually got several options ranging in cost from relatively inexpensive to very costly.

If you've never been inside a major recording studio, you're probably thinking that you don't have a reference point for what a professional studio might sound like. Actually, you do! But it's far more important to know what a professional studio *doesn't* sound like: It doesn't sound like your living room, your kitchen, your garage, your bedroom, your office, or your bathroom. Although it certainly is possible to locate your home studio in any of these areas, in most cases they simply won't do if you expect to produce high-quality recordings. So where will you put your home studio?

All of these rooms have inherent acoustic "problems" in most homes, the most common of which is simply referred to as room reverberation—the common echo that is the result of sound waves bouncing off parallel surfaces like walls, windows and hard surfaces. A typical bath room is probably the best example of room echo. Solid hard walls, mirrors, and tile surfaces cause sound waves to bounce all over the place. A less offensive, but still unacceptable, form of reverberation is known as *slap echo*. This form of reverberation is what you'll usually hear when you clap your hands in an empty bedroom. The resulting reflections of sound waves bouncing off the walls creates a relatively short echo that may not be obvious until you actually listen for it.

The one place in your home that most closely resembles the "sound" of a professional recording studio is your bedroom closet stuffed full of clothes. The soft fabrics absorb sound waves, preventing them from bouncing off the walls. A room with overstuffed furniture, thick carpet, heavy drapes, and lots of wall hangings will have the same effect of killing reflected sound waves, resulting in a relatively "dead" sound.

Acoustically treating the walls of a room will achieve a similar result. Acoustic treatments can be anything from hanging a few heavy, absorbent blankets on the walls to applying specially designed acoustic foam to the walls. The general idea is that you want to create an acoustically non-reflective environment, one in which your voice will be recorded without any reflections from objects or walls.

More expensive options include purchasing a preconstructed booth to serve as your recording space or building an acoustically isolated room within a room. An Internet search for "voiceover booth" will reveal numerous manufacturers of prefabricated booths and even a few sites with instructions on how to build your own.

If you are going to try to "fix" the acoustics of an existing room, be careful not to overlook the corners. Corners where walls meet the ceiling or floor are notorious for reflecting low frequency sound waves often resulting in a "boominess" that can seriously affect your recordings. The right-angle corners where walls meet can also create a similar bass reflector. These areas can be dealt with in many ways through the use of acoustic foam, corner bass traps, mounting absorbent material, or even using an overstuffed pillow to effectively eliminate the right angles of the corners.

Although fabric wall hangings, blankets, pillows, overstuffed furniture, and other inexpensive acoustic treatments may appear to "fix" many acoustic issues in a room, they still may not create an environment that is ideal for your voiceover recordings.

Creating an acoustically dead area surrounding your microphone may work better than trying to treat an entire room. A few inexpensive three-fold screens or custom-built wooden or PVC frames draped with heavy blankets can serve as a portable recording area. There are even "portable" voiceover booths and stand-mounted acoustic baffles available that create a sound absorbent area around the microphone.

If you are using one of these portable acoustic environments, be sure to aim your mic toward the acoustic material if at all possible. Any room reflections will first hit the back of the mic, which is a dead zone for cardioid microphones, and then be absorbed by the acoustic material, thus minimizing being picked up by the mic.

Chances are you will need to do some additional research and spend some money to make your recording environment as quiet and "dead" as possible. If you're just getting started in voiceover, you should strive for the best room acoustics you can achieve that is within your budget. I would not recommend purchasing a prefabricated voiceover booth or building your own isolation room without first doing a lot of study on the subject of acoustics. Explore other options first. The time to upgrade your home studio environment is when you are generating a substantial income from your voiceover work—not when you are just getting started.

Advanced Home Studio Technology

If you are just getting started in voiceover, you don't need anything more than the basic equipment discussed earlier in this chapter. However, as you begin to get booked, you'll soon discover that some clients might request *ISDN* or *phone patch* services. You may never need either of these, but it is still worth knowing what they are and how to use them.

ISDN

ISDN stands for *Integrated Services Digital Network*. This is basically a hard-wire digital phone line provided by your local phone company that will connect your home studio to any other ISDN studio in the world. It's the next best thing to the client actually being at your home studio. To connect your studio to another studio requires a codec (coder-decoder) that will convert your audio to an appropriate digital signal that can be transmitted over phone lines. The receiving studio must have a compatible codec in order to receive your audio.

Most ISDN set-ups use two sets of phone lines—one for transmitting audio and the other for receiving. Depending on location, installation of the phone lines can cost up to several hundred dollars per line and the codec alone can cost up to about $3,000. Add to that the monthly fees for maintaining the phone service and the price of ISDN can add up quickly.

For many years, ISDN was the preferred method for digital audio transmission because it allowed for real-time, high-quality remote audio recording. With the rapid improvements in Internet technologies, ISDN has steadily been losing favor and, in fact, many telephone companies are reducing or discontinuing ISDN service in their service areas.

A number of alternatives to ISDN have made their appearance, most of which use advances in Internet technology to provide real-time remote recording similar to ISDN. Two of the more popular systems are Source Connect, **www.source-elements.com**, and AudioTX, **www.audiotx.com**.

Do you need ISDN—or even one of the alternatives? Probably not. Because any digital remote recording scheme will require matching equipment at both ends and the potential for monthly service fees, I do not recommend investing in any technology of this sort until you have a client base that will support the initial costs and produce a return on your investment. And when you do get that first ISDN booking, there's no need to rush out and install expensive equipment. Most cities have at least one ISDN studio where you can book the session.

PHONE PATCH

Many voiceover talent buyers and producers prefer to direct or supervise their VO sessions from the comfort of their office. In order for them to do this, they need to be able to clearly hear you as you record their script. At the same time, you need to be able to clearly hear them as they direct you through the session. Using a phone patch to your studio is the next best thing to ISDN or its alternatives, and it's a lot less expensive.

Although holding a telephone to your ear will do the job, this is not practical for a voiceover session as the mere action of holding the phone will restrict your ability to move as you perform. A Bluetooth headset or some other telephone headset will do the job nicely and may be the simplest, most cost-effective way to go. Many VO professional use a Bluetooth headset with great success. However, ideally, you should be able to connect your home studio directly to your phone system.

A properly configured phone patch system will require an analog or digital mixer with the ability to send your audio both to your computer and to an external telephone hybrid that will interface your audio output with your telephone system. Once installed, your client will call your phone number (or you call them) and you push a button on your hybrid to connect your studio to the phone line. Connecting a phone patch digital hybrid to your equipment can be a bit tricky and is fully explained in *The Voice Actor's Guide to Professional Home Recording ebook*, available through **www.voiceacting.com**.

EQUIPMENT UPGRADES

At some point in your voiceover career you may want to upgrade your home studio. By the time you are ready for this you will have plenty of experience and you'll have a better knowledge of what you might want to upgrade, what equipment you might want to add and, more importantly, why. Equipment that is installed outside of your mixer or computer is generally referred to as *outboard equipment*. There are literally dozens of audio equipment manufacturers and equipment to fit any budget.

A few of the possible equipment upgrades or additions are listed below:

- **Outboard signal processor**—A signal processor can be any device that modifies or adjusts an audio signal. The most common outboard processors include a compressor/limiter, equalizer, de-esser, noise reduction, signal enhancer, or mic preamp. Some devices include multiple functions while others are dedicated to a single purpose.

- **Outboard microphone preamp**—Your mixer or USB interface already includes a mic preamp. A microphone produces an extremely low electrical output. A mic preamp boosts this audio signal to a level that can be used by a mixer or other device. An external mic preamp will generally be of a higher quality than the built-in preamp in your mixer or USB interface.

- **Powered speakers**—Advances in speaker design have produced new speaker systems with the power amplifier built in. This results in extremely high quality audio from relatively small speakers.

- **Additional microphones**—Different microphones can produce different results from your voice recordings. You or your client might want a certain sound for a specific project or type of voiceover work. Careful selection of your mic can help you achieve the desired results and allow you to offer greater versatility with your recording services.

Upgrading your home studio should only be a consideration when you have either a specific need for the upgrade or you have generated enough income to justify the expense.

Managing Your Computer

Regardless of your computer or its operating system, you will need to devise a method for managing your files. There is no one correct or ideal file management system, so this is something you'll need to work out for yourself. However, I do have some suggestions that you might like to keep in mind as you work out your personal computer management system.

- **Devise a folder system that is easy to work with**—Think of your computer as a file cabinet. You might have a primary folder for each client and subfolders for each project you work on for that client. Within each project's subfolder there might be additional subfolders that hold files for various aspects of that project. The key to a successful and efficient file management system is to thoroughly think it through, even mapping it out on paper, before you start creating folders on your hard drive. This advance planning may reveal potential structural issues or result in some better organizational ideas that you might otherwise miss.

- **Include an "Upload" or "Deliver" folder**—Rendering your MP3 or other deliverable files to a separate folder will make it easier to locate them when burning to a CD-ROM, attaching to an email, or uploading to an FTP site.

- **Use a separate external hard drive for your voiceover projects**— If you keep all of your VO recordings on your C Drive, you stand a chance of losing everything when your hard drive crashes… and eventually, it will! By using a separate hard drive for your voiceover work, you remove a considerable amount of stress from your C Drive, allowing it to work more efficiently for running your system's programs. Using an external USB drive can give you the additional flexibility of being able to move to a different computer to work on your projects. Large capacity portable and network USB hard drives are inexpensive, and the investment will be well worth it.

- **Master your computer software**—If you are going to operate as a professional business person, you must be able to handle any computerized business functions quickly and efficiently. Learning how to use your software now will pay big dividends later on.

- **Know in advance how to deliver files to your clients**—There are many ways to deliver audio files: CD-ROM, FTP, email, and third-party delivery uploads. Any of these may be used to deliver MP3, .wav, .aiff, or other files. Learn which delivery methods work best for which file formats. CD-ROM may require software to properly burn the files to a disk. Email cannot handle a file larger than about 8–10 MB, and realistically, anything much larger than 4 MB will tend to slow things down. FTP (*File Transfer Protocol)* can be challenging to work with, but is often the preferred delivery method for very large files. There are many third-party delivery methods that allow you to upload your files to a server and then notify your client via email with a link for them to download the files. An Internet search for "file delivery service" will reveal dozens of options, many of which are free and some of which can handle files up to several GB in size.

Wearing the Hat of the Audio Engineer

Having a home studio where you can record your auditions and voice tracks does not necessarily qualify you as a recording engineer, nor does it mean that you should even consider handling extensive editing or complete audio production with music and sound effects. If you know nothing about production, and this is something you would like to offer your clients, you will need to learn yet another set of skills. But that's another book!

At the very least, you will need to master your chosen audio recording software. Be prepared to spend some time learning how to record pristine audio, edit out breaths, adjust spacing between words and phrases, normalize, adjust the quality of a recording using equalization (EQ), properly utilize the many features of the software, and render high-quality audio in different formats. There are many ways to learn audio production and postproduction skills, and many excellent books on the subject.

As a voice actor, you really only need to know how to record your voice at the highest quality possible. Chances are you may never be asked to provide any extensive postproduction services.

Your home studio is critical to your voiceover business. By mastering the technical skills necessary to build and operate your home studio, you will be in a much better position to build a successful career as a voice actor.

24

Auditions

The Audition Process

Auditions may seem frustrating and nerve-wracking, but they are an essential part of the voiceover business. Without auditions, it would be very difficult for performers to get exposure to producers and other talent buyers.

Over the past several years, the audition process has changed from one where the voice talent would go to a studio or other site for a live audition, to one where, more and more often, the talent will now audition via the Internet or email by recording their audition in their home studio. Agents in larger cities will often handle auditions at their office.

No matter how it's done, the audition process is still the most efficient way a producer or advertiser has of choosing the best performer for a project. Once a script is written, copies are sent out to talent agents, casting directors, and online audition services. Specific performers or character types may be requested for an audition, and in some cases voiceover talent may be cast directly from a demo or prior work.

Talent agents and casting directors will select performers from their talent pool who they feel will work best for the project being submitted. If a specific voice actor is requested, the talent agent will attempt to book that performer. You, the *talent*, are then contacted and scheduled for an audition.

If you are just starting out, you may get the audition call from one of your contacts, through classes, online sites, friends, networking, or sending out your demo. You may receive the call several days in advance, the day before, or even the day of an audition. And it may come in an email.

ONSITE LIVE AUDITIONS

Onsite live auditions can be held anywhere. Some ad agencies have a recording booth for handling voice auditions. Sometimes auditions are held at a recording studio, the client's office, a radio or TV station, or even at a hotel conference room. In Los Angeles, many agents ask their talent to record auditions at the agency. In this way the agent has some degree of control over the quality of the audition for the talent they are representing.

You will be given a time and location for your audition, but usually you will not be asked when you are available, although you often can arrange a mutually agreeable time. Auditions are generally scheduled over one or two days, every 10 to 20 minutes and, depending on the scope of the project, there may be dozens of performers auditioning for the same roles.

You may or may not be told something about the project, and you may or may not receive the copy ahead of time. I've actually done auditions while on my cell phone after the casting person dictated the copy.

Once scheduled for an audition it is your responsibility to arrive at, or prior to, your scheduled time, prepared to perform. Only if you absolutely cannot make the scheduled appointment should you call the casting agent to let him or her know. The agent may, or may not, be able to reschedule you.

In this day of electronic communication, a variety of other types of auditions are becoming popular. You might receive a script by email or fax with instructions to call a phone number to leave your audition. Or, if you have your own recording equipment, you might be asked to record your audition and email it as an MP3 file.

ONLINE AUDITIONS

The Internet has spawned a number of specialized websites that exist for the sole purpose of providing talent buyers the opportunity to reach hundreds of voice talent for auditions, and talent the opportunity to audition for hundreds of talent buyers.

There are currently four types of websites where voice talent can post their information and demos in the search for voiceover work. The first is a website operated by an Internet talent agent or casting agent. To be listed on these sites, a voice actor must be accepted by the talent agency for representation. These online agents operate like any other talent agent—it's just that their talent pool is spread across the Internet.

The second is a free listing site that will usually offer to post the talent's name, a brief description of the talent's services, and a link to the talent's own website. As with most things in life, you get what you pay for, so these free listing sites are often not very well promoted and don't result in much work—if any. The real benefit of listing on these free sites is not in the listing, because there simply won't be that may people visiting the site. The benefit is the fact that Internet search engines are constantly scanning websites to catalog names and links. The more frequently your name is found by these search engines, the easier it will be for someone to find you.

The third type of listing website is one where voice talent can receive requests to submit auditions for a fee. These sites are very clear in stating that they do not act as a talent agent, but instead, serve as an intermediary between the voiceover talent and those seeking voiceover performers. They do not take a commission for work obtained by the talent, but they do charge a membership fee for voice talent to gain access to audition requests. Many voiceover professionals refer to these as *pay-to-play* sites.

A fourth type of online audition site has begun to appear over the past few years. These sites are usually operated by legitimate talent agents and, although they operate in a similar manner to the other online audition sites in that they charge a fee for audition delivery services, these sites will also take a commission from any work obtained through the agency. Although some of these sites are quite popular and can produce impressive results for their members, the fact that they are both charging a fee and taking a commission may put the ethics of their business model in question.

This act of taking two income streams from the same source is commonly known as *double-dipping*, and is illegal in many States. Owners of these sites will say that the membership fee they charge is for services that are in addition to representation, such as web pages, consultation services, and management services. However, most of these services have traditionally been considered a cost of doing business for an agent and many of the online talent agents do include these services as part of their representation. The fact that an online agent might charge a membership fee for services beyond those of representing their clients should not be taken as an implication that they are operating illegally. It is only mentioned here so you are aware of how online agents and audition services operate so you can make an educated decision when considering their services.

The vast majority of membership audition websites are legitimate, do not act as a talent agent, and truly do their keep their promise to market and promote the website (and thus the talent) to producers, ad agencies, and other talent buyers. Two of the largest voiceover membership sites in the U.S. and Canada are **www.voices.com** and **www.voice123.com**. At the time of this writing, the main European counterpart is **www.bodalgo.com**.

On many of these sites a voice actor can, as part of their membership, receive a personalized web page with their bio, performing styles, and several demos ready to be heard. Members also receive a continuous stream of audition requests. The web page is a huge benefit for a beginner who doesn't have a site of their own. A website is a powerful marketing tool, and this feature alone may make a membership worth the price, but for an experienced voice actor who has their own site, this may be of little value, primarily because the web page is actually a *sub domain* of the hosting site with little or no direct marketing value for the member.

Another benefit for beginning voice talent is the constant stream of audition requests. Each audition is an opportunity to practice performing skills and experiment with voiceover technique with real-world copy without risking anything. For this reason alone, it can be well worth the membership fee just to get access to a wide variety of scripts. Experienced voiceover performers don't need the practice, and dealing with the vast number of unqualified auditions can be very time consuming.

Although most online audition sites truly have the best interests of their members in mind, there are some that seem to cater to the talent buyers who do not pay for the service, rather than to their voice talent members who pay a fee. This has created a great deal of controversy in the voiceover community primarily due to the large quantity of low-end, inexperienced talent buyers; poorly written copy; and low-ball talent fees.

Some audition requests are submitted by experienced producers who know what it takes to be a qualified voice actor; and who appreciate the time, energy, and money that a serious voice talent has spent developing their skills and business. Their audition requests are clear, specific, and informative. They know what the work will entails, what it is worth, and they offer a reasonable fee, expecting professional results.

Then there are the talent buyers who don't know what they're doing, or who simply want the most work for the lowest price. These producers do not appear to have even the slightest comprehension of what it takes to create a voiceover performance that will actually get the desired results. Their auditions are often fragmented, incomplete, or excessively demanding and unrealistic. Here's an example of a typical "low-ball" audition request:

> I need this VO done ASAP. My script is only about a page, so it shouldn't take longer than about 10 or 15 minutes. What you send should be finished with the VO, music, and sound effects tracks. If this works out, I'll have a lot more work for you. I would prefer if you do the spot spec. I will pay on completion, just prior to delivery. I have a budget in mind for this, but I'd like you to send me your prices so I know who will work within my budget range.

One major problem with audition requests like this is that the talent buyer is obviously looking for the lowest price he can get for a lot of work. This places the auditioning voice actor in a position of trying to come up with the lowest bid for the best performance, which can only serve to lower the voice actor's credibility. This producer is probably inexperienced and knows very little, if anything, about what it takes to record a high-quality voiceover performance, let alone one that has music and sound effects.

One of my students submitted an audition through one of these member sites and was awarded a job to provide the voiceover work for a radio commercial. It was only after she was hired, and had committed to a fee, that the producer told her they also needed music and sound effects for a completely produced commercial. She was not prepared for this, she is not a production engineer, and she did not have access to the music and sound effects libraries she needed. Yet, she was under contract to provide her services, and because she was eager to please her new client, she was placed in the very uncomfortable position of having to deliver a complex job for a minimal fee. Had she only recorded and delivered the voice track that she had originally agreed to provide, she considered her compensation would have been adequate, because this was one of her first jobs. However, by the time she completed the production, she had put in many hours more than necessary and had to spend her own money for production elements, all of which resulted in her actual compensation equaling far less than minimum wage. Her mistake was that she agreed to the additional demands of the client, and did not have a clear agreement about what she would provide. At her choice, she will never work with that client again.

Many experienced voice talent have commented that it appears as though some of these online audition website operators do not screen audition requests, and that they provide no education or training to those seeking voiceover talent or to their voice talent members. Although the site operators may truly be attempting to serve their paid members by providing some form of ongoing training through podcasts, articles, and webinars, the outward appearance is that they really don't care—or worse, that they care more about the free-riding talent buyers than their paying members.

Some pay-to-play audition sites may actually put members in a bidding war by requiring them to state what they would charge for their work, often without having enough information from which to make an intelligent estimate. The result of this is that the experienced voice talent will not submit auditions, and the beginning voice talent end up setting their talent fees to unrealistically low levels. This does a disservice to everyone concerned. The talent are not fairly compensated for their work, and the client ends up receiving a voice track that, although it might fit his budget, is less than ideal or effective. Do your research and ask for comments from current members before joining any pay-to-play audition site.

Internet audition websites can, however, be a useful resource and a valuable tool for obtaining work for both beginners and professionals. It may take some time, but it is quite possible to land a single job that could pay for several years of membership. If your performing skills are good enough; if your recording quality meets the standards of the producer; if your talent fee is comfortable for both you and the producer; and if you don't mind submitting at least 40 to 60 auditions, or more, before you get a good lead, joining a membership audition website may be a good choice for developing your skills and marketing your talent.

Auditioning from Your Home Studio

The advent of home studios for recording voiceover has made it easy to provide high-quality auditions on a moment's notice. The key to a great sounding audition from your home studio is in two parts: 1) your studio design and equipment, and 2) your performing skills.

If your studio isn't designed properly, your recordings may contain excessive room echo or unwanted outside noise. These must be addressed because there is always the possibility that you may be expected to deliver a high-quality voice track if you get the job. Many voice talent believe that the sound quality of an audition recording is not important because producers are using the audition only to hear how you interpret their script.

I disagree! You have only one chance to make a good first impression. Most auditions will be the first time a producer will experience what you sound like and your performing abilities. If your audition is full of room echo or there's a lawnmower, dog barking, or baby crying in the background, their first impression of you will likely be considerably less than desirable. If you have given so little care to the quality of your audition, should a producer reasonably expect anything more from your work if they hire you? Think of each audition as a customized demo.

RECORDING A "KILLER" AUDITION

To submit auditions that stand a chance of getting you work, you must know how to properly use your equipment, you must know how to work the microphone (mic technique), you must know how to deal with adverse noise conditions, and you must know what you are doing as a performer. In short, you must know how to produce a "killer" audition.

Let's assume you've taken care of all equipment and acoustic issues and that you've got the expertise to record excellent voice tracks. Now what?

The first thing you need to know is that just because you think a script is a perfect match for you, it doesn't mean your performance is what the producer is looking for. All you can do for any audition is to perform to the best of your abilities, using what you consider to be the best choices for your performance. Then, let it go.

Most talent buyers request auditions be sent as MP3 files without any production, music, or effects. In other words, they want to hear only your *dry voice* at the best possible quality. An audition is not the place for you to demonstrate your production skills or your talent for choosing music.

For most auditions, you simply don't have enough information and have absolutely no idea what the producer is looking for. You are effectively second-guessing the producer in an attempt to come up with a performance that you think will meet their needs. Sometimes, they don't really know what they're listening for, so it may be worth sending two, or at most three, different interpretations of their script.

One of the most important things to keep in mind when sending out an audition from your home studio is to follow instructions to the letter, especially if the audition request came from a talent agent. If you are asked to *slate*, or identify, your audition in a certain way, do it exactly as requested! If you are asked to name your MP3 file a certain way, name it that way. If you are asked to send only one track, don't send two. If you are asked to send your MP3 file at a specific sample rate, you had better know how to do it. If you are asked to upload your audition to an FTP website, don't email it. Producers want to know that you have the ability to follow their instructions and take their direction. There may be a specific reason for their request, they may simply be a very controlling producer, or they may be testing you. Read audition instructions carefully, and follow them. If you don't, there's an excellent chance your audition will be never be heard.

WHO WROTE THIS COPY?

It is an unfortunate aspect of the voiceover business that many auditions will arrive as very poorly written scripts. Grammatical and punctuation errors, misspellings, nonsensical syntax, poor sentence structure, and confusing phrasing are commonplace. When you are auditioning a script with any of these issues, your gut instinct may be to change a word here or there, move a sentence, or even rewrite the script to make it "better."

There are several problems with this idea. To begin with, no matter how "bad" a script might seem, you really don't know that the way it is written might be exactly the way the writer intended. If the errors are obvious and extensive, it may be that the script was written in a hurry or perhaps the copywriter is inexperienced, not qualified, or may not be conversationally fluent in writing in English. Or, it could be that the errors are just simple mistakes. It really doesn't matter because your job as a voice actor is to deliver the best audition you can with the script you are given. Your services as a copywriter are not on the table at this stage of the project.

Many inexperienced voice actors make the mistake of thinking that by taking the time to "fix" the copy, they are showing their prospective client that they can work as part of the team, or that their copywriting skills can be an additional benefit to the project. This is flawed thinking because at that stage of the process the client isn't interested in anything you can do beyond your performance of their script. But that's not the worst part.

Suppose you do take the time to massage the script, and you actually do improve upon the original copy. You record a brilliant performance of your revised audition script and submit it. The talent buyer is under no obligation to book your voiceover services, and aside from the fact that you were never asked to rewrite the copy, you have just given them an improved version of their script. If they like your revisions, they are under no obligation to pay you for it, and will very likely use it when they book someone else. Because you provided a free rewrite of their original script, you don't even have any copyright to your new version. All your effort will have been to no avail.

Attempting to "fix" an audition script is essentially a waste of your time and energy. Of course, minor corrections might be appropriate, but the time to discuss major copy corrections is after you are booked for the recording session. You may be surprised that the final script has been miraculously fixed. Some talent buyers have been known to send out a flawed script as a sort of test to see (or hear) what auditioning voice actors do to handle the errors. It's sneaky, but it does happen from time to time.

As a general rule of thumb, when you get a "bad" script, just do the best audition you possibly can and leave the copywriting to the client.

UNCOMPENSATED USE OF YOUR AUDITION

So, you've followed all the instructions and you've sent out a very good dry voice track recording as your audition. At this point, you don't really know who you're sending your audition to, and you certainly don't have any sort of agreement for compensation should you be chosen for the job. If you've sent out a clean recording, the only thing preventing a producer from using your work without telling you is their personal morals and ethics. The vast majority of producers maintain ethical standards and will not use a performer's work without compensation. But there are those unscrupulous producers who will take advantage of a situation. How can you protect yourself so your work will not be used without compensation?

There are several ways to do this, all with the goal of making the audition unusable as a final recording but still providing a good representation of your work. One is to simply send only a partial performance as your audition, leaving out a few critical lines. A similar approach is to change the client name, product name, or phone number. This is not a substantial rewrite, but it does serve to make the audition track unusable while giving the producer a good idea of your performance.

Yet another approach to protecting your audition is to use a *watermark* or drop out the audio at certain key words. In its simplest form, a watermark is a beep, tone, or click, usually on a second audio track, that is strategically placed to interfere with certain words, thus making the track unusable. At first glance this might make perfect sense as a way to protect your performance, and some of the pay-to-play audition sites encourage its use. However, there is a definite down side to using a watermark.

The only reason you would use a watermark is if you suspect that the recipient might steal your work. Producers know this! And many consider use of a watermark to be a reflection of the voice actor's level of experience and professionalism. The logical thought is this: "If this voice actor doesn't trust me to handle their audition with integrity, why should I hire them for this job?" This is simply not a good way to start a professional relationship.

Aside from this negative affect, a watermark can easily be a serious distraction for a realistic evaluation of an audition. No matter how low the volume is set for the watermark, the beep, click, or tone can be extremely distracting and annoying. The result of a watermarked audition can easily be loss of the job. A far better approach to protecting your audition performance is to simply change a few key words of the script.

Preparing for Your Audition

You most likely will feel a rush of excitement when you get the call for your first live audition or you get that email audition that is a perfect fit for your skills. That excitement could quickly turn to panic if you let it. Don't let yourself get caught up in the excitement. Whether you're going out for a live audition, or you're recording in your home studio, focus on the job before you and keep breathing. Approach the audition with a professional commitment to do your best.

If you're auditioning from your home studio, you can take your time and keep recording until you have recorded something you feel comfortable in sending out. If you get the audition script early enough, you can record your audition, let it sit for an hour or so, then listen to it to see if its really something you want to send out. If not, you have the luxury of redoing it or editing out the bad takes.

Live onsite auditions are becoming more and more rare, but when they do happen, they are a completely different experience.

WHERE DID THOSE BUTTERFLIES COME FROM?

As soon as you get the call for a live audition, you will probably begin to feel butterflies in your stomach. This is a good time for you to practice some relaxation exercises. You need to prepare yourself mentally and physically for the audition. Just the fact that you were called to audition is a good sign, so keep a positive mental attitude. After all, you have been invited to be there, and the client wants you to succeed.

THE DAY HAS ARRIVED

On the day of the audition, loosen up with your daily stretches and voice exercises. Dress comfortably, yet professionally. Be careful not to wear clothing or jewelry that will make noise when you are on-mic. If your audition is close to a meal, eat lightly and avoid foods that you know cause problems with your performance.

Plan to arrive at your audition about 15 to 20 minutes before your scheduled time. Make sure you leave enough time to allow for any traffic problems and for parking. If you do not plan ahead, you may arrive too late to read for your part, especially for multiple-voice auditions. When in your car, continue with some warm-up exercises and listen to music that will put you in a positive frame of mind. Sing to songs on the radio to loosen up your voice and relax your inhibitions, but don't overdo it. Use your cork.

Always bring several sharpened pencils for making copy notes and changes, and a bottle of water. A briefcase or tote bag containing your supplies, business cards, a bottle of water, and several copies of your demo can add that extra touch of professionalism to your image. Don't plan on giving your demo or business cards to the people you are auditioning for, unless they request them—they already know who you are. These are for other people you might meet whom you did not expect to be there.

Act as if you know what you are doing, even if this is your first audition. Watch others, follow their lead, and keep a positive attitude.

KEEP TRACK OF THINGS

Under current tax laws, any expenses you incur that directly relate to earning income are deductible, including travel expenses to and from auditions and parking fees, whether or not you get the job. It's a good idea to keep a journal with you so that you can itemize your mileage and expenses. You also may want to keep a record of auditions you are sent on, who the casting people are, where the audition was held, and how you felt about it. You might include names, addresses, and phone numbers to add to your follow-up mailing list. If you are using computer software or an online service to keep track of your expenses, be certain to keep it up to date.

What to Expect at a Live Audition

When you arrive at the audition, you may find several other performers already there. Also, you may find that several auditions are being conducted at the same time, with different copy for a variety of projects. Find the correct audition and pick up your copy. If the audition is for a large account, someone may be "checking-in" the scheduled performers. In most cases, there will simply be a sign-in sheet at the door and a pile of scripts. Once signed in, you are considered available to audition and may be called at any time. If you are early and want to take some time to study the copy, wait a few minutes before signing in. If you are scheduled for a specific time, be sure you are in the waiting area a few minutes before the appointed time.

In many cases, you will see the copy for the first time only after you have arrived on site. However with email, fax, and online casting services, it is becoming more and more common for audition scripts to be delivered ahead of time. On some occasions, for reasons only the producer can understand, you will have to wait until you are in the booth before you know what you are doing. You may even experience an audition where there is no script and the you are simply asked to improvise on lines or props provided by the producer. Fortunately, this is rare, but it does happen.

BE PREPARED TO WAIT

Even if the audition starts on schedule, chances are that within a short time, the producers will be running late. Have something to read or do while you wait for your turn. Stay relaxed and calm, and keep breathing. This is a good opportunity to get to know some of the other performers who are there, if they are willing to talk to you. Many performers prefer to keep to themselves at an audition in order to stay focused or prepare themselves. Always respect the other people who are auditioning. You may end up working with them some day. If the opportunity arises to get to know someone new, it might be in your best interest to take advantage of it.

Remember, networking can be a valuable tool when used properly—it's often not what you know, but who you know that gets you work. Even though these people may be your direct competition, you may make a connection for future jobs that would have otherwise passed you by.

If the copy is for a dialogue spot, you may find another performer willing to *run lines*, or practice the copy with you. This can be an advantage for both of you, even if you do not do the audition together. However, do keep in mind that interaction with the competition can often be distracting.

EXPECT TO BE NERVOUS

Nervous energy is only natural, but it is something you need to control. You must be able to convert your nervous energy into productive energy for your performance. Focus on your acting rather than on the words. You know you are nervous and so do the casting people. Don't waste time trying to suppress or conceal your nervousness. Breathe through it and focus on converting the nervous energy into positive energy. Many of the top stage and screen actors become very nervous before a performance. It's a common condition of all performing arts. Bob Hope, one of the top comedians of the twentieth century, was known to be incredibly nervous before going on stage. When asked about it, Mr. Hope said he valued his nervousness because he felt it gave him an edge while performing. Adjust for your nervousness by taking a long breath deep down through your body to center yourself and focus your vocal awareness. Chapter 5, "Using Your Instrument," explains how to do this.

EXPECT TO BE TREATED LIKE JUST ANOTHER VOICE

At most auditions, the people there really want you to be the right person for the job. However, if the audition is for a major account in a major city, expect the possibility of being treated rudely by people who just don't care and are trying to rush as many performers through the audition as possible in a limited amount of time. If anything other than this happens, consider yourself lucky. Many times, the people handling the audition are just there to record your performance and have little or nothing to do with the client who will eventually be hiring the actors.

PREPARE YOURSELF

Use your waiting time for woodshedding: to study it for your character, key words, target audience, and for anything that is unclear—especially words you don't understand or don't know how to pronounce. Try to get a feel for what they are looking for. What attitude? What sort of delivery? Most of the time, your choices will be clear. Sometimes, there will be a character description on the copy, or some notes as to what the producers are after. If there is a graphic or sketch of the character you are to play, make note of any physical features, body language or other characteristics that might be used to develop your performance choices. Note the important words or phrases, the advertiser and product name, where to add drama or emotion, where to pull back. Mark your copy in advance so that you will know what you need to do to achieve the delivery you want. Rehearse out loud and time yourself. Don't rehearse silently by merely reading and saying the words in your mind. In order to get an accurate timing and believable delivery, you must vocalize the copy. Make sure you know how you will deliver the copy in the allotted time.

Be careful not to overanalyze. Read the copy enough times to become familiar with it, then put it aside. Overanalyzing can cause you to lose your spontaneity. Decide on the initial choices for your delivery, and commit to them. But be prepared to give several different variations. Also, be prepared for the director to ask for something completely unexpected.

Auditions for a TV spot may or may not have a storyboard available. This may be attached to the script, or posted on a wall. It may be legible or it may be a poor copy. A *storyboard* is a series of drawings, similar to a cartoon strip, that describes the visual elements of a TV commercial or film that correspond to the copy. If there is a storyboard for your audition, study it thoroughly. Instead of a storyboard, many TV-commercial scripts have a description of the visuals on the left side of the page with the voiceover copy on the right side. The storyboard or visual description is the best tool you have to gain an understanding of a video or film project. If you only focus on the words in the script, you will be overlooking valuable information that could give you the inspiration you need to create the performance that gets you the job.

MAKE A GOOD FIRST IMPRESSION

Greet the producer or host, introduce yourself, shake hands, be spontaneous, be sincere, and be friendly. If you are auditioning near the end of a long day, the people in the room may not be in the best of moods. You still need to be friendly and professional as long as you are in that room. Remember, first impressions are important. Your first impression of them might not be very good, but you need to make sure that their first impression of you is as good as possible. Your personality and willingness to meet their needs will go a long way.

Answer any questions the casting producer, agency rep, or engineer ask of you. They will show you where the mic is and let you know when they are ready for you to begin. Do not touch any equipment—especially the mic. Let the engineer or someone from the audition staff handle the equipment, unless you are specifically asked to make an adjustment.

There will probably be a music stand near the microphone. Put your copy here. If there is no stand, you will have to hold the copy, which may restrict your physicalization. If headphones are available, put them on—this may be the only way you will hear cues and direction from the control room. In some cases, you may be asked to read along with a *scratch track* for timing purposes and you will need the headphones to hear it.

A *scratch track* is a preliminary test recording that is usually used as a guide for video editing or as a sample for the client. Sometimes, you might be lucky enough to actually have a music track to listen to as you perform. This can be very helpful, because music is often used to help set the mood and tone for a commercial and can provide clues about the target audience. If you don't have anything to work against, you might ask the producer or director to give you an idea of the rhythm and pacing for the project.

Before you start, the engineer or producer will ask you for a *level*. This is so he can set the proper record volume. When giving a level, read your copy exactly the way you plan to perform it. Many people make the mistake of just saying their name or counting 1, 2, 3,... or speaking in a softer voice than when they read for the audition. Use this as an opportunity to rehearse your performance with all the emotion and dynamics you will use when the engineer starts recording. In fact, many times, the engineer will actually record your level test—and occasionally, that take, or portions of it, may end up in the final product.

MAKE THE COPY YOUR OWN

Your best bet for getting a job from an audition is to discover the character in the copy and allow that character to be revealed through your performance. Play with the words! Have fun with them! Put your personal spin on the copy! Do not change words, but rather add your own unique twist to the delivery. Don't focus on technique or over-analyze the script. Use the skills of voice acting you have mastered to make the copy your own. If they want something else, they will tell you.

Making the copy your own is an acquired acting skill. It may take you a while to find your unique style, but the search will be worthwhile. Chapter 10, "The Character in the Copy," discusses this aspect of voiceover work.

INTRODUCE YOURSELF WITH A SLATE AND DO YOUR BEST

You will have only a few moments to deliver your best performance. Remember, you are auditioning as a professional, and those holding the audition expect a certain level of competency. When asked to begin, start by slating your name, then perform as you have planned.

To *slate*, clearly give both your first and last name, your agent (or contact info), and the title of your audition. Your agent or the instructions may request that you slate in a specific order, add additional information, or leave an item out. Many talent buyers only want your name and your agent's name in the slate. The following is a typical audition slate:

"Reina Bolles with Cameron Ross Agency."

There are two schools of thought on slates: One is to slate with your natural voice. The other is to slate in character or in a manner consistent with the copy. Slating in your natural voice may be like a second audition by giving the casting person a taste of who you really are and what your voice is like. This may result in a future booking based only on your slate. Slating in character provides continuity for the audition. Neither approach is correct in all situations. If you have an agent, ask how they would prefer you handle the slate. For some auditions, you may be asked to *not* include a slate, but you may be asked, instead, to include your name as part of the MP3 file name if you are auditioning from your home studio. Use your best judgment when slating, but always keep it short.

After your slate, wait a few beats as you prepare yourself mentally with a visualization of the scene, and physically with a good diaphragmatic breath, then begin your performance. Don't just jump in and start reading. If you are auditioning from home, you have the luxury of editing later on.

At a live audition, you may, or may not, receive direction or coaching from the casting person. If you are given direction, it may be completely different from your interpretation. You may be asked to give several different reads, and you need to be flexible enough to give the producer what he or she wants, regardless of whether you think it is the right way. You may, or may not, be able to ask questions. It depends entirely on the producer. There may be one, two, three, or more people in attendance. Stay focused and don't let yourself get distracted by the people in the room.

Many auditions are simply intended to narrow down possible voices and the performance is secondary. The copy used in some auditions may be an early draft, while other auditions may provide a final script. Either way, you are expected to perform to the best of your abilities. Do your best interpretation first, and let the producer ask for changes after that. It may be that your interpretation gives the producer an idea he or she had not thought of, which could be the detail that gets you the job. In some cases, you might be asked to simply improvise something, and won't even have a script.

Offering your opinions is usually not a good idea at an audition, but it is something you can do if it feels appropriate. Some producers may be open to suggestions or a different interpretation, while others are totally set in their ways. If the producer is not open to it, he or she will tell you. These are not shy people. At other times, the audition producer will be doing little more than simply giving slate instructions and recording your performance.

The casting person will let you know when they have what they want. Two or three reads of the copy may be all the opportunity you have to do your best work. They may, or may not, play back your audition before you leave—usually not. If you do get a playback, this is a good opportunity for you to study your performance. Do not ask if you can do another take unless you honestly believe you can do a much better performance, or unless the producer asks if you can do something different. When you are done, thank them, and then leave. Your audition is over. If you like, take the script with you, unless you are asked to return it.

After the Audition

Most of the time, after leaving a live audition, you will simply wait for a call. If you do not hear anything within 72 hours, you can safely assume that you did not get the job. As a general rule, agents call only if you get the booking or are requested for a callback.

While you are waiting for that call, don't allow yourself to become worried about whether or not you will get the job. Write your follow-up letter and continue doing what you usually do. Remember that voice acting is a numbers game, and that if you don't get this job, there is another opportunity coming just down the road.

WHEN THE ACTORS ARE GONE

At the end of the day, the audition staff takes all the auditions and returns to their office. There, they listen to the recordings and narrow down the candidates. They may choose the voice they want right away, or they may ask for a second audition—called a *callback*—to further narrow the candidates. The audition producer will contact the appropriate talent agents to book talent for a session or callback, or may call independent performers directly. Voiceover audition callbacks are fairly rare, but when they occur, they are usually for a major regional or national account.

If you are scheduled for a callback, you may find there is less pressure, the attitude of the people involved may have changed, and the script may be different. At a callback, the producer may say that they really liked what you did on take 3 of the first audition. Chances are, unless they have a recording of that audition to play for you, or unless you have an exceptionally good memory, you will not remember what you did on take 3, or any of the takes for that matter. When this happens, all you can do is go for your best interpretation of the copy (which probably changed since the original audition), and use any direction from the producer to guide you.

The simple fact that you are called back for a second audition shows that there is something about your performance that the producer likes. Try to find out what it was that got you the callback. Do whatever you can to stay on the producer's good side and make friends. If for some reason you do not get this job, the producer may remember you next week or next month when another voiceover performer is needed for another project.

After the callback, the audition staff once again takes their collection of auditions (much smaller this time), and returns to their office. This cycle may be repeated several times until the producer or client is satisfied that the right voice is chosen.

BE GOOD TO YOURSELF

When you left the audition, you probably came up with dozens of things you could have done differently or "better." You might even feel like going to your car, winding up the windows, and screaming real loud. Second-guessing yourself is self-defeating and counterproductive. Instead of beating yourself up with negatives, do something positive and be good to yourself. You've done a good job! You have survived your audition. Now you deserve a treat. Take yourself out to lunch, buy that hot new DVD you've been wanting, or simply do something nice for yourself. It doesn't really matter what you do—just do something special. Then let it go.

Agency Demos

Not all auditions are held for the purpose of casting a final project. In some cases, you will be auditioning for a demo. This type of *demo* is a spec spot that is produced by an ad agency as a potential commercial to sell an idea to their client. Often, the entire concept of an advertising campaign is changed between the demo and the time the final spot is produced.

You may be told that the audition is for a demo at audition, or at some time later. Either way, the recording from your audition normally will not be the recording used for the demo spot. If it is to be used for the demo, you will be compensated for your time at the audition.

AFTRA has a separate rate for demo sessions, which is different from their commercial scale. Demos are usually paid for on a one-time-only fee basis. However, a demo can be upgraded to a commercial if the client decides to use it. In this case, if you are a union member, your fee would also be upgraded to the commercial rate. Independent voice actors need to negotiate their own fee for a demo, or let their agent handle it.

The Voiceover Survival Kit

As you begin to work voiceover auditions and sessions you will find there are certain things you will want to always have with you. Here are some common items for a voiceover "survival kit." Feel free to add more.

- Water
- Pencil
- White-out pen
- Small photos
- A small travel pack of tissue
- Throat lozenges
- Dry mouth spray
- Chap stick or lip balm
- A green apple or dry mouth spray
- A wine bottle cork
- Business cards
- Demo CD
- Blank invoice or agreement
- Other items you'll think of later

25

You're Hired!
The Session

Congratulations, you've got the job! Your audition was the first step—and on average, you'll have submitted about 40□60 auditions for every job you land. The client likes your audition better than anyone else's. You already know the details about the project, they've agreed to hire you at a fee you or your agent has negotiated, and you have a signed booking agreement.

Now, you need to know the details of the session: When does the session start (your *call time)*? Where will the session be recorded, or are you expected to provide the recordings from your own equipment? If you're going to be recording in your home studio, you may or may not get direction from your client. For the purpose of this chapter, your session will be at a local recording studio in your city. Either way, the process will be basically the same. It's just that if you're recording voice tracks in your own studio, you'll be wearing many hats.

A Journey through the Creative Process

The recording session is where your voice is recorded and all the pieces of the puzzle are put together to create a final commercial or soundtrack. Besides your voice, the project may include music, sound effects, other voices, recordings of interviews, or other *sound bites*, and digitally processed audio. It is the job of the recording engineer to assemble these various puzzle pieces to form the picture originally created in the mind of the producer or writer. It can be a challenging and time-consuming process.

If you are recording from your personal home studio, you are your own engineer, producer, and director. You may have your client on the phone or on a *phone patch* connection to your audio mixer, but you are ultimately in

345

control of both the recording process and your performance. Doing both can be a challenge, and although the purpose of this book is not about audio production, it is important that you at least have a basic idea of what's happening on the other side of the glass.

Much of the creative process involves a lot of technology and a high level of creativity from the engineer. As a voiceover performer, only a small portion of the recording process involves you. To give you a better idea of how your performance fits within the whole process, the rest of this chapter will be devoted to walking you through a typical production.

THE PRODUCTION PROCESS

It all begins with an idea! That idea is put into words on a script, which may go through many revisions and changes. At some point during the script's development, thoughts turn to casting the roles in the script. In some cases, a role may be written with a specific performer in mind, but this is usually the exception to the rule. To cast the various roles, the producers listen to demos and hold auditions. The audition process (Chapter 24) narrows the playing field to select the most appropriate voice talent for the project at hand. If your voice is right for the part, and your demo or audition was heard by the right person, you could be hired for a role.

Be absolutely certain you arrive *before* your scheduled session time. It is much better to be early and have to wait a few minutes than for you to be late and hold up the session. Recording studios book by the hour, and they are not cheap. Basic voiceover session time can be in the range of $100 an hour or more, depending on the studio. Some Hollywood and New York studios book out for $300 to $500 per hour. You do not want to be the person responsible for costing the client more money than necessary.

Time is also of the essence when you are in the studio. Things can happen very fast once you are on-mic and recording begins. You need to be able to deliver your best performance within a few takes. If the producer or director gives you instructions, you need to understand them quickly and adapt your delivery as needed.

If you are working a dialogue script with a performer you have never met before, you both need to be able to give a performance that creates the illusion that your separate characters are spontaneous and natural. This is where your character analysis and acting skills really come into play.

SESSION DELAYS

Studio time is a valuable commodity. The producer will want your best performance as quickly as possible. In reality, it may take a while to get it. A voiceover session for a :60 radio commercial can take as little as 5 minutes to as much as an hour or longer. A long session for a seemingly simple spot can be the result of one or more of the following factors:

- There may be several voices speaking (dialogue or multiple-voice copy), and it may take some time to get the characters right.
- Microphone placement may need to be adjusted or the mic may need to be changed.
- The copy may require major changes or rewrites during the session.
- A session being done to a video playback may require numerous takes to get the timing right.
- There may be technical problems with the equipment.
- The voice tracks may need to be inserted into a rough spot for client approval before the performers can be released.
- The session may be a *phone patch* and the client may request changes that need to be relayed through the producer or engineer.
- Your client may not know what he or she really wants.
- There may be several would-be directors trying to offer their ideas, creating unnecessary delays.
- The voiceover performer may lack experience, and may not be able to give the producer the desired reading without extensive directing.
- An earlier session may have run overtime, causing all subsequent sessions to start late.

Regardless of how long you are in the studio, you are an employee of the ad agency, producer, or client. Present yourself professionally and remain calm. Above all, do your best to enjoy the experience. Keep breathing, stay relaxed, and keep a positive attitude.

WORKING WITH PRODUCERS, DIRECTORS, WRITERS, AND CLIENTS

A voice-actor friend of mine once described a producer/director as "headphones with an attitude." Regardless of the producer's attitude, you need to be able to perform effectively. You must be able to adapt your character and delivery to give the producer what he or she asks for. And you need to be able to do this quickly with an attitude of cooperation.

It is common for a producer, after doing many takes, to decide to go back to the kind of read you did at the beginning. You need to be able to do it! It is also common for a producer to focus on getting exactly the right inflection for a single word in the copy. You might do 15 or 20 takes on just one sentence or a single word, and then a producer will change his mind and you will have to start all over.

Every producer has a unique technique for directing talent. You must not let a producer frustrate you. Occasionally, you will work for a producer or writer who is incredibly demanding, or simply does not know what he or she wants. When working for this type of person, just do your best and when you are done, leave quietly and politely. When you are alone in your car, with the windows rolled up, you can scream as loud as you like.

There are some producers who operate on a principle of never accepting anything the first time—no matter how good it might be. Your first take might be wonderful—you hit all the key words, get just the right inflection, and nail the attitude. Yet, the producer may have you do another 10 takes, looking for something better, all the while drifting off target. When all is said and done, that first good take may be the one that's used.

WHO ARE THOSE PEOPLE?

Some sessions may be crowded with many people deeply involved with the project you are working on. Of course, the studio engineer will be present, and there will usually be someone who is the obvious producer/ director. But the client or storeowner may also be there, as well as his wife, their best friend, the agency rep from their ad agency, the person who wrote the copy, and maybe even an account executive from a radio or TV station. All these people have an opinion about what you are doing, and may want to offer suggestions about what you can do to improve your performance. It's a nice thought, but too many directors will make you crazy.

You may actually find yourself getting direction from more than one person. One of the obvious problems with this is conflicting direction. As a performer, you must choose one person in the control room to whom you will listen for direction and coaching. Most of the time this should be the producer handling the session. However, if it is obvious that the producer cannot control the session, you might choose someone else, if you feel the person is a better director. Most studios that record voiceover have engineers who are very experienced in directing voice actors. It is not uncommon for an engineer to "take over" the session if he recognizes that the client or producer is not getting an effective performance or the desired results.

Once you have made your choice, you must stick with that person for the duration of the session. Changing directors in mid-session will only make your performance more difficult. Simply focus your attention on the person you picked and direct your questions and thoughts to only that person, mentioning him or her by name when necessary. There's a way of doing this that won't offend anyone.

When someone else presses the talkback button and gives you some direction, you need to bring control back to the person you chose. Allow the interruption to happen, and then refer to your chosen director for confirmation or further comment. After this happens a few times, the would -be director will usually get the hint and let the person in charge handle the session. Future comments will then be routed to you via your chosen producer or director—as they should be.

Types of Sessions, Setups, and Script Formats

There are many different types of voiceover projects, and recording sessions come in all shapes and sizes, with a variety of format styles.

DEMOS

A demo session is for a project that has not yet been sold to the client. It will be a demonstration of what the ad agency is recommending. The client may or may not like it. The ad agency may or may not get the account. A demo is a commercial on spec (speculation).

Mel Blanc, one of the great animation character voices of the 1950s and 1960s, once gave the following definition of working on spec:

> *Working on spec is doing something now for free, on the promise you will be paid more than you are worth later on. Spec is also a small piece of dirt!* (Mel Blanc, from *Visual Radio*, 1972, Southern California Broadcasters Association)

Ad agencies, television stations, and radio stations often do projects on spec when they are attempting to get an advertiser's business. The potential profit from a successful advertising campaign far outweighs the cost of producing a spec commercial—provided the agency lands the account.

Demos will not air (unless they are upgraded by the client), and are paid at a lower scale than regular commercials. In some cases, the demo serves as an audition for the ad agency. They may have several different voiceover performers booked to do the demo session. It is not technically an audition, since completed spots will be produced. Instead, demos are intended to give the advertiser a choice of performers for the final commercial. If the demo is simply upgraded, your agent will be contacted and you will be paid an additional fee. If a separate session is booked, you will be contacted, scheduled, and paid an additional fee.

SCRATCH TRACKS

A *scratch track* is similar to a demo in the sense that it is the preliminary form of a commercial. The major difference is that a scratch track is used as a reference for a commercial that is already in the process of being produced. Scratch tracks are most often used for TV commercials and other video productions, and serve as a reference track for the video editor before the final voice track is recorded. A scratch track will often be voiced by the producer, director, or sometimes the editor or audio engineer, and the music, sound effects, and other elements of the spot may or may not be in their final placement.

As a voiceover performer, you may be providing the original voice for a scratch track, or you may be providing the final voice that replaces an earlier recorded voice used on an already-assembled scratch track. Either way, your job will be to perform as accurately as possible to the existing timing. The process is similar to *ADR* (Automated Dialogue Replacement) used in the film industry, except that you are working to an audio track instead of lip-syncing to a picture.

Just as for a demo session, your performance for a scratch track may be good enough for use in the final spot. You or your agent will know if the scratch track session is for a demo or a final commercial, and you will be paid accordingly.

REGULAR SESSION

This is a session for production of a final commercial. Many engineers refer to *regular sessions,* to differentiate them from demos, tags, scratch tracks and so on. The only difference between this type of session and all the others is that it is for a complete production.

SESSION SETUPS

There are two basic session setups: *single session* and *group session.* At a *single session*, you are the only person in the studio, but this does not mean you are the only voice that will appear in the final project. Other performers, to be recorded at another time, may be scheduled for different sections of the project, or for the tag. There will be only one microphone, a music stand, a stool, and a pair of headsets. Many recording studios also have monitor speakers in the studio, so you can choose to wear the headset or not. Let the engineer make all adjustments to the mic. You can adjust the stool and music stand to your comfort.

Multiple-voice, or *group sessions*, are often the most fun of all types of sessions simply because of the ensemble. Each performer normally has his or her own mic, music stand, and headset. Depending on the studio, two performers may be set up facing each other, working off the same mic, or on separate mics in different areas of the studio. A group session is like a small play, only without sets. Looping is almost always done as a group session with from a few to a few dozen voice actors in the studio.

SCRIPT FORMATS

There are a variety of script formats used in the business of voiceover. Radio, television, film, multimedia, video game, and corporate scripts all have slight differences. Regardless of the format, all scripts include the words you will be delivering and important clues you can use to uncover the building blocks of any effective performance.

The Session: Step-by-Step

Let's walk through a session from the moment you enter the studio, until you walk out the door. Much of this is reviewed from other parts of this book; however, this will give you a complete picture of a studio session. After reading this section, you will know what to expect and should be able to act as if you have done it all before. Although the studio session process is very consistent, there are many variables that may result in variations on the following scenario. Just "go with the flow" and you will be fine.

Once you enter the studio lobby, your first contact will be the receptionist. Introduce yourself, and tell her which session you are attending. If the studio is in an office building and you paid to park in the building's parking structure, don't forget to ask if the studio validates.

The receptionist will let the producer know you are there. If you don't already have the script, you might be given your copy at this time, or you might have to wait until the producer comes out of the control room. Depending on how the session is going, you may have to wait awhile.

The producer or engineer will come out to get you when they are ready, or the receptionist will let you know that you can go back to the control room. Or, someone might come out to let you know that the session is running late. There are many things that can put a session behind schedule. Remember, this is a hurry-up-and-wait kind of business.

When you enter the control room, introduce yourself to the producer, the engineer, and anyone else you have not yet met. You can be certain that anyone in the control room is important, so be friendly and polite.

If you did not receive the copy earlier, it will be given to you here. This is your last, and sometimes only, opportunity to do a quick "woodshed," or script analysis, set your character and ask any questions you might have about the copy. Get as much information as you need now, because once you are in the studio, you will be expected to perform. Get a good idea of the target audience and correct pronunciation of the product's and client's names. Make notes as to attitude, mood, and key words. Mark your script to map your performance so that you will know what you are doing when you are in the studio. The producer or engineer may want you to read through the copy while in the control room for timing or to go over key points. When the engineer is ready, you will be escorted to the studio.

In the studio, you will usually find a music stand, a stool, headphones, and the microphone. Practice good studio etiquette and let the engineer handle any adjustments to the mic. Feel free to adjust the music stand to your comfort. If a stool is there, it is for your convenience, and you may choose not to use it if you feel more comfortable standing. Some studios will give you the option of performing without having to wear headphones, but for most you will need to wear them to hear the director. Find out where the volume control is before you put on your headphones.

Make sure your cell phone is turned off, or better yet, leave it in the control room.

The microphone may have a *pop stopper* in front of it, or it may be covered with a foam *wind screen*. The purpose of both of these devices is two-fold: first, to minimize popping sounds caused by your breath hitting the microphone and second, to minimize condensation of breath moisture on the microphone's diaphragm. Popping can be a problem with words containing plosives such as "P," "B," "K," "Q," and "T." If the wind screen needs to be adjusted, let the engineer know. If the mic is properly positioned, the pop screen may not be needed.

When the engineer is ready to record, you will be asked for a *level* or to *read for levels*. He needs to set his audio controls for your voice. Consider this a rehearsal, so perform your lines exactly the way you intend to once recording begins. You may do several reads for levels, none of which will likely be recorded. However, the producer or engineer may give you some direction to get you on the right track once recording begins. Some engineers will record your rehearsals, which occasionally are the best takes.

The engineer will *slate* each take as you go. You will hear all direction and slates in your headphones. This is not the same as slating your name for an audition. The engineer may use an audio slate or identify the project or section you are working on, followed by "take 1," "take 2," and so on. Or he may simply use flag markers inserted into the digital project. Before or after an audio slate, you may receive some additional direction.

Do not begin reading until the engineer has finished his slate and all direction is finished. You will know when you can start by listening for the sound of the control room mic being turned off. If you speak too soon, your first few words might be unusable. Wait a second or two after the slate, get a good supporting breath of air, begin moving, then begin speaking.

As you are reading your lines, the engineer will be watching your levels and listening to the sound of your voice. He will also be keeping a log sheet and will time each take with a stopwatch. He may also be discussing your delivery or possible copy changes with the producer or client.

Common Direction Terms

After each take, expect to receive some direction from the producer. Do not change your attitude or character, unless requested by the producer. Do not comment about things you feel you are doing wrong, or ask how you are doing. Let the producer guide you into the read he or she is after.

Marc Cashman (**www.cashmancommercials.com**) has compiled a list of common direction terms from some of the top voiceover resources available, including prior editions of this book. Here's his list:[1]

Accent it: Emphasize or stress a syllable, word or phrase.

Add life to it: Your reading is flat. One expert advises: *"Give it C.P.R.: Concentration, Punch, Revive it!"*

Add some smile: Simply put, smile when you're reading. It makes you sound friendly and adds more energy to your read.

Be authoritative: Make it sound like you know what you're talking about. Be informative.

Be real: Add sincerity to your read. Similar to *"make it conversational."* Be genuine and true-to-life in your delivery.

Billboard it: Emphasize a word or phrase, most always done with the name of the product or service.

Bring it up/down: Increase or decrease the intensity or volume of your read. This may refer to a specific section or the overall script.

Button it: Put an ad-lib at the end of a spot.

Color it: Give a script various shades of meaning. Look at a script as a black and white outline of a picture that you have to color, with shading and texture.

Don't sell me: Throw out the "announcer" voice, relax; the read is sounding too hard-sell.

Fade in/fade out: Turning your head toward or away from the microphone as you are speaking, or actually turning your entire body and walking away. This is done to simulate the "approach" or "exit" of the character in the spot.

False start: You begin and make a mistake. You stop, the engineer refers to this as a *false start* and either goes over the first slate or begins a new slate.

Fix it in the mix: What is done in postproduction, usually after the talent leaves. This involves fixing levels, editing mouth noises, etc.

Good read: You're getting closer to what they want, but it's not there yet.

Hit the copy points: Emphasize the product/service benefits more.

In the can: All recorded takes. The engineer and producer refer to this as having accomplished all the takes they need to put the spot together.

In the clear: Delivering your line without *stepping* on other actors' lines.

In the pocket: You've given the producer exactly what they want.

Intimate read: Close in on the mic more, speak with more breath, and make believe you're talking into someone's ear.

Keep it fresh: Giving the energy of your first take, even though you may be on your twentieth.

Let's lay one down: Let's start recording.

Less sell/More sell: De-emphasizing/stressing the client name/benefits.

Let's do a take: The recording of a piece of copy. Each take starts with #1 and ascends until the director has the one(s) they like. Also heard: *Let's lay it/one down.*

Let's get a level: The director or engineer is asking you to speak in the volume you're going to use for the session. Take advantage of this time to rehearse the copy. Any shouts or yelling will require you to turn your head 45☐90 degrees away from the mic. If the mic needs to be adjusted, the engineer will come into the booth. Do not move the mic unless instructed to do so.

Make it conversational: Just like it sounds, make your read more natural. Throw out the "announcer" in your read, and take the "read" out of your delivery. If it sounds like you're reading, you won't be believable. Pretend you're telling a story, talking to one person. Believe in what you're saying.

Make it flow: Also heard as: **Smooth it out.** Avoid choppy, staccato reads, unless the character calls for it.

More/less energy: Add more or less excitement to your read. Use your body to either pump yourself up or calm yourself down. Check with the engineer (i.e., do a level) to make sure you are not too loud or soft.

Mouth noise: The pops and clicks made by your mouth, tongue, teeth, saliva and more. Most mouth noises can be digitally excised, but make sure that you don't have excess mouth noise, because too much is an editing nightmare and will affect your work. Water with lemon or pieces of green apple can help reduce or eliminate most mouth noise.

One more time for protection: The director wants you to do exactly what you just did on the previous take. This is similar to "that was perfect, do it again." This gives the director and engineer a bit more selections to play with, should they need them in post-production.

Over the top: Pushing the character into caricature.

Pick it up: Start at a specific place in the copy where you made a mistake, as in: *Pick it up from the top of paragraph two,* or *Let's do a pick-up at the top of the second block.*

Pick up your cue: Come in faster on a particular line.

Pick up the pace: Pace is the speed at which you read the copy. Read faster, but keep the same character and attitude.

Play with it: Have fun with the copy, change your pace and delivery a bit, try different inflections.

Popping: Noise resulting from hard consonants spoken into the mic. Plosives, which sound like short bursts from a gun, are most evident in consonants like B, K, P, Q and T.

Punch-in: The process of recording your copy at an edit point in real time. In a punch-in, as opposed to a "pick-up," the engineer will play back part of the copy you recorded and expect you to continue reading your copy at a certain point. The director will give you explicit directions as to where in the script you will be "punched in," and you will read along with your prerecorded track until your punch-in point. From there, you'll continue recording at the same level and tone you originally laid down.

Read against the text: Reading a line with an emotion opposite of what it would normally be read.

Romance it: Also "Warm up the copy." Make it more intimate.

Run it down: Read the entire script for level, time, and one more rehearsal before you start recording.

Shave it by…: Take a specific amount of time off your read. Also heard as "shave a hair." If your read times out at :61, the director might ask you to "shave it by 1.5 seconds."

Skoche more/less: A little bit, just a touch more or less. This can refer to volume, emphasis, inflection, timing, attitude, etc.

Split the difference: Do a take that's "between" the last two you just did. For example, if your first take comes out at :58, and your second take comes out at :60, and the director asks you to "split the difference," adjust your pacing so the third take should be in at :59. Or, if your first take is monotone-ish and your second one is very "smiley," and the director asks you to "split the difference," adjust your read so that the third take will be somewhat in-between the first two.

Stay in character: Your performance is inconsistent. Whatever character and voice you commit to, you have to maintain from beginning to end, take after take after take. Focus. Be consistent with your character and voice.

Stepping on lines: Starting your line before another actor finishes theirs. Sometimes the director wants actors to "overlap" their lines, or interrupt. Others want each line "in the clear," where there is no overlapping or stepping.

Stretch it/Tighten it: Make it longer/shorter.

Take a beat: Pause for about a second. You may be asked to do this during a specific part of the script, like in-between paragraphs, or inside of a sentence or in a music bed. A good sense of comic timing is particularly helpful.

Take it from the top: Recording from the beginning of a script.

That's a buy/keeper: The take that everyone loves—at least the director loves. If the client loves it, then it's accepted.

That was perfect—do it again: An inside joke, but a compliment. Usually the producer wants you to reprise your take "for safety" (i.e., to have another great alternate take).

This is a :15/:30/:60: Refers to the exact length of the spot in seconds, also known as a read or take.

Three in a row: Reading the same word, phrase, sentence or tag three times, with variations. Each read should have a slightly different approach, but all should be read in the same amount of time. The engineer will slate three in a row "a, b, and c."

Throw it away: Don't put any emphasis or stress on a certain phrase, or possibly the whole script.

Too much air: Noise resulting from soft consonants spoken into the mic. Most evident in consonants like F, G, H, and W, and word beginnings and endings like CH, PH, SH, and WH.

Under/over: Less or more than the time amount needed. If you were *"under or over"* you need to either shorten or lengthen your delivery and *"bring it in"* to the exact time.

Warm it up a little: Make your delivery more friendly and personal. Whatever makes you feel warm and fuzzy is the feeling you should inject into your delivery.

Woodshed: To practice or rehearse a script, reading out loud. From the old days of theater where actors would rehearse in a wood shed before going on stage.

Wrap: The end—as in "that's a wrap!"

You will hear many other directions. Do your best to perform as the director requests. There is a reason why he or she is asking you to make adjustments, although that reason will sometimes not be clear to you.

One of my favorite directing stories is one that Harlan Hogan tells about a session he once voiced. He had just completed a delivery that the producer said was extremely good, but wasn't quite where he wanted it. In the producer's words, "...that last take was a bit burgundy, I'm looking for something a little more mauve." With direction like that, what could Harlan do? So he delivered the script exactly the way he had just done, and the producer's response was "...now that's what I'm looking for!" Go figure.

Producers usually have an idea of what they want, and may or may not be receptive to your suggestions. Find out what the producer is looking for when you first read the script. Once in the studio, you should be pretty much on track for the entire session. If you get a great idea, or if it appears that the producer is having a hard time making a copy change, by all means speak up. You are part of a team, and part of your job is to help build an effective product. If your idea is not welcome, the producer will tell you.

Recording studio equipment sometimes has a mind of its own. There are times when the engineer may stop you in the middle of a take because of a technical problem, and you may have to wait awhile until it is corrected. Once corrected, you need to be ready to pick up where you left off, with the same character and delivery.

If you left your water in the control room, let the engineer know and it will be brought in for you. If you need to visit the restroom, let them know. If you need a pencil, let them know. If you need *anything*, let them know. Once your position is set in front of the microphone (on-mic), the engineer will prefer that you not leave the studio, or change your position. If your mic position changes, you can sound very different on different takes, which can be a performance continuity problem for the engineer if he needs to assemble several takes to build the final commercial. This can be a problem when doing long scripts or lots of takes. If you must move off-mic, try to keep your original mic position in mind when you return to the mic.

Be consistent throughout your session. Changes in dynamics may be useful for certain dramatic effects, but, generally, you will want to keep your voice at a constant volume or in a range consistent with your character. If your performance does call for sudden changes in volume, try to make sure they occur at the same place for each take. This becomes important later on, when the engineer edits different takes together. If your levels are erratic, the changes in volume may become noticeable in the final edit.

You know what the producer wants. You stay in character. Your timing and pacing are perfect. Your enunciation and inflection are on track. Your performance is wonderful. The producer is happy. The engineer is happy. And, most important, the client is happy. That's it! You're done, right?

Not quite.

Wrap It Up

Before leaving the studio, make sure you sign the contract for your services. If you are a union member, the producer will probably have a contract already filled out for you. Read the parts of the contract that apply to your session before signing. If you were booked for one commercial (spot announcement), and the producer had you do three spots plus tags, make sure the changes are made on the contract. Also make sure you call your agent and let her know about the changes. If you are unsure of anything on the contract, call your agent *before* signing the contract.

For union work, send your AFTRA form to the union within 48 hours of the session to avoid any penalties. The union form is the only way AFTRA has of tracking your work, and making sure you are paid in a timely manner. If you are working freelance, make sure you are paid before you leave the studio, or that you have a signed invoice or deal memo—and make certain you have the contact address and phone number of your client. If you have a merchant account, you can take a credit card number to be processed, or to hold as a guarantee until your check arrives. You've completed your part of the agreement, and you are entitled to be paid. It's up to you if you agree to have your payment sent to you, but keep in mind that you take a risk of delays or not being paid if you do this.

It's good form to thank the producer, engineer, client, and anyone else involved in the session before you leave. Keep the script for your files, if you like. If you think your performance was especially good, you can ask the producer for a copy of the spot when it is finished. You can ask for a CD but don't be surprised if you only get an MP3 file emailed to you. If the project is a TV commercial, there may be a charge for you to receive a copy. In this digital age, finished commercials are increasingly being distributed to stations via ISDN networks directly from the studio's computer, emailed as MP3 files, uploaded to a website, or sometimes mailed as a one-off CD. One way to ensure that you get a copy is to include a clause to that effect in your agreement. However, even with that, you may find yourself waiting several weeks, or even months, before you get it.

Once your session is over and the paperwork is done, you are free to leave. Your job is done, so don't stick around for the rest of the session or to talk. The producer and engineer have lots of work to do and your presence can cause delays, costing time and money. After you are gone, the process of assembling all the pieces of the puzzle begins. It may take from several hours to several days before the final audio track is complete.

If your session is for a TV commercial, the completed audio will often be sent to a video postproduction house where the video will be edited to your track to create a final TV spot. In some cases, just the opposite occurs—the video may have been edited to a scratch track, and the purpose of your session would have been to place your voiceover against the preproduced video. Once mastered, a number of copies are made and distributed to the radio and TV stations scheduled to air the spot.

Follow up your session with a thank-you note to the producer. Thank him or her for good directing or mention something you talked about at the session. Be honest and sincere, but don't overdo it. A simple note or postcard is often all that's necessary to keep you in the mind of the producer or director and get you hired again. If you haven't already, be sure to add their names to your mailing list for future promotions you send out.

---

1 Adapted and compiled by Marc Cashman from the following sources:

Alburger, J. R. (2006). *The Art of Voice-Acting* (3rd ed.). Focal Press.

Blu, S. & Mullin, M. A. (1996). *Word of Mouth.* (revised edition). Pomegranate Press.

Apple, T. (1999). *Making Money in Voice-Overs.* Lone Eagle Publishing Company.

Whitfield, A. (1992). *Take It From the Top.* Ring-U-Turkey Press.

Thomas, S. (1999). *So You Want to Be a Voice-over Star.* Clubhouse Publishing.

Berland, T. & Ouellette, D. (1997). *Breaking into Commercials.* Plume Publishing.

Douthitt, C. & Wiecks, T. (1996). *Putting Your Mouth Where the Money Is.* Grey Heron Books.

Jones, C. (1996). *Making Your Voice Heard.* Back Stage Books.

Clark, E. A. (2000). *There's Money Where Your Mouth Is.* Back Stage Books.

26

Tips, Tricks
and Studio Stories

The world of voiceover is not only a unique area of show business as a performing craft, but it is also unique in that professionals in this business are, for the most part, far more supportive and generous than those working in many other areas of the entertainment industry. While some performers may appear to be protective of their processes and techniques, most voice actors are more than willing to help those who sincerely desire to learn more about this craft. Perhaps it is because these professionals are very confident and self-assured in their work, or perhaps it is simply a reflection of their passion for what they do. Whatever it may be, I am proud to be a part of the voiceover community and to consider so many of these professionals as friends.

This chapter is a gift from some of the many professionals around the world whom I've come to know over the past several years. Each has many years of experience from which they have gained unique insights and wisdom. Please join me in thanking them for sharing their knowledge, and use this chapter to learn from their experience.

BEAU WEAVER (Los Angeles, CA)
www.spokenword.com

If you watch TV, you've heard Beau Weaver's voice. Considered by his peers as one of the top voiceover professionals in the world, Beau is among the handful of voice actors who regularly voices national network promos, movie trailers, and major documentaries. Coming from a radio background, he knows what it takes to make the transition to voiceover. Beau is also one of the nicest people you'll ever meet and I'm honored that he not only wrote the Forward for this edition, but has shared what he considers to be one of the secrets to his success.

The Beginner's Mind

One of the most important tools you need to cultivate is a way of being what the Buddhists call, "beginner's mind." If you already know everything, there is no room to learn.

Radio was my first love. I weaseled my way into my first on-air job at age 15, after having already spent years hanging around radio stations learning the arts and sciences of broadcasting. In just a few years, I managed to make my way to the biggest of the big time: the Number One pop music station in Los Angeles, the legendary 93/KHJ. It was 1975, and my dream had literally come true. I was working with cats who had been my radio idols at the station everyone in America wanted to work for. Our pictures were constantly in the trade press, and the record companies were only too happy to give us front-row tickets to all the hottest rock concerts and take us backstage to meet the band. I got fan letters from deejays in big cities, asking for tips on making it to the big time. Pretty heady stuff for a 24-year-old.

One night, I came in to the station, located on the Paramount lot in Hollywood, and noticed a light in one of the production studios, which should have been dark and locked. I stuck my head in there, and saw one of the other KHJ air personalities (fancy word for disc-jockey) at the console, plugging in patch cords.

Something was very wrong with this picture.

At KHJ, and most other L.A. stations in those days, the air personalities (AFTRA members), could not even so much as touch the equipment. That was the exclusive territory of our NABET union engineers. We could be fined for even leaning up against a tape machine. But there he was, in flagrant violation, my friend Dave Sebastian Williams (today, the owner of Dave & Dave, Inc., publishers of the *Voice Over Resource Guide* and the **www.everythingvo.com** website). "Oh my God! Dave, what are you doing in here, man?" said I in a panicked tone. "You could get in big trouble for this!"

He nodded at me to close the door. He had small spools of tape cued up, running dubs of what sounded like a string of commercials. Dave: "It's cool, man, I gave a twenty-spot to the shop steward… I'm making copies of my voiceover demo." Me: "Voiceover demo? What the hell is that?" Isn't that just a bunch of spots?" Dave patiently explained: "Beau, don't you realize that the commercials are the *real* business. What we do as disc-jockeys is just filler. *This* is where the real money is." A moment of transformation occurred. The scales fell from my eyes, and in that moment, as Dave went on about the world of opportunities available the freelance voice actor… my life changed.

"Voiceover demo, huh? I gotta get me one of those things!"

Not so fast. Dave informed me that radio guys were *persona non grata* in the advertising agency world. They think we have a "radio accent." If you sound like a radio guy, you are sunk before you even start. I was stunned. We were at the very pinnacle of our industry. Everyone wanted to be at KHJ in Los Angeles. And now he was telling me that everything I knew was wrong? You don't get to the top station in the country without a healthy ego, so this was more than a little hard to hear.

"But, there is a kind of therapy for the problem we have," Dave reassured me. "There is this workshop for voice actors taught by Joan Gerber, where you can practice reading copy, and taking direction. I go every week. You can go with me." So we made a plan to drive together up into the Hollywood Hills that Wednesday night.

Joan Gerber, known to many as Gary Owen's "Story Lady," was the voiceover field's first true female superstar. In her spacious living room, she had set up a copy stand with a reading light and a script for a television commercial. As a warm-up exercise, each actor would come up to the front and "cold read" the script with no rehearsal. In my on-air job at KHJ, I was required to deliver a lot of live commercial copy. While I was nervous in this new environment with "actors" (not radio jocks), I was completely confident (cocky perhaps?) in my ability to nail it, first time. So, I strode up to the copy stand and got half way through the first line, and Joanie interrupted my performance: "Disk Jockey! Right? Ha! I can smell you guys a mile away." A bit of a ball-buster, yes? But, remarkably, I did not bolt. (Partly because Dave drove!) And, I learned some things that night.

The fact that there was a whole set of skills that I did not have did not insult or offend me, but rather I felt intrigued, challenged, and invigorated. Sure, I had been recording and delivering live commercials professionally since I was 15. But, it was dawning on me that there was a completely different way to approach commercials. And, for some reason, I was willing to stay and see what possibilities there might be that had not yet occurred to me. Against all of my big-time radio guy instincts, I was willing to be... a beginner... at something I had been doing all my life.

In the months ahead, I continued the weekly workouts at Joan's, and I found two other voiceover workshops. I was soon almost obsessed with mastering this new skill set. While I continued to work in radio, for the next decade my main focus was on becoming a voice actor. Since then, I have had the privilege of working in almost every genre that uses voice talent: national television and radio commercials, documentary narration, television promos, movie trailers, and animated cartoon characters.

All because I was willing to be a beginner.

MIKE HARRISON (New Jersey)
www.mike-harrison.com

Mike Harrison has, since 1973, been writing, voicing and producing radio commercials, plus narrating and/or producing audio tracks for many Fortune 500 corporate/industrial clients. He was a two-time co-finalist, for copy and production, in the 1985 International Radio Festival of New York, and his voice is currently heard in various markets across the U.S. as well as in the United Kingdom.

It Ain't Just Talking
The Sugar-Free Truth about Pursuing a Voiceover Career

There are some folks interested in becoming a voiceover talent who have become discouraged—a few even angry—after receiving the honest feedback they solicited from those of us who've been around the block a few times. What I'll attempt to explain here is that our goal is NOT to take the wind out of your sails, but only to make you aware of… reality.

What probably makes a career in voiceover so appealing to some is that it has a lot of glamour attached to it. It's very much like a career in film acting. The growing popularity of entertainment publications and television shows reflect the public's fascination with people who make six-figure (and higher) salaries for what appears to the uninitiated to be easy work. After all, those who speak and act for a living make it seem so effortless.

Here are a couple of facts: Anyone with aspirations for success in any field first has to realize that we don't often get from point A to point G without first having gone through points B, C, D, E and F. As we've seen in science documentaries, even laboratory animals learn that. Nothing in life that is worth achieving comes easily or for free. Success is relative: you get out of it only what you put into it.

For as long as Hollywood has been making motion pictures, there have been countless stories of everyday folks with stars in their eyes who have gone there to find fame and fortune. Trouble is, there are only so many roles to play and there are many more actors already there, waiting to play those roles. Most of them wind up waiting tables or doing some other work to pay the bills until their big break comes along. Voiceover, just like film acting, is hugely competitive.

This doesn't mean you should give up. What it does mean, though, is that you'll have to work very hard. And, that's assuming there is a foundation of basic talent on which to build. It's much more than having a "nice" or even "great" voice. Do you read well? Aloud? Are you able to properly INTERPRET what you're reading so that you can sound as though you're NOT reading… and be

CONVINCING? Can you read well, interpret copy well, and take direction? How about read well, interpret copy well, take direction well, and finish within the allotted time? And then do it all again but make it sound different? Are you able to do all that and still retain your composure and professional demeanor? Take after take?

Contrary to what some may think about the craft of voiceover, it ain't just talking. What I mean by that is not everyone is cut out to be in voiceover. And an honest coach will tell you if you truly have potential, or if you are about to waste your money. If you've just been told you're ready for a demo—after only a few lessons and/or workshops—you should clarify whether you are ready to have a COMPETITVE demo produced. Because, when marketing yourself, you want to put your absolute best foot forward. Your demo will be in the pile with those of others who have every bit the talent, experience, and drive that you have... and perhaps more. The agents and producers you send your demo to will know immediately whether you are a contender or not.

Sure, you can have a "quickie" demo produced; one that might get you some work doing local cable TV spots. And there's nothing wrong with that. But if you want more, be prepared to roll up your sleeves, spend some good money on proper training with qualified coaches (know their background and experience), take the time to read as many books on voiceover as you can, and practice, practice, practice. And, while buying recording equipment may help you to practice, do keep in mind that simply owning a stove does not make one a chef. In the beginning, instead of spending money on professional recording gear, consider whether your money would be best spent on training. And don't forget about all the marketing you'll have to do after your demo is produced. There's the 80/20 rule: even established talent spend roughly 80% of their time marketing themselves, with the remaining 20% on actual voiceover jobs. Included in the 80% is auditioning. And with auditioning, comes... rejection. You'll do lots of auditions but will land only a few jobs. And you're never told why you don't get the part; in fact, you'll never hear ANYTHING unless you DO get the gig. Finally, you'll also want to keep your full-time job as you slowly get the wheels turning.

To sum up, you'll need to spend lots of time working hard, and you'll need to spend some money. But you will also need thick skin to weather some negative feedback and rejection along the way, and the tenacity to keep moving forward. This helps us learn and become better. When asking for advice, feedback, and/or opinions, we all hope to hear only positive things. But we all learn that we have to eat our vegetables before we get any dessert.

BILL RATNER (Hollywood, CA)
www.billratner.com

Bill Ratner is one of America's most heard and most versatile voices. He's voiced characters on hit shows like *G.I. Joe, Family Guy, Robot Chicken* and dozens more. He's voiced movie trailers for just about every major film company in Hollywood, including DreamWorks, Paramount, Nickelodeon, Disney, MGM/UA, Warner Bros., and Sony Pictures. Bill has narrated programs for Discovery, History, A&E, and Travel Channel, and handles daily promo duties for networks and TV station affiliates across the U.S. All of this is, of course, in addition to his work in commercials for many of the top ad agencies and television advertisers in the country. Let's just say, Bill Ratner knows voiceover!

Do It Like You Talk

Just a few days after I signed with my first Los Angeles voiceover agent I booked a set of radio spots for Mexicana Airlines. I was thrilled; my agent would finally see that she'd made the right decision by signing me, and I wouldn't have to return to selling pens and toner over the telephone!

The gig was booked at SSI in Hollywood, right across Santa Monica Boulevard from the Formosa Bar and the old Samuel Goldwyn Studio lot. The writer/director for the session was a disarmingly lovely and intelligent woman from the ad agency in San Francisco. We chatted amiably, and then Gary the mixer beckoned us into the dark leather luxury of his room, Audio A. The AKG 414 gleamed in the moody pin-point-spot lighting of the voiceover booth. "Let's have a level," Gary said cheerily into the talkback.

The energy which on-camera actors often experience as stage-fright, voiceover performers feel as euphoria—they're actually going to pay me to do what I love best: say words into a microphone. Gary slated it, "All right, let's have a go at this… For Mexicana Air radio—Bill Ratner voiceover—call this take one." My head swollen with pride to the size of Gibraltar, I purred into the microphone for twenty-nine-and-a-half seconds.

"H-m-m-m," came the director's voice into my headset. "O-k-a-y," she said indecisively. My mood plummeted from celebratory to self-doubt. Why wasn't she saying, "That was great," or "I loved that?"

Sitting behind the dark California oak client's desk she put her chin in her hand and stared at me through the glass. Was she disappointed? Had she made the wrong choice in hiring me? Had she drunk too much Merlot at lunch? She cradled the talkback switch in her hand, running her beautifully manicured fingers absent-mindedly over the button. Finally she pressed it, sighed, and said, "You know how, a few minutes ago, you and I were talking together out in the lobby? You have such a beautiful voice when you just talk, but when you got in the booth and read just now you sounded... I don't know... "

"Announcery," I said.

"Exactly. Don't do that." And she smiled at me and said in a near-whisper, "Read this copy exactly the way you talked to me out there... *Do it like you talk.*"

I was being profoundly seduced by this woman, and not in an inappropriate way; I was getting the best voiceover direction I would ever receive in this business, delivered in the most elegant, diplomatic manner I could ever hope for. Now the real work would begin—to speak into the microphone to effectively sell a pleasant airplane ride to a far more pleasant vacation destination, and *not* sound like I am delivering time and temp on an FM radio station.

After a few false starts, requiring her to repeat her simple direction, the session went swimmingly; the client and the director were happy with the result, I was extremely happy with my work, and I had learned an invaluable lesson. Up to this point in my career I had been able to skate by on my pipes—that funny little pink flap of skin behind my Adam's apple that allowed me to sound like The Real Don Steele at age 14. But neither the advertiser, nor the TV promo producer, nor the movie trailer business is interested in just a sonorous delivery of their copy. They want the *real* person—the one they hired because you actually had a hint in your audition of what the client is looking for—the silliness or the petulance or the seductiveness or the aloofness or the anger or the quiet authority they are hoping to hear in their spot.

Each one of us is saddled with our own unique set of emotions and personality, separate and distinct from our voice-print. And that's who they want to hear—the *real person*—not the announcer. Shedding our habitual announcer skins and discovering that person behind the microphone is the most difficult and most rewarding part of our job.

JANET AULT (Phoenix, AZ)
www.janetault.net

Janet Ault, "The Voice of Choice," is a voiceover professional based in Scottsdale, AZ. With over a decade of experience, she has been providing her voiceover services in commercial, narration, interactive, e-learning, and imaging genres, to name a few. Her projects have ranged from Delta Airlines, Disney/Hasbro and Fisher-Price to thousands of local and regional spots.

Leads From Biz Journals Work!
This Cold Call Got *Hot* Fast...

As a professional voiceover talent in what is considered a "not-major" market (even though Phoenix is the fifth largest city in the country), I scour my local Business Journal, the Business Section in the local newspaper, trade magazines, etc., to find new clients.

And I have had great success obtaining new clients this way. However, one in particular became what I call my "Christmas miracle." Even I am having a hard time comprehending it! Here's what happened...

December 26:

In the Business Section of the local newspaper, I noticed that a new grocery chain is expanding in the Valley. Grocery chains, especially new ones, do a LOT of radio and TV spots and, of course, need VO talents.

I Googled the company, found their website, and discovered that I already knew the marketing person mentioned in the newspaper article.

Because of the Christmas holiday, rather than call, I sent that person an email with a voice sample, with the intention of calling on the 28th.

December 28:

No need to call!

The marketing person forwarded my sample to the ad agency, which emailed me to request a hard-copy CD since they loved the sample!

I sent the hard copy overnight (along with a couple of wooden Post-it note holders with my logo on it. Hey, it couldn't hurt!)

December 31:

Happy New Year! Here is the response I got:

"Thank you so much for sending me your packet. We listened to your demos and we really do love your voice. Your timing was great. We just finished scripts for the January campaign. We are, in fact, looking for a new voiceover talent!"

The ad agency continued: "Our primary demo is women. We think this would be a great fit for both of us. We are so excited! Scripts to follow!"

January 4:

I received the scripts and have learned that this is an ongoing job for a minimum of six months in several regions, with a lovely monthly income.

Like the lady said in her email: "Your timing was great!"

In a little over a week, from one newspaper article, a repeat-customer was born!

As voiceover professionals, we can't just sit and wait for the phone to ring. We've got to create opportunities. Once in a while, it really does pay off!

PHILIP BANKS (Portgordon, Scotland)
www.philipbanks.com

A former Investment Director for a Swiss merchant bank, at the ripe old age of 30 Philip Banks found himself sitting in front of a mic doing a "voiceover" for a friend who worked as a radio producer. From that point on it was all downhill. In 1992 Philip moved from investment banking to being what he likes to call "a Voice Overist".

Philip spends most of his time in a VO booth about 100 yards from the beach in the little Scottish village of Portgordon. ISDN connected to the world, his credits are global CNBC, CNN, BBC, Zone Horror (guess what type of movies they show) and he's even been used as a voice on *The Queen* starring Helen Mirren and *Volver* starring Penelope Cruz.

Of his success Philip says. "Were I to move to L.A. I'd be one of the top ten voices in town so I'm staying in Portgordon because here I'm the number one!" Out of how many Philip?

The Ten Top Traps Voice Overists Love To Put Their Feet in… and The Real-World Truth You Need to Know:

Trap #1: One size fits all.

The Truth: The demo or demos are designed to get any kind of work out there as opposed to the jobs for which they are best suited. What are your strengths? Play to them on your demos other opportunities are for auditions.

Trap #2: Demo critique.

The Truth: When you get an honest critique, you don't like it, so you FIGHT BACK. Be honest, you really didn't want a demo critique… you wanted demo approval. OH YES YOU DID! Cut and paste this – "Hey your demo is great!"

Trap #3: Once I'm signed by the big VO agent, that's when the money arrives by the sack full.

The Truth: If you ask most of the VOs with top agents how much they earned last year they will not tell you, not because they believe it's confidential but because the truth is embarrassing. If you're interested last year I earned well into six figures and all I do is voiceover work. Having said that I am happy to report I am the highest earning Voice Overist in Portgordon.

Trap #4: Your movie trailer sounds like everyone else.

The Truth: At best most movie trailer demos are an amalgam of the men who are making a living doing movie trailers. If you know a man with a Porsche, try to sell him a fake Porsche and see how far it gets you. Your movie trailer demo should not sound like anything other than *your* movie trailer demo.

Trap #5: Any Female VO can do sexy.

The Truth: Most can't! Maybe when you're sounding sexy you really ARE bored but I doubt it. In order to deliver a smoldering read you need to be at your most comfortable and to a degree, a little vulnerable. The eyes have to say "Yes, I REALLY like you" and the voice will follow. It is extremely difficult.

Trap #6: Gear makes you great.

The Truth: You need the best gear possible in order to deliver quality from your own studio; once people start buying what you're selling at the top pro level then you can stop buying, unless it's a hobby. No one will exclude you as an option for movie trailer, TV promo work because you don't use an MKH 416 but they will exclude you if you suck.

Trap #7: BS factor 10.

The Truth: This is the highest level of BS and most VOs never work on reducing it. No one will ever get to BS factor 1 but you should aim to get as low as possible. I've lost count of the number of great jobs someone has told me they've been signed to do but they can't tell me what it is because of an NDA (non-disclosure agreement). In 20 odd years of VO work I've signed no more than five and none of the big gigs required any kind of signature on anything. If you got a job, the people who matter are proud of you so no need to BS.

Trap #8: Location, location, location.

The Truth: I know VOs who have moved to get their careers on track only to discover they have moved to a location full of out of work VOs. In order to move up a level, a change of location may help, but like everything else in this business it has a 95% chance of achieving nothing.

Trap #9: Anyone can be taught how to be a Voice Overist.

The Truth: NO, THEY CAN'T. Just in case you're unsure I'll give you a little more detail – NO, THEY CAN'T, Girl Scout's honour. The ability to do voice work is an intuitive skill and all any teacher/coach can do is identify that skill in a few minutes and then offer to teach you how to master it. You cannot be taught to act, but you can be taught to be a better actor.

Trap #10: Pay to play (subscription based voiceover sites) work on a number of levels.

The Truth: If you want to be taken seriously as a Voice Overist, the best advice is to avoid such sites at all costs. Should you believe that the site makes sense then you should test it for financial sense. The purpose of these sites is to get you work and the value of that work should exceed your membership fee by a factor of around 10. No "buts," this is business. The site made $300 from you, you had to earn around $500 to end up with $300 to spend. What did you get back?

SYLVIA AIMERITO (Los Angeles, CA)
www.audiogirlproductions.com

As co-owner of AudioGirl Productions, Sylvia Aimerito is a busy voice actor, on-camera actor, producer, coach, and casting director with a client list that includes some of the biggest companies and advertising agencies in the U.S. Sylvia Aimerito is an accomplished voiceover actress, business owner, and on-air radio personality at K-EARTH 101, Los Angeles.

What's the Big Deal?

I teach a voiceover workshop about three times a year and one of the first questions I ask the aspiring actors is: "What is your primary motivation for taking this class?" And what is one of the most common answers I receive? "Well, I hear these people on TV and radio and say to myself, 'Hey! I can do that! What's the big deal?'"

Then, after a couple of weeks into the class they understand what the "big deal" is. They discover that it's more than just reading copy into a microphone; it takes a lot of talent, skill, and preparation to do what we do. And like any profession, the better we are at our job, the more we work. And how do we get better at our job? By continuing to raise the bar and acquire, improve or add skills to our voiceover tool belt.

Which brings us to my area of focus: The skill of giving a fast-paced performance while maintaining the proper attitude, emotion, and intention. Or in voiceover language, the skill of fitting 40 seconds of copy into a 30-second spot.

Many of you know that copy time constraints are a common demand of clients and producers so it's an important skill to master. A great place to start is with your preparation. As my old music teacher used to say, "Failing to prepare is preparing to fail," and it's a philosophy that also works well in the field of voiceovers.

Preparation suggestion number one: *Forget about time.* Anyone can read the copy fast enough to fit into a time constraint.

To quote the aforementioned student, "What's the big deal?" The big deal is maintaining your attitude and emotion while making the copy fit. But first you must approach the copy as if you had all the time in the world.

Now, with time limits out of the way, you're ready to study the copy in some depth and work out the technical, academic details such as word pronunciation or where to take breaths, if that's an issue.

After that part of your technique is worked out, you're free to get creative and embrace the next level of focus: bringing the right attitude, emotion, and intent to the piece. Throughout your preparation it's a good idea to remind yourself that your primary goal as a voiceover artist is to sell the attitude and emotion—not the product. Again, without any consideration of a time limit, practice the copy until you feel comfortable.

Now you're ready to work on speed.

While maintaining the emotion and relaxed delivery, start increasing your speed—little by little, a second at a time. Keep applying this technique until the goal is reached. Be patient with yourself and you'll be fitting 40 seconds of copy in a 30-second spot with your emotion, attitude, and intent wonderfully intact!

Having a rock solid technique in working within time parameters is an important skill to master because it comes up often on the job. And, as with all skills, the only way to master it is with disciplined practice and patience. This skill, is one among many tools in our professional belt that keeps us sharp and confident and consequently working.

One final thought. A great tool for every voiceover artist to have is a convenient way to record your practice sessions and play back for critical listening. And now, more than ever, there are a plethora of terrific handheld digital recorders on the market. They're also handy for the particular focus of time constraints since most, if not all, automatically give the record time making it easy to determine your progress.

Well, that's my story and I'm stickin' to it. Stay positive and fierce.

BETTYE ZOLLER (Dallas, TX)
www.voicesvoices.com

Bettye Zoller is among the best-known voiceover coaches in the U.S. She is an international voiceover talent with an impressive collection of Golden Radios, ADDYs, and CLIOs, as well as an audio engineer/producer/director and recording studio owner in Dallas, Texas. For more than 30 years, Bettye has been performing and coaching voiceover talent, including some who have gone on to the Fox Channel, MSNBC, Broadway, Las Vegas, New York, and L.A. She's served on several college faculties, including the University of Texas and Southern Methodist University. Bettye knows this business from both sides of the microphone and, as with most pros in this business, she is eager to share her vast experience with her students. Her coaching and voiceover work are in demand worldwide.

It Used to Be So Easy…
(Wearing Two Hats Is Challenging but Fun)

It was 1995. I checked my voicemail: "This is your agent," the voicemail message said. "Be at Fuller Recording Studio at 2 PM August 4th. You'll voice a commercial for Rainbo Breads. The contract's at the date. Bring it to our offices after the session."

Easy. All I had to do was drive to a studio, go inside, and after a short conference with the producer or client, read a script into a microphone. The audio engineer did all the rest.

My voiceover agents (six of them in various cities) all began urging me to refresh my skills as an audio engineer. "Laura Stevens already has a recording studio in her home. So does Jeb Brown," one taunted me.

"My voice talents who have studios are getting more jobs," said another. "Being an audio engineer as well as a voice talent will mean more money for us both and you already have the skills." I turned a deaf ear.

Now, I look back with regret for procrastinating putting in my studio but I first had to build a studio of my own. Where to start? When a student wanted a voiceover demo, I produced the demo in someone else's studio. I lost thousands upon thousands in revenue not to mention jobs. I was bewildered. Live and learn.

I first became an audio engineer about 20 years ago as the Creative Director of a large audio production house. We created jingles, songs, and libraries used worldwide by radio and TV stations. Once, I created and engineered a 38-CD library of jingles and voiceovers! The firm had three cushy recording studios that looked like spaceships! There were hundreds of knobs and sliders on the engineering console. Sometimes, we used two engineers because one person couldn't manipulate all of the knobs at the same time. In the four years without a recording studio in which to work, my engineering skills were dormant. What's more, I knew nothing of the digital recording revolution.

With advice from audio engineer friends, I purchased digital recording equipment, a new computer, soundproofing materials, and prepared a room in my home. Since then, I've spent many thousands more. A studio is always *a work in progress*! We even replaced windows in our home with triple pane glass and hung multiple pairs of heavily lined draperies.

Now, I also have a state-of-the-art portable recording studio on my laptop. My Internet "stick" provides instant Wi-Fi. I travel worldwide serving clients, teaching at universities, presenting workshops and symposiums. Not only can I record at home in my wonderful studio, I can record in a tiny Swiss village or in the rain forests of Panama. I've done both traveling the globe! I'm wired for sound!

So, I wear two hats… one as a voiceover talent/studio singer, the other, as an audio engineer. Do I get paid double? Not often. Can I ask a client for my full audio engineering rate when I also am the talent? Today, most clients expect a one-price package. Competition is fierce. Too many people are eager to record lowball jobs for practically *no wage*! That hurts us all!

Everything's changed.

That's not to say I don't love—and I mean, really, really love—being an audio engineer. I do. I am proud of my skills. It's fun to add effects, aircraft engine noises that move from one speaker to another giving the effect of motion, animal sounds, African drums, night sounds on the farm, creating what radio buffs used to call, "theatre of the mind." Postproduction is so creative. It's magical. For me, being an audio engineer, especially creative postproduction, is the most fun I have as a voiceover talent or singer.

Quite some time ago, I began urging online "pay to play" websites to separate experienced audio engineers from those who were beginners or who did not record at all. Being capable of recording oneself voicing an audition is not the same as calling oneself a professional audio engineer! Maybe someday this important distinction will be observed by online casting sites. At this writing, that has not happened. Audio engineers with vast expertise are not set apart.

Some projects are extremely complicated. When I see these types of jobs on an online casting site, I wonder how many are auditioning whose recording skills are inadequate. The client chooses a voice, assuming that that person can also engineer it. Stories abound concerning botched projects, time wasted, nerves jangled, because someone took on a project for which they were not equipped!

What to charge?

When you audition for (or accept) a voiceover project requiring audio engineering, you must first decide if it is worth your time. If the price is too low and the hours too great, pass it up (unless you like working for nothing or consider it a free learning experience!). If it is "workable," name your fee. Be sure to mention that changes (redos) cost extra. Specify what constitutes a redo for which you expect to be paid. Of course, you agree to fix a mistake if you're at fault.

What is your time worth? Estimate how many hours a project might require to both voice and engineer it. Every job should have a "bottom line" below which you will not go. Some engineers feel that one hour of finished (playable) audio requires about two-and-a -half hours of recording and engineering. Many websites contain budget guides. Check it out.

Be cautious about voiceover projects with unrealistic (or undoable) expectations. That is, read the specs carefully. Does the project require recording multiple voice talents who come to your studio? That's time-consuming. Does the project need special music or effects you may have to find or buy? Is the project long-form audio (an audio book or other lengthy piece)? Must you divide the work into segments over a month-long period or more, turning down other jobs that come your way? Do you have a day job that precludes spending long hours in your studio?

Never hesitate to ask advice from those more advanced than you. Budgeting projects is a skill unto itself. You may want to offer to pay an advisor to help you with a workable budget. It will be well worth it.

If you're so inclined, I encourage you to hone your audio engineering skills and keep improving them over-time. But know your capabilities and don't over-promise or take on projects you shouldn't until those skills are well in place.

Maybe you'll stay at a rudimentary audio engineering level, someone capable of recording a voice audition, but incapable of postproduction. That's OK. Not everyone will become a skilled audio engineer. Not everyone will succeed at voiceovers either. We're all unique. It's a big world and all of us can contribute. Keep up with new technology. Investigate new recording software and hardware. Enroll in a class at your local community college or a recording studio. Many cities are offering audio engineering instruction of various types. There are "support groups" and community groups in some locales.

Pay for a few private lessons from an audio engineer at a local studio. Ask an engineer if you can observe a session or two. Perhaps you can even offer to be an unpaid recording studio apprentice or helper to soak up the studio environment and sit in on some sessions. Several students of mine became office workers at recording studios. They later emerged as engineers! On -the-job training's always best.

We live in an "instant world" and becoming a really accomplished audio engineer is not "instant." Like playing the piano and tennis, it takes time for experience to grow. Go for it!

PAUL M. DIAZ (South Florida)
TV Promo Producer

Paul M. Diaz, is a seasoned television promo producer/editor based in South Florida. Since 1994 he has worked in news, entertainment, branded entertainment, music videos, commercials and special events. Some of his most refined work has been done on television projects like, the *Latin Billboards HD* (Telemundo), *Miss Universe HD* (Telemundo), and *Beijing Olympics 2008* (Telemundo). In 1998 he and his crew of four were awarded a Suncoast Regional Emmy for Best Entertainment Program. Most of his work involves casting, recording, and directing voice talent. I asked Paul to share some of his experiences of working as a promo producer.

Caring Makes a Difference!

I've worked in video production and broadcast for a little over 15 years and I've hit the record button with a variety of people facing a mic; professional VO talent, show hosts, reporters, and even producers who love to hear themselves speak. Overall the one thing I love about recording or listening to voices is how unique every voice is and how each voice talent handles their work differently.

Some of the talent I've recorded would voice a piece and then want to hear it back before recording their next take, while some would just voice it once and walk away.

Some just seemed to care while others didn't. I once worked with this guy who would have me press play on a video tape and he would just ad-lib a 2- to 3-minute segment on the fly. This other girl, whom everyone thought was a diva, was very receptive when I gave her some direction when no one else bothered to tell her that her delivery was too flat. I have no real experience teaching or coaching anyone, but I've been producing long enough to tell her how *not* to do it.

My best work always comes from those recordings where the voice talent and I are working as a team, and it is always a pleasure working with a professional who takes their work seriously and truly cares about doing the best job they can. When I'm recording someone who just doesn't care… the quality of the whole production suffers, and that's not good for any of us.

MARC CASHMAN (Los Angeles, CA)
www.cashmancommercials.com

Marc Cashman creates and produces copy and music advertising for radio and television. Winner of over 150 advertising awards, and named one of the Best Voices of the Year by *AudioFile Magazine*, he also instructs voice acting of all levels through his classes, The Cashman Cache of Voice Acting Techniques in Los Angeles, and offers one-on-one coaching via email or phone.

Coloring Our Words

It's hard to remember exactly when we got our first coloring book, but we do remember it was fun. At first, we sprayed crayon colors all over the page, without a care as to whether we stayed inside the lines or not. As we became toddlers, our coloring got more refined. We learned boundaries, we assigned certain colors to certain objects, and were more discerning in our choice of colors. A few years later found us drawing with colored pencils or markers. Later still, we marveled at the results of paint-by-numbers, and then on to watercolors and pastels.

The coloring books were random, assorted pictures, themed pictures or page-by-page pictures that laid out a story. And each one of these pages had the same format: a black outline on a white page that showed a picture. At a glance, we could see a cabin on a lake, with smoke rising from its chimney; a boat tethered to a pier, fishing poles jutting out at its end; a winding road leading up toward the cabin, and a big, broad apple tree on its front lawn, with majestic mountains towering behind the cabin, backed by a sky full of puffy clouds and a bright sun.

As children, we looked at this black and white tableau and made some decisions: the sky would be blue, we'd leave the clouds white; the cabin would be brown and the lawn would be green; we'd apply the same colors to the apple tree, but add some red for the apples; the road might be charcoal; the lake would be blue, the mountains might be gray, and the sun would be yellow. We colored in the outline of a story.

As adolescents, we got better at drawing. Our sky might be bluish-purplish. The clouds might have shades of gray and green; the water on the lake would be a mixture of many colors, possibly reflecting the boat that floated on it; the road might be a mixture of brown, dusty tan or beige to signify dirt, with black rocks and pebbles strewn about; the cabin would have a different colored roof, and, like the apple tree, cast shadows from the light of the sun.

We'd use different coloring tools in our teens: pastels, colored pencils, watercolors and markers. And we'd start adding depth and shading, because we could discern perspective and light better. And we'd spend much more time at our task; we were more exacting and meticulous.

Printed words are groupings of black symbols on white paper. Strung together intelligently and creatively, they tell a story, just like the outline of a picture in a coloring book. It's our job as voice actors to color words, to give them depth, shading and perspective. Our tools: our voice, vocal techniques and acting abilities. And it's our acting that has to come to the fore through our voice. And through our voice needs to pour conversationality and emotion in order for us to sound believable.

The reason that most great stage and screen actors are believable is because we can see their characters. We see their body language, their movements, their gesticulations, and their eyes. We see them embody characters through their actions. But people can't see voice actors—they can only hear us. So all the color and emotions we bring to a script has to come out of just one place—our mouth.

The nuances of the human voice are extraordinary. Millions of years of human evolution have made the sound of the human voice a wonder to behold and something no machine will ever duplicate. Oh, they've tried. At first, people thought that developing speech recognition would be a simple matter of replicating phonemes, and they've had some success in transplanting those basic sounds into myriad applications. But like astronomers exploring the universe—the more they peer into the vastness of space, the more they realize how complex it is—in their quest to simulate real speech with a machine, scientists have found that the more they try to perfect speech recognition, they realize they can't. Because the human voice is so incredibly unique. Our vocal cords hold a powerful gift: the power to paint pictures, with an infinite variety of colored shades, textures, depth, patterns and mixtures. We have the innate ability, through our voice, to convey meaning without even uttering a word! No machine could do that.

Many of us refer to ourselves as voice artists as well as voice actors. If we're artists, then we have to take out our palette of vocal colors and brush those words, wash, tint and dab them. We have to channel impressionism, cubism, pointillism, abstract art, op art and realism into our phrasing. We have to apply the endless color combinations of emotions and infuse them into words. When we're presented with text that cries out for coloring, take out your 120-count box of vocal Crayolas® with all their wonderful hues and shades and create a masterpiece!

We're blessed with the ability to lift words off the page effortlessly and to articulate them clearly. But if we don't inject emotional depth and real meaning into them, if we don't artistically color in the outlines of those pictures, we'll never do justice to beautifully crafted text or capture a listener's attention. And we'll waste a great opportunity.

DAVID KERSTEN (UK)
www.davidkersten.com

David Kersten has a delivery style that is truly international. His client list includes dozens of the highest-profile, major-name advertisers in the world. Even with his many years of experience, David—just as every other voice actor—is constantly confronted with the issue of getting paid for his work. Economic challenges over the past several years have become a common argument for clients who attempt to delay paying for voiceover services. David has some specific ideas on the subject of getting paid that is food for thought for all voice talent.

On Getting Paid...

There is a counter argument to using the economic crisis as an excuse to delay payment, and that is to insist on earlier payment as you don't even know if they'll still be in business next week or the week after! The last thing you want to do is take a number and stand in line for your payment should your client become insolvent.

Seriously though, I don't haggle much on price, if at all. My rate is generally not negotiable... seriously. I've found that most will just try it out because they have a certain perception of voiceover artists being there with their freelance, itinerant, gypsy caps in hand, desperately pleadingly in need of work. A polite but very firm, if at times awkward, "no" soon puts an end to that and is also able to invert that perception. Suddenly you become something... special. Perhaps even, heaven forbid... good. A logical explanation as to how that rate was arrived at also helps.

I'm of the belief it's the quality of the client that matters most. Both rates and terms can be adjusted according to the relationship I have with the client, the frequency of the work and the kind of income that work generates for me. Net 30 is my standard, Net 60 seldom happens and I have generally refused work when told this is their payment policy.

The excuse of not paying me until they get paid is heard all too often, coming mostly from agents and production houses, but again, it's from the perception of the voiceover needing to be grateful for having got the work in the first place. I find this stance arrogant and patronizing, as they have forgotten how essential you are to their business. A little gratitude and respect would be appreciated rather than their matronly dictating the terms of how they intend to pay. Don't forget, without you, and others like you, they're out of business. My response generally is to make it clear that if they can't afford to pay me they shouldn't book me. Credit terms are for me to determine, not for them.

Still, I will at times agree to Net 60 for certain clients if only for the reasons already mentioned—frequency of work, how well it pays, not needing to pursue payment, and how well I get along with them.

With new clients my credit terms are the standard 30 days if they are local as it's unwise to offer different terms to different people who may perhaps meet or know each other and have the opportunity to compare notes. I'll insist on payment up front, usually before I step in the booth, from clients I've had complications with in the past, especially as regards payment.

All online work is delivery after payment. I don't let anything go until I'm paid first. The client has the opportunity to hear the recording by way of a watermarked audio link but that's as far as it goes. Once it's approved and paid for I then fire off the file.

Oh, and I won't be offering discounts for earlier payments. I'd be surprised if it actually works as an incentive and besides, it's really going to mess with paperwork that I'm loathe to do, especially toward the end of the financial year.

JEFFREY KAFER (Seattle, WA)
www.jeffreykafer.com; www.voice-overload.com

Jeffrey is a Seattle-based voice actor, with "actor" being the key word here! He has more than 20 years of theatre and improv experience that he applies to every voiceover job ranging from commercials, to corporate narration, to audio books. He is also quite the creative artist! The cartoons sprinkled throughout this book are Jeffrey's creation. Visit his website, subscribe to his blog, and study his demos. You'll learn a lot!

Do Your Voiceover Clients Know
What Kind of Salsa They Want?

I'm on the search for perfect restaurant salsa. There's a nearby Mexican food joint that has darn near perfect salsa. Thin (but not watery), deep red, with some cilantro, a few pieces of onion and some spices. Delicious! No carrots, no fruit, no beans.

And I can't find anything remotely close in the grocery store. I can't even find a lesser-quality homogenized, pasteurized, factory-created replica of this style of salsa. Instead, I get bombarded with thick 'n chunky, peach, mango, pineapple, corn, or picante salsa. I don't want giant pieces of food in my salsa.

I know exactly what I want, but all of these companies offer me what they think I want. Or worse, what *they* want me to want.

This got me thinking. Are we doing the same thing with our clients? Are we delivering what our clients want? Do we know what our clients want? Do they? To answer that, we need to break down our clients into three different types:

1. Clients who know what they want, how they want it, and don't want anything but that. This is me as it relates to salsa.
2. Clients who know what they want, but are open to other things.
3. Clients who don't really know what they want and expect you to wow them.

There's nothing wrong with any of these kinds of clients. But as a voice actor you need to be able to recognize the kind of client you're dealing with. You can't treat a #3 as a #1 or they'll walk away wondering why they weren't wowed. And if a client is a #1 and you give them something they don't want, they're going to walk.

And this is where the newbie lowballers on the P2P site du jour can't compete. This is where your value add comes in as a seasoned pro. Let your experience be your guide. Understand your clients and figure out what they need. Don't be afraid to ask questions. The client will be more than happy to give you an idea of what they want, if they know. Then you can deliver the results.

AMY TAYLOR (Los Angeles, CA)
www.amytaylorvoice.net

Amy Taylor is a bilingual voiceover artist whose clients include Verizon Wireless, Dannon Activia, Sony, South Beach Diet, SOAPnet, ESPN, MSN, and Nintendo Wii. She played Detective Natalie Evans on the CSI NY virtual game and won the 2009 Best Female Voice Voicey award from **www.voices.com**. She offers voiceovers in English and Spanish.

The Announcer Vortex

Those who've made crossover from radio to voiceover may know what I'm talking about. If you are a disc jockey, newscaster or otherwise employed by a radio station with on-air duties and are trying to start your own voiceover career, you'll need to be aware of the "Announcer vortex" and learn how to get around it.

Many on-air personalities want to make the shift into a voiceover career. But, just because you've voiced hundreds or even thousands of commercials, don't assume that voiceover work will be the same as in-house radio station ads. Sure, you have good microphone techniques, you're used to the sound of your own voice and you probably know your way around a production studio. But you may not be ready to step into the national—or these days "global" arena just yet.

A former radio personality myself, I've found that radio attracts some really fun-loving, exceptionally talented people. The same skills that helped you land a job in radio are tremendously helpful tools in the realm of voiceover. You'll need to pack your impeccable sense of timing, vocal stamina, and most of all your sense of humor when you go to auditions.

When I made the decision to leave radio and dedicate myself to voiceover, I thought it would be a smooth and easy transition. It wasn't. And if I hadn't been guided by some great voiceover coaches, I wound not have enjoyed the success that I've been blessed to experience.

Through my 11 years in radio, I had developed a default read that I used on almost all of my in-house productions. It was a cerebral template. I read the same way every time. It didn't matter what the copy was; a spot for the local furniture store, car dealer or restaurant. It was always the same energy and cadence. I'll admit it now. I could finish the take and walk away not even remembering what I just read. Now how could I expect the listener to believe my words if I didn't even think about what I was saying?

I had fallen into the "announcer vortex." I call it the announcer vortex because even if I'd start off with a great read, as I'd start listening to myself, the announcer vortex would begin to suck me in. Inevitably, my read would sound like an announcer and not a real person.

The announcer vortex is particularly important to avoid these days. Ad agencies are shelling out big bucks to sell products and services to people who are bombarded by scores of commercials a day. Agencies want their commercial to stand out from the sea of announcer spots on the air without sounding like they're selling anything. It's a trick successful voice talents pull off and they get paid very well to do it.

Almost daily, I see these directions on audition scripts:

> "Not announcery."
> "Not looking for announcer read."
> "Looking for conversational read."
> "Real person. No announcers, please."

The reason voice seekers say that is because many of us give announcer reads on auditions. We think that's what they want. The old time announcer read is woven into the very fabric of radio. It was virtually the only read that was heard from the dawn of radio until late in the twentieth century. Now, don't get me wrong. We should all have that announcer read in our back pockets. Sometimes the client will ask for it. But we need to hone better voicing skills in order to read the majority of today's scripts.

The first thing I suggest to anyone serious about voiceover is to study with a reputable voiceover coach. They can identify areas where you need to improve to sound less "announcery" and more "conversational." Enrolling in an acting class can also do wonders for your voicing skills.

Those who are still working in radio may be tempted to make a voiceover demo at the station using clips from local spots they've done. Please don't! It'll sound like a demo you made out of local spots you've done! It is much wiser to invest in a well-produced, genre-specific demo. There are several reputable producers who specialize in making voiceover demos. Your demo is the single most important thing you'll need for your voiceover career. You'll want it to be the best showcase of your work that you can make. It needs to be made by a professional *after* you've studied with a voice coach.

With the right training and preparation, voiceover can be quite profitable for the radio crossover artist.

DAN HURST (Lees Summit, MO)
www.danhurst.com

Dan Hurst (Daniel Eduardo), was raised in Honduras, the son of missionary parents, and is considered one of America's most experienced and versatile bilingual voice talents. He grew up speaking Spanish and English equally and is fluent in both languages. His client list includes numerous international clients, such as Ford, Sprint, Macromedia, Walmart, and Butler Manufacturing.

SHUT UP Already!!!

Today I speak for producers.

Those guys (men and women) who often decide which voice to use… which talent moves ahead in his/her career… who is actually the best voice for their product or project.

And here's what they want us voice talents to know: It's not about us.

I was recently in a session with an engineer from one of the production houses I work with. He asked me point blank, "Let me ask you something. Yesterday I was in a session with a talent… the client was on the line… and the talent just went off on how I was handling the session. Am I doing something wrong?"

He actually caught me by surprise because I think he's one of the best producers I've worked with, and I certainly didn't think the way he ran a session was inappropriate. But what is even more disconcerting is that the VO talent mouthed off in front of the client.

WHAT??? Shut up already!!! It's not about you!

The more I thought about that the more agitated I got. On the one hand, it just made me look better to the producer and the client, but on the other hand it was an insult to our industry. No matter what you may think, how you represent yourself represents the rest of us. When a voice talent leaves a good or bad impression on a producer or client it affects how they think about voice talent in general.

So with that in mind, here are five things to consider:

1. The most important thing in the world at that time to your client is that project. Not you. Not what you think about the copy. Not what you think about the production. And not what you think about the direction.

2. Piss off the client or the producer and you'll never work for them again. Do me a favor would you? When you decide that the producer or the client is a moron, would you please let me know? I figure they won't be hiring you again and I might as well get my name in there. I could use the work.

3. The producer and/or the client generally know what they want. They've been working on that project for a long time—much longer than you've been a part of the project. They've got the sound they want in their head. They thought you could deliver it. It's your job to figure out what they want, not theirs to convince you.

4. Along with being a voice talent comes a responsibility for professionalism. Unfortunately too many talents have not lived up to that. There is no union, association, or universal criteria that guarantees that. It's no wonder producers and clients are so hesitant and guarded with us.

5. Voicing your unrequested opinion… or even showing an attitude in a session is not only arrogant but ignorant. It just proves how out of touch you are with the process. Shut up already!!! It's not about you!

The good news is that the VO talents who violate these principles are few and far between. To you producers and clients who may read this, know that the vast majority of voice talents get it—we're just part of the process; one of the tools in your toolbox.

Guaranteed satisfaction. It's a creed most of us voice talents live by. And it can only happen when we really understand how we fit into the scheme of things, and then make sure the people we work for are happy they used us. It is, after all, about them.

MICHAEL MINETREE—CD/26
www.minewurx.com; www.thevoiceovercoach.com

Michael Minetree is the owner of MineWurx Studio, a voiceover production studio in Washington D.C. As a coach, he specializes in training new voice talent. He also works in the industry as a performer and producer, provides studio construction consultation and is an avid technician and studio engineer.

Timing Is Everything!

In comedy, you're dead in the water without it. In drama, you've broken the illusion if you lose it. You've heard it before… Timing is everything. Along with comedians who blow though a seven minute set in four minutes because they were nervous, to stage actors completely stomping all over another actors lines, voice talent seem to struggle with timing quite a bit when faced with unfamiliar material. Anyone who has ever booked voice actors or engineered studio sessions has seen it happen.

For some reason, voice actors new and old seem to lose touch with their timing. It happens more in the early days of one's career, but I've seen it from national talent, experienced actors and professional broadcasters. No one is immune, but veterans encounter it much less often. The biggest issue I encounter with timing is speed, or pace. Everyone, from time to time simply goes too fast.

Whereas new voice talent will likely always go too fast when they first start learning, a veteran talent will do it more when the copy is unfamiliar. Here I hope to give you a few rehearsal ideas about how to slow things down so that when it's crunch time you'll be well-versed in hitting the brakes.

Much of voiceover is about creative imagination and the development of the ability over time, to bring your imaginative creations to life very quickly while reading through a script. Practice and training are really nothing more than exercising and becoming familiar with one's ability to tell an audience of 1 or 1,000 what is going on inside their head. By rehearsing, we are developing the ability to do it more fluidly and believably. Practice enough and you will eventually develop a predictable, repeatable pattern of delivery; one that is familiar to you and if you're lucky, to your audience.

Those of us who do not rehearse a lot, or rehearse with bad habits in place, tend to fly past or glaze over conversational opportunities that pop up in a script. When a writer or producer tells us to slow down what they are really saying is, "Hey, take a moment to ingest the next line and consider how you really want to share it before you present it." What we need to remember as talent, is that just because the copy is laying there before us waiting to be read, it doesn't mean we have to give it to them right away.

Hold back a little. Savor the flavor of the words and in much the same way you might wait for your favorite wine to open on your palate, hold fast for a moment and try to grasp the entire intention of the writer. I absolutely propose that voice talent who more accurately capture the voice inside the writer's head at the time the copy was written will land more auditions than any other talent in the waiting room, hands down. While so much of voiceover still has everything to do with your voice, there are many jobs won solely because of someone's given interpretive ability.

Here is a small exercise which I have passed on to many new talents over the years, and even a few people who thought they knew everything, including myself.

When you encounter any piece of copy, commercial or narrative, do not look at the content as a continuous piece that should be read from A to B, top to bottom as one collective thought. Rather, look at it as a collection of small fragments on a page, each one possessing its own set of circumstances and at times, limited connectivity to the others. Each fragment needs and deserves its own space and time. It has a life and meaning all its own.

Some fragments will need to be read a little faster, some much slower and some may even need to be fragmented further once they're give a second look. Though it's not a universal rule and at times impossible to apply completely, there will never be a time where a line in a piece of copy wasn't put there for a reason. Something may have happened along the way to eliminate the need for a particular line in a script, but as talent our job is to give every piece of the script the proper attention. Notice I didn't say equal, just proper. Not every fragment in a piece of copy deserves an equal amount of attention.

Just take your script and break it down to what's between the periods and go over your initial observations of fragments. From there, look for lines that can be broken down into even smaller fragments, usually lines with commas and colons. From those fragments the copy can sometimes be broken down even further, especially when it is first person narrative or highly conversational in feel. Now try to read your fragments out loud, one at a time. During your first few run-throughs on a given piece of copy, you will feel like you're trying to play tennis in lead shoes. It's very clunky and your recorded auditions will sound choppy and lack continuity. During rehearsal this is of little importance.

After you feel confident you have deciphered the intention of the writer and more, the true intended tone and inflection of the words and fragments, begin to read each fragment collectively, with ever smaller pauses between them. Keep a keen ear out for areas where two fragments need more connectivity. You will find that very little copy carries a great deal of connectivity from line to line and certainly very little from paragraph to paragraph. Though the script as a whole was written with one overall intention and meaning, the small parts of which a script is constructed often reveal more to the reader when focused on individually. You just have to look for them. Rehearse better and you will audition better. Audition better and you will more frequently get what all of us as working talent are after: the job.

MAXINE DUNN (Denver, CO)
www.maxinedunn.com

Maxine Dunn is a top voiceover artist and on-camera spokesperson who has been seen and heard in hundreds of commercials, documentaries, corporate narrations, voicemail systems and websites. She is a British native and her ability to also deliver a perfect American accent gives her business a tremendously wide range. She enjoys working out of her professionally equipped home recording studio with Fortune 500 companies and award-winning creative teams, and maintains an extensive clientele, locally, nationally, and internationally. While Maxine is best known for her voiceover and spokesperson expertise, she is also an avid writer who enjoys bringing stimulating and motivating material to her readers.

The Importance of Marketing Materials

If you're interested in a career in voiceover or if you're already working regularly as a voice actor, one of the keys to your success will be your marketing materials.

I think of marketing materials as anything that your clients or prospective clients see, hear, or use to learn about you or to interact with you. They represent you to your potential buyers, which is why it's crucial that you pay attention to their importance right from day one. I want you to realize that your marketing materials encompass much more than just a website.

They include:

- Your online reputation and credibility
- Your appearance and demeanor on auditions and jobs
- Your outgoing voicemail message on your phone
- Your email correspondence
- Your business cards
- Your voiceover demos
- Your website
- Your thank you cards and invoices

In other words, ANYTHING that creates an image of you in a prospective client's mind is marketing.

If you're interested in making your voiceover career the best it can be, then I strongly suggest you make your marketing materials the best they can be BEFORE you start actively marketing yourself.

Let me briefly outline how you can best use each of these vital tools to your advantage.

- **Your online reputation and credibility:** Don't discount the importance of your reputation and your online presence. If you contribute to forums, Facebook, etc., never use profanity or denigrating verbiage. Be aware of what pictures of you are available online. Remember, clients can see all that too so make sure your online image is one you're proud of.

- **Your appearance and demeanor on auditions and jobs:** Many voice actors think because it's only their voice that's being heard and not their face that's being seen, it's permissible to show up at auditions and jobs in torn jeans and tee shirts and flip-flops. I couldn't disagree more. You don't have to get dressed up to the nines, but I believe a professional appearance and a patient, cooperative demeanor on auditions and jobs are vitally important. It shows respect for your clients and portrays you as a professional with high personal standards.

- **Your outgoing voicemail message on your phone:** Your outgoing message should be clearly spoken, professional, short, to the point, and gracious. No music, background sounds, slang, or sarcasm.

- **Your email correspondence:** Your emails should be professional, error-free, and contain opening and closing salutations. As your relationship with a client progresses, you may become more informal but NEVER cross the line of becoming too personal, emotional, or using profanity. Even if your client's emails are all lower case, contain no opening or closing salutations and are filled with typos and slang, take the high-road and keep it professional. Trust me on this one. Create a code of conduct for your email correspondence and stick to it.

- **Your business cards:** You should have professionally printed business cards, ready to go at all times. They don't need to be fancy but they should be clear, easy-to-read (no tiny fonts or wild graphics) and contain all the contact information you can supply. As you secure your domain name, email, and website, you can add information as you go. If you're just getting started and don't even have a demo yet, at least get a business card with your name and phone number. But DO have a business card, no ifs ands, or buts on this one. No business card = unprofessional.

- **Your voiceover demos:** Depending upon your market, you should have a demo for each genre of voice acting that you do: animation, corporate, commercials, etc. They should be current, fast-moving, have superior sound quality and be professionally produced. You must also be able to perform in the studio at the level demonstrated on your demo. Have your demos available on CD, as individual MP3 files on your computer, and also available for download on your website.

- **Your website:** As soon as you have your demo or demos produced, you'll need a professional website. Your website should be current, easy to navigate, and well-optimized. When you're first starting out a simple, one-page website is just fine. Any call-to-action or important information should be "above the fold" (the top part of the page) so no scrolling is necessary. Check your website regularly to make sure everything is working as it should be.

- **Your thank-you cards and invoices:** Thank-you cards are, of course, a given—after jobs, referrals, assistance from others. Make them prompt, brief, and specifically mention the reason for your gratitude. Avoid including any unrelated information, questions, or concerns. Your invoices should be easy-to-read, clear, and detailed. They should contain your contact information and address, payment information and terms, job details, as well as a thank-you comment ON the invoice, thanking your client for their business.

This is just a brief overview of some of the important aspects of your marketing arsenal, but if you pay close attention to these from day one, (or get them back on track if you're mid-career), you'll be way ahead of those that don't.

Wishing you success!

CONNIE TERWILLIGER (San Diego, CA)
www.voiceover-talent.com

For close to four decades, Connie has been working professionally in the many facets of broadcasting, as a writer, producer, director, and of course, voice actor. Her client base is a "who's who" of prestigious companies that provide her the opportunity to work in many areas of voiceover. For more than two decades she has also taught a college-level course on voiceover in San Diego.

Auditioning in The 2000s

Technology has changed the way we do business as voice talent in many ways, but one thing that stands out is auditioning. I have never auditioned so much in my life as I have since Internet casting and remote recording came in vogue. Back in the 80's I rarely auditioned for anything. More often than not, voice talent were booked off their demos. In some areas where the stakes were a bit higher than my little pond, the initial casting was done off the demos and the talent brought in for call backs.

But nothing stays the same. After about a ten year hiatus from VO work while wearing lots of hats for a major corporation (producer, writer, on and off-camera talent), I started to ease myself back into voiceovers—and boy oh boy had the world changed!

It seems that the moment I had the ability to record out of my home studio and send voice files over the Internet I suddenly had to prove that I could read the client's specific words. If you are competing for pretty much any kind of voiceover work, no longer is the artfully prepared demo enough. And the competition is fierce.

After many years in this business, I am starting to have the opportunity to go vocal cord to vocal cord with some of the best, most well-known talent in the industry. Several times a day, I hear spots on the radio and TV that I auditioned for and didn't get. I have returned to my audio files to compare what ended up on the air with what I submitted. And I know that my audition was truly competitive—but being the subjective business this is—I was just not the voice in the head of the people making the decisions that day. And it is a mystery for sure sometimes as to why one voice is selected out of all the choices submitted.

So what makes a winning audition when the process is so subjective? Your audition has to catch someone's ear. They do have a sound in mind – but may not have been able to express it. Or they fall into the "know it when they hear it" group. Either way, there are some basics to help your audition stand out.

1. It must be technically clean – no background noise, volume not too loud or too low. It is expected that you have your own professional studio, or have access to one. Your competition does.

2. You must know that you are "right" for the copy you are reading. And yes, "right" is subjective. But none of us are good at everything. Know your strengths and pass on

auditions that are not "right." There is still some debate over this point. But auditioning does take time, so choosing auditions that are up your alley instead of going for long shots may help your audition to booking ratio.

3. Be ready to audition. Warmed up, focused, relaxed, confident.

4. Understand the script. Just what the heck is going on!

5. Understand who you are talking to. In the case of a commercial, who is the target buyer for the product? In the case of a corporate piece, are you talking to a peer? Are you talking to someone who really doesn't want to hear what you have to say?

6. Understand what you are saying. Tell the story that lies beneath the words. Don't just read the words. Dig under, around, above each word to communicate the meaning and the emotion of the copy.

7. Know that what you have recorded is actually competitive. You are the one hitting the send button, so you must be the one to make that final decision.

Hard to tell what the magic sound will be for any specific client, but if you submit high-quality auditions technically and creatively, your audition-to-booking ratio should pay off.

BEVERLY BREMERS
www.beverlybremers.com

Beverly Bremers entered the world of voiceover backwards. Her first VO job was the result of being a singer in the right place at the right time—with zero voiceover experience. She has since gone on to voice commercials and other projects for national clients like Nissan, Shell, Sony and hundreds more. She is also an excellent voice-acting coach and independent producer.

Helping Hands

A successful voice actor once quipped that if he didn't have hands, he couldn't do voiceovers. How true! Second to the ability to read well, the use of the hands is the most important physical tool a voice actor has. Use your hands to point to a person or thing, caress an adjective, or simply wave them around for emphasis. Voila! It's like magic.

You can experiment with this concept easily. Record yourself saying "He searched all over the world," first without using your hands, and then again, waving your hand in the air from left to right on the word "all". You'll notice immediately that the sentence sounds more descriptive and engaging.

How can something so simple be so effective? I firmly believe that how you place your body reflects your emotions, moods, and attitudes. It's a "mind extension." That's why you must always be aware of your body stance when doing a voiceover. Standing tall versus hunched over gives opposite emotional results. Continue that concept with your hand gestures, and you can communicate different emotional results. If you cross your arms, you reflect a guarded, doubting attitude toward your listener and subject. Hands on your hips show hostility, hands in pockets reflect casual cool, etc. Try reading something that's supposed to be enthusiastic without using your hands and you'll see how difficult it is to show that emotion. The hands help!

So remember to channel your inner Italian and talk with your hands. If you're talking about a list of items, point them out. If you're describing scenery, wave your hand in a sweeping motion. Hold the objects, demonstrate the verbs, caress the adjectives. Just like the old Yellow Pages slogan, "Let your fingers do the walking," I recommend that you "Let your hands lend a hand."

PENNY ABSHIRE (San Diego, CA)—CD/10, CD/25
www.pennyabshire.com; www.voiceacting.com

It has become a tradition that my business partner, Penny Abshire, closes this book with her words of wisdom. Penny has been a performer since the age of 7. She entered the world of voiceover at the tender age of 47 and never looked back. Since then, she has gained a reputation for her exceptional copywriting skills, her outstanding performance coaching and directing, her positive and motivational perspective on the business, and her skill in creating believable "real-people" characters.

If You Think You Can... or You Can't... You're Right!

I studied journalism and theater arts in college. My dream was a career in the theater. I was also a highly trained classical pianist and had aspirations of going on tour. But, alas, it was my heart that swayed me from that direction. I fell in love and put homemaking and motherhood ahead of those dreams. Mind you, I have never regretted that decision! I have been married to the same great guy for 41 years—have two amazing sons, two equally amazing daughters-in-law, one very handsome grandson, and three beautiful granddaughters! But as one does, as the years passed I often wondered what "could have been."

When my kids got older, I went back into the work force as a secretary and then later as a paralegal. It was good, steady money, it was safe and I was very good at it. But for the most part, I hated it! For a creative person, that kind of repetitive work is just a slow, lingering death. Please don't misunderstand and think I don't admire those who work as secretaries or paralegals—I certainly do! It just wasn't where I was supposed to be. But because it was "safe," I continued working and justified my unhappiness by saying to myself I was doing the responsible thing.

I went to my very first voiceover class at the ripe old age of 47. In fact, I went kicking and screaming! I didn't want to go and a friend practically had to shove me through the door! Mostly, I think, it was because I was afraid of looking foolish or failing. I am so glad I did it anyway! I left that first three-hour class (taught by James Alburger) knowing that voice acting was what I would be doing the rest of my life. I can't explain it, I just knew. This short introductory class literally changed the direction of my life! It seemed I had been asking God for so many years what my "real" purpose was in life and that night I got my answer.

Then, the "voices" started talking to me… "You're too old, Penny! What's wrong with you? This isn't the type of career you start when you're almost 50 years old! You idiot! You're going to look ridiculous in class. Everyone there will be younger and more talented than you!" And on and on and on and on.

It was a continuous flow of negative thoughts. Like most people, those voices in my head were VERY loud and for almost a year, I let them win. My family would ask, "Are you ever going to take those voice classes, mom?" And I would reply, "Oh, yes… definitely. I'm definitely going to take them… someday!" One day, my youngest son asked me that question and when I replied with, "Someday…" he said, "Mom, admit it, if you weren't scared to do it, you would have already done it!"

Out of the mouths of babes! OK… Well, he was 22 at the time, but still…

Anyway, I called Jim that day and signed up for classes. And it was hard to make that call. My "voices" were extremely loud! But because of my son, I finally had the courage to tell them to shut up! I finally found the courage to follow my dream. One of the things that got me "off my duff" was this thought: "In 10 years I will be 57. Do I want to look back in 10 years and say, 'Darn, I wish I'd done that?' Or do I want to look back and see all I have accomplished in the last 10 years doing what I love?"

It was pretty much a no-brainer in my book. But imagine my very conservative husband's face when I announced I was quitting a good paying job with benefits to work as an actor! Fortunately, after 29 years of marriage, he understood me pretty well. But I'm sure at the time he thought I'd lost my mind! Now he can't tell enough people what I do and he brags about me and my accomplishments constantly. I love that!

That was 1998. Today I work full-time as a busy voice actor, director, copywriter and producer. I have written my first book. I am the co-executive producer of the only Voiceover Convention in the world—VOICE (VoiceOver International Creative Experience, **www.voice-international.com**) and I have hundreds of students who I now call dear friends! I have traveled all over the world and met some incredible people while teaching. My days are filled with excitement and anticipation—something I never experienced as a secretary! It's something new every day. New projects, new scripts, new techniques, new teachers, and best of all, new people! And I work with the nicest guy on the planet, my first teacher and now my partner and best friend, James Alburger.

Yes, the years since I made that decision have been absolutely wonderful and I believe I'm only getting started. So, what is my goal for the next several years?

I want to help as many people as I possibly can to realize *their* dreams!

When another decade has passed, I know I will again look back and be excited about what's been accomplished!

So, for any of you out there who think you're too old (or any other excuse you're currently using) to pursue your voice acting career, please don't let your age be an obstacle. Age is only a number. I know I don't feel like I'm 59 even though, as I write this, that's what the calendar says. Actually, I'm darn proud of being my age and to be doing what I'm doing. Sadly, I've seen too many people hold themselves back from accomplishing their dreams because of what other people might think. At some point in your life, you just have to have the courage to say, "Ah, to heck with them!"

If you have a dream (and it is truly your heart's desire) the ONLY person who can stop you from pursuing and attaining it is **you**.

You only have one life. Please choose to enjoy it while doing something you love!

Index